Reverse Osmosis and Ultrafiltration

ACS SYMPOSIUM SERIES **281**

Reverse Osmosis and Ultrafiltration

S. Sourirajan, EDITOR
National Research Council of Canada—Ottawa

Takeshi Matsuura, EDITOR
National Research Council of Canada—Ottawa

Based on a symposium sponsored by
the Division of Industrial and Engineering Chemistry
at the 188th Meeting
of the American Chemical Society,
Philadelphia, Pennsylvania,
August 26–31, 1984

American Chemical Society, Washington, D.C. 1985

Library of Congress Cataloging in Publication Data

Reverse osmosis and ultrafiltration.
(ACS symposium series, ISSN 0097-6156; 281)

"Based on a symposium sponsored by the Division of Industrial and Engineering Chemistry at the 188th Meeting of the American Chemical Society, Philadelphia, Pennsylvania, August 26-31, 1984."

Includes bibliographies and indexes.

1. Reverse osmosis—Congresses. 2. Filters and filtration—Congresses.

I. Sourirajan, S. II. Matsuura, Takeshi, 1936–
III. American Chemical Society. Division of Industrial and Engineering Chemistry. IV. American Chemical Society. Meeting (188th: 1984: Philadelphia, Pa.)
V. Series.

TP156.07R48 1985 660.2′8424 85-9155
ISBN 0-8412-0921-9

PRINTED IN THE UNITED STATES OF AMERICA

ACS Symposium Series

M. Joan Comstock, *Series Editor*

FOREWORD

The ACS SYMPOSIUM SERIES was founded in 1974 to provide a medium for publishing symposia quickly in book form. The format of the Series parallels that of the continuing ADVANCES IN CHEMISTRY SERIES except that, in order to save time, the papers are not typeset but are reproduced as they are submitted by the authors in camera-ready form. Papers are reviewed under the supervision of the Editors with the assistance of the Series Advisory Board and are selected to maintain the integrity of the symposia; however, verbatim reproductions of previously published papers are not accepted. Both reviews and reports of research are acceptable, because symposia may embrace both types of presentation.

CONTENTS

PREFACE

RESEARCH ENDEAVORS in the areas of technological improvement and fundamental understanding of reverse-osmosis–ultrafiltration membranes and processes constitute the central theme of this book. The two main features of the symposium upon which this book is based were reports of some significant advances in the field with respect to both fundamentals and applications and major contributions from overseas. More than sixty percent of the papers presented came from ten countries other than the United States.

The research and development reports contained in this book offer a valuable aid to the further development of the science, engineering, and technology of reverse-osmosis–ultrafiltration membranes and processes.

S. SOURIRAJAN
National Research Council of Canada—Ottawa

TAKESHI MATSUURA
National Research Council of Canada—Ottawa

February 22, 1985

1

Materials Science of Reverse-Osmosis–Ultrafiltration Membranes

TAKESHI MATSUURA and S. SOURIRAJAN

Division of Chemistry, National Research Council of Canada, Ottawa, Ontario, Canada K1A 0R9

A new field of materials science has been evolving as a means of designing RO/UF membranes for different applications. This science is based on the interfacial properties of membrane materials at membrane material-solution interfaces, origin and development of pores on the surface of asymmetric porous membranes, method of determining the average pore size and pore size distribution on the surface of RO/UF membranes, and the relationship between the above interfacial properties and the porous structure of membrane surface, and the RO/UF performance of the membrane on the basis of the surface force-pore flow model for RO/UF transport. This science is broadly outlined and illustrated.

In the process of manufacturing membranes which are to be used for particular reverse osmosis (RO)/ultrafiltration (UF) applications, consideration has to be given to the choice of the membrane material (whether polymeric or nonpolymeric) and to the formation of appropriate physical structures in the membrane. The preferential sorption-capillary flow mechanism for reverse osmosis states that the solute concentration gradient induced by the interfacial interaction force and the pore size and its distribution on the membrane surface together govern the solute separation as well as the product rate of the membrane permeated solution (1). The membrane material considered being the only factor which affects the interfacial interaction force for a given solute-solvent system and the physical structure of the membrane surface being expressed by the average pore size and the pore size distribution on the membrane surface, the problem set above is reduced to the search of the membrane material which demonstrates the most appropriate interfacial interaction forces to the given solute-solvent system and also the pore size distribution on the membrane surface. This approach to membrane design constitutes a new branch of the materials science with respect

0097–6156/85/0281–0001$06.00/0
Published 1985, American Chemical Society

to RO/UF membranes; this paper is concerned with such materials science involving polymeric materials and membranes.

For the development of such science the determination of the interfacial interaction forces and of the pore size and its distribution are of vital importance. The above two factors (interfacial interaction forces and pore size distributions) have to be further incorporated into appropriate transport equations so that the membrane performance data (solute separation and product permeation rate) which are experimentally obtainable can be calculated. Such transport equations have already been developed on the basis of the surface force-pore flow model (2,3) which is a quantitative expression of the preferential sorption capillary flow mechanism.

As for the determination of the interfacial interaction forces, the liquid chromatography (LC) method, which simulates those forces working in the membrane polymer material-solvent-solute system, has proved particularly useful (4). Furthermore, in the course of establishing the method of determining the average pore size and the pore size distribution on the membrane surface, it was found that, in general, the bimodal normal distribution satisfies RO and UF data most satisfactorily (5-9). The origin of such bimodal pore size distribution is related to the structure of the film casting solutions (10b); viz. the polymer network pores which constitute the smaller pore element of the bimodal distribution arise from the spaces among the polymer segments forming the polymer network in each supermolecular polymer aggregate in the film casting solution, and the polymer aggregate pores which constitute the larger pore element of the bimodal distribution arise from the spaces created among neighboring such supermolecular aggregates themselves.

The origin of the bimodal pore size distribution having thus been conceptually established, the method of determining such pore size distributions is briefly outlined. Further, the methods of controlling the average pore sizes involved in the bimodal distribution and the ratio of numbers of pores belonging to each distribution are discussed. The effect of the bimodal distribution on the performance of RO/UF membranes is also illustrated in the paper.

Analysis of Casting Solution Structure and the Origin of Bimodal Pore Size Distribution

It is generally accepted that the membrane structure is determined incipiently by the structure of the polymer in the casting solution (11-13); therefore, the study of the polymer solution structure is necessary to discuss the origin of the membrane pore structure.

Aggregate Pores. Aggregate pores are formed from spaces surrounded by neighboring polymer aggregates when they are closely packed in a space occupied by the polymer solution, called "solution space" hereafter, or from spaces which are devoid of polymer aggregates. Hence, the size of the polymer aggregate and its distribution in the solution space should be known in order to calculate the size and number of aggregate pores. The measurement of the size of the polymer aggregate was made with respect to the solution of aromatic polyamide polymer of different molecular weights and calcium chloride in dimethylacetamide solvent, maintaining polymer-$CaCl_2$ weight ratio in the solution constant (14). The structure of the polymer used in this study is the same as that given in (15). Polymers of different molecular weights were laboratory synthesized by the method of Gan et al. (16). By measuring the intrinsic viscosity of aromatic polyamide-$CaCl_2$ complex (weight ratio 3.3:1) in dimethylacetamide solvent, it was found that the number of polymer chains in one aggregate was 2.14, when polymer molecular weight was 10,500 (this polymer is called PA-10,500 hereafter), thus magnifying the apparent molecular weight of polymer aggregate to 22,500. On the other hand, the number of the polymer chains in one aggregate was 4.12 when the molecular weight of the polymer was 31,300 (this polymer is called PA-31,300 hereafter), the apparent molecular weight of the aggregate being 129,090 (14). Further, by applying the method of Rudin and Johnston (17) the polymer aggregate sizes in the casting solutions used (15) were determined. The results are 31.6×10^{-10} m and 57.2×10^{-10} m for PA-10,500 and PA-31,300, respectively, at 20°C. The effect of temperature on the aggregate size is almost negligible up to 95°C. The above numerical values are applicable for the specific casting compositions tested, and they could vary for different casting compositions.

On the basis of the above experimental data we are going to examine the distribution of polymer aggregates in a given solution space more closely with the aid of statistical thermodynamics. For the statistical calculations, we need to specify a unit cell, either occupied or unoccupied by a polymer aggregate. It is convenient to consider a number of cubic cells whose sizes are equivalent to those of polymer aggregates and which are distributed three-dimensionally in the solution space. Therefore, in the following discussion, a spherical polymer aggregate is approximated by a cube, the length of whose edge is the same as the diameter of the spherical aggregate under consideration. Furthermore, from the polymer content in the casting solution, it was calculated that 3.93×10^{21} total cubes exist in one kg of the PA-10,500 solution of casting composition while 0.58×10^{21} cubes of equal size are filled by solvent only. Likewise, there exist a total 0.741×10^{21} cubes in one kg of PA-31,300 solution of casting composition, among which 0.079×10^{21} cubes are filled by solvent only.

Distribution of Polymer Aggregates in the Bulk Polymer Solution. Let us now assume a large cube which is composed of m^3 number of unit cubes. Such a large cube can be regarded as a solution space. Each unit cube is either filled with a polymer aggregate or unfilled. Let us call hereafter the unit cubes filled with a polymer aggregate "filled cubes" and those filled by solvent only "vacant cubes". Let us then count the number of ways by which ℓ^3 number of vacant unit cubes can be distributed, separately, in the aforementioned large cube of m^3 unit cubes. When $m^3 >> \ell^3$ the number of such ways is approximated by the number of combination of m^3 things ℓ^3 at a time

$$w_1 = {}_{m^3}C_{\ell^3} = (m^3)!/(m^3 - \ell^3)!(\ell^3)! \qquad (1)$$

When ℓ^3 vacant cubes are put together into a single cube which consists of ℓ^3 unit cubes, the number of ways of positioning such a cube into the large cube (of m^3 unit cubes) is

$$w_2 = (m - \ell + 1)^3 \tag{2}$$

The entropy increase resulting from splitting an assembly of ℓ^3 vacant unit cubes into isolated vacant unit cubes, which is designated as ΔS_{cube}, is therefore

$$\begin{aligned}\Delta S_{cube} &= \underset{\sim}{k}\ \ell n\ \frac{w_1}{w_2} \\ &= \underset{\sim}{k}\ \ell n\ \left[\frac{(m^3)!}{(m^3 - \ell^3)!(\ell^3)!} / (m - \ell + 1)^3\right]\end{aligned} \tag{3}$$

and when $m^3 >> \ell^3 >> 1$, it is approximated by

$$\Delta S_{cube} = 3\ \underset{\sim}{k}\ \ell^3\ \ell n\ (m/\ell) \tag{4}$$

Splitting a large vacant cube into a number of small vacant unit cubes is also accompanied by an increase in the surface area of contact between filled and vacant cubes, resulting in increase of the surface energy which can be regarded as increase in enthalpy, ΔH_{cube}. The latter can be calculated by,

$$\Delta H_{cube} = 6\ \ell^3\ L^2\ \sigma - 6\ (\ell L)^2\ \sigma \tag{5}$$

where L and σ denote the length of the edge of unit cube and the interfacial tension between polymer aggregate and solvent. Combining Equation 4 and 5 the free energy change involved in the splitting of a large vacant cube is written as

$$\Delta G_{cube} = \Delta H_{cube} - T\Delta S_{cube} \tag{6}$$

$$= 6\ (\ell^3 - \ell^2)\ L^2\ \sigma - 3\ \underset{\sim}{k}\ T\ \ell^3\ \ell n\ (m/\ell) \tag{7}$$

Assuming $\sigma = 0.202 \times 10^{-3}$ N/m between aromatic polyamide polymer and dimethylacetamide solvent, and using the size and number of cubes (either filled or vacant) given earlier ΔG_{cube} was calculated to be 23.42 J/kg and 11.84 J/kg for polymers PA-10,500 and PA-31,300, respectively. The positive free energy changes resulting from a large increase in the surface energy, ΔH_{cube}, which could not be compensated by the entropy increase, indicate that splitting of vacant cubes into small unit cubes is thermodynamically unfavorable. Therefore, vacant cubes (filled by solvent only) tend to gather in the bulk solution phase.

Distribution of Polymer Aggregates in the Solution Surface and the Size of Aggregate Pores. The situation at the surface of the solution differs from that in the bulk. Because of surface tension, the polymer aggregate at the air-solution interface is flattened (18), and two dimensional distributions of polymer aggregates have to be considered rather than three dimensional ones. Let us now assume that there is a large square which consists of $(2\mu)^2$ number of unit squares. This large square corresponds to a two-dimensional solution space. The length of the side of the unit square is assumed to be Λ. Again, we define the unit square occupied by a flattened polymer aggregate as "filled square" and the unit square occupied by solvent only as "vacant square". Let us now distribute total λ^2 number of vacant unit squares in such a way that λ_1^2 of them are distributed in a completely isolated manner, while the rest is grouped into a number of squares, each one of which is composed of four unit squares. Supposing the number of latter squares to be λ_4^2,

$$\lambda^2 = \lambda_1^2 + (2\lambda_4)^2 \tag{8}$$

Following the same principle as in the case of three dimensional distribution, the free energy change involved in the splitting of a large vacant square (which consists of λ^2 vacant unit squares) can be calculated approximately under the condition $\mu^2 >> \lambda^2 >> 1$ as,

$$\Delta G_{square} = \left[4\ (2\lambda_4^2 + \lambda_1^2)\ \Lambda\ \tau\sigma\right] - k\ T\ \left[\lambda_4^2\ \ell n\ \left(\frac{\mu}{\lambda_4}\right)^2 + \lambda_1^2\ \ell n\ \left(\frac{2\mu}{\lambda_1}\right)^2\right] \tag{9}$$

where τ indicates the thickness of the flat polymer aggregate. It was already mentioned that the most thermodynamically stable state in the bulk solution is where the vacant spaces are all assembled in one place. The above thermodynamic state considered being the reference point for the free energy change on the surface, ΔG_{square}, the equilibrium between the bulk and surface phase solutions requires the condition

$$\Delta G_{square} = 0 \tag{10}$$

As for the numerical value of Λ, the spherical polymer aggregates are assumed to be flattened to circles of two polymer layers and their diameters are set equal to Λ. Then $\Lambda = \sqrt{2}\ L$ and therefore $\Lambda/2 = 44.65 \times 10^{-10}$ m and 80.88×10^{-10} m for PA-10,500 and PA-31,300, respectively. Considering the molecular structure of aromatic polyamide molecule, 6.8×10^{-10} m was used for τ. Then, assuming the ratio of vacant squares to the total square $(\lambda^2/(2\mu)^2)$ is the same as that of vacant cubes to the total cubes in the bulk solution the ratios (λ_1^2/λ^2) and $\{(2\lambda_4)^2/\lambda^2\}$ were calculated from Equations 8, 9 and 10. The results are $\lambda_1^2/\lambda^2 = 0.029$ and 0.980 for PA-10,500 and PA-31,300, respectively, which means that 97.1% of vacant unit squares exist as isolated assemblies of four unit squares with respect to PA-10,500 polymer, while 98% of vacant unit squares exist as isolated single squares with respect to PA-31,300 polymer. (The possibility of the formation of larger vacant cells than those with four unit squares was rejected by the preliminary calculation which showed that Equation 10 could not be satisfied by such distribution modes.) The above results indicate that the space unfilled by the polymer aggregate tends to be split into small cells at the polymer solution surface. This may offer a thermodynamic reason for the asymmetricity of RO/UF membranes, though the asymmetric structure is usually attributed to kinetic effects (19). These vacant cells may be considered as incipient aggregate pores. The above results also indicate that the size of vacant cell is $2 \times 44.65 \times 10^{-10} = 89.3 \times 10^{-10}$ m for PA-10,500 polymer (radius of polymer aggregate = 31.6×10^{-10} m) and 80.9×10^{-10} m for PA-31,300 polymer (radius of polymer aggregate = 57.2×10^{-10} m). From these data one may conclude that the size of the aggregate does not affect very much the size of vacant cells and consequently the size of the aggregate pores. Furthermore, the aggregate pore size may be expected to become smaller as solvent evaporation proceeds, as indicated in our earlier paper on aromatic polyamide hydrazide membrane (8).

The Size of the Network Pore. While the aggregate pores are associated inherently with the polymer structure in the casting solution, the final network pores are associated with the structure of polymer aggregate in equilibrium with water environment. For calculating the size of such network pores, let us now consider a polymer aggregate which contains N number of water molecules. Let us also consider that most of polymer chains are in crystalline form and there are several loose noncrystalline regions which connect neighboring crystalline regions. Let us assume that ν number of water molecules intrude into one noncrystalline region and form a pore; thereby the length of the polymer segment is slightly stretched. Since the polymer aggregate is flat on the surface, we regard this problem again as that of two-dimensional distribution of water molecules. Then, the entropy increase accompanying the dispersion of water molecule in a polymer aggregate in the aforementioned manner may be approximated by using Stirling's formula as

$$\Delta S_{disp} = \underset{\sim}{k}\, N \,\ell n \left(\frac{N}{\nu}\right) \tag{11}$$

The surface energy increase involved in the generation of the interface at the water-polymer boundary can be considered as the enthalpy change and can be written as

$$\Delta H_{interface} = \frac{N}{\nu} \cdot 2\pi \sqrt{\nu}\, r_{water}\, \tau\sigma \tag{12}$$

where $\sqrt{\nu}\, r_{water}$ is the radius of a circle which is formed when ν water molecules are two-dimensionally fused. Besides, we have to consider the work stored in the polymer segment when it is stretched by the intrusion of water molecules into the noncrystalline region. When a polymer segment of cross-sectional area S and of initial length L_o is stretched by ΔL, the work required for the stretching, which may be regarded as the further addition to enthalpy increase in the system, can be written as

$$\Delta H_{stretch} = \frac{1}{2} \frac{N}{\nu} \frac{ES}{L_o} \Delta L^2 \tag{13}$$

where E is Young's modulus of the polymer. Then the free energy change involved in the process of two-dimensional dispersion of water molecules can be written as,

$$\Delta G_{disp} = \Delta H_{interface} + \Delta H_{stretch} - T\Delta S_{disp} \tag{14}$$

$$= \frac{N}{\nu} \cdot 2\pi \sqrt{\nu}\ r_{water}\ \tau\sigma + \frac{1}{2}\frac{N}{\nu}\frac{ES}{L_o}\Delta L^2 - k\ T\ N\ \ln\left(\frac{N}{\nu}\right) \qquad (15)$$

When the system in which water molecules are dispersed in the polymer is in equilibrium with that where water and polymer exist separately, ΔG_{disp} should be equal to zero. Furthermore, the first term of the right side of Equation 15 is negligible as compared with the rest of the equation, and in the second term ΔL can be approximated by the area occupied by intruding water molecules divided by the distance between two polymer segments d_s. Then,

$$\Delta L = \nu\ \pi\ r^2_{water}/d_s \qquad (16)$$

Setting $\Delta G_{disp} = 0$ in Equation 15, and inserting Equation 16

$$\frac{1}{2}\frac{ES}{L_o}\frac{\nu\ \pi^2\ r^4_{water}}{d_s^2} = k\ T\ \ln\left(\frac{N}{\nu}\right) \qquad (17)$$

Considering that the water content in the aromatic polyamide material as 0.39 (20) we can calculate the number of water molecules in one aggregate, N. As for the quantity E, Young's modulus of polyhexamethylene terephthalamide (4.84×10^9 N/m^2) reported in reference (21) is used. The values for S, L_o and d_s are taken from the structure of the poly-m-phenylene isophthalamide given by Herlinger et al. (22) as 22.1×10^{-20} m^2, 16.5×10^{-10} m and 6.7×10^{-10} m, respectively, assuming that the meta-structure in the polymer forms the noncrystalline region, while para-structure forms the crystalline region. Then, from Equation 17 ν is calculated to be 24.1 and 38.5 for PA-10,500 and PA-31,300 polymers, respectively. When water molecules are two dimensionally fused they form pores with radii 4.3 $\times 10^{-10}$ m and 5.4×10^{-10} m, respectively.

The above calculation is very enlightening in its scope and direction. It indicates that there are indeed two distinct kinds of membrane pores. The first one, called aggregate pores, originates from the spaces in the polymer solution surface, which are devoid of polymer aggregates. The radius of such pores is less than 90×10^{-10} m. Obviously, the sizes and numbers of these pores are strongly related to the polymer structure in the casting solution, and they are of transient nature. The other kind, called network pores, are related more to the structure of polymer aggregate itself which is in thermodynamic equilibrium with water environment. The radius of such pores ranges from 4 to 5×10^{-10} m. These pores are of such a nature that the polymer ultimately acquires the structure in water environment. The transition from aggregate pores to network pores is possible as will be shown later. It is also interesting to note that the network pore radius depends on the intrinsic property of the polymer as well as its disposition in the aggregate. According to Equation 17 the number of water molecules in the network pore, ν, increases when N increases. Obviously, increase in water content in the polymer increases N as well as ν. The value of ν decreases with increase in Young's modulus E, and with decrease in L_o. Therefore, crosslinking of the polymer to shorten the segment length L_o in the noncrystalline region is an effective method to reduce ν. Thus, Equation 17 contains properties of polymer materials which govern the radius of the network pore. The bimodal concept described above is schematically represented in Figure 1.

It should be noted, however, that all the numerical values obtained above refer to the particular polymer solution composition studied. They are subject to change with the change in the nature of solvent, swelling agent and the casting solution composition.

Experimental Detection of Bimodal Distributions

The basis for the determination of the pore size distribution is to apply transport equations developed on the basis of surface force pore flow model (2,3) to the RO/UF separation data of chosen reference solutes and find the best pore size distribution to optimize the agreement of calculated and experimental separation data. The details of the above procedure are described in the literature (5). For the clarification of symbols used in this paper the frame work of the theory is outlined below.

In the transport equations, the pore size distribution is expressed in terms of one or more Gaussian normal distributions. For describing such pore size distributions the distribution function of the ith component given as

$$Y_i(R_b) = \frac{1}{\sigma_i\sqrt{2\pi}} \exp\left\{-\frac{(R_b - \bar{R}_{b,i})^2}{2\sigma_i^2}\right\} \quad (18)$$

and a quantity defined as

$$h_i = n_i/n_1 \quad (19)$$

are necessary, where $\bar{R}_{b,i}$, σ_i and n_i denote the average pore size, standard deviation and the number of pores which belong to the ith distribution (5). We also define that $\bar{R}_{b,i}$ becomes progressively larger as the number i increases.

With respect to nonionized organic solutes in aqueous solutions, the interfacial interaction force constants are expressed as constants which define the interfacial potential function by

$$\phi(\underline{d}) = \begin{cases} \text{very large} & \text{when } \underline{d} \leq \underset{\sim}{D} \\ -\dfrac{\underset{\sim}{B}}{\underline{d}^3}\underset{\sim}{R}T & \text{when } \underline{d} > \underset{\sim}{D} \end{cases} \quad (20)$$

where $\underline{d}$ is the distance between the polymer surface and the center of the solute molecule, $\underset{\sim}{D}$ is a constant associated with the steric hindrance (distance of steric repulsion), and $\underset{\sim}{B}$ expresses the nature and the magnitude of the van der Waals force (3). The quantity $\underset{\sim}{D}$ is always positive and when the solute shape is assumed spherical, it can be approximated by the molecular radius such as the Stokes law radius, while $\underset{\sim}{B}$ may be either positive (corresponding to an attractive force) or negative (corresponding to a repulsive force). The parameters associated with the pore size distribution, i.e. $\bar{R}_{b,i}$, σ_i and h_i and the interfacial interaction force parameters $\underset{\sim}{B}$ and $\underset{\sim}{D}$

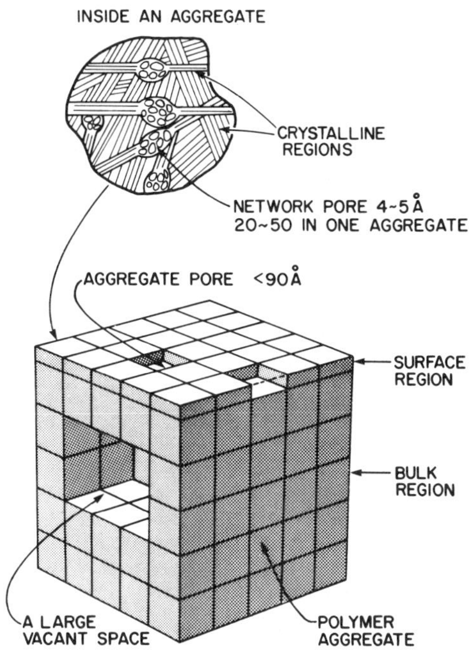

Figure 1. Schematic representation of aggregate pores and network pores.

are related to the specific surface excess, $\Gamma_A/c_{A,b}$, obtainable from the chromatographic retention volume data, and the solute separation f, obtainable from RO experiments by

$$(\Gamma_A/c_{A,b})_j = \{g\ (\underset{\sim}{B},\ \underset{\sim}{D})\}_j \tag{21}$$

$$f_j = \{h\ (\bar{R}_{b,i},\ \sigma_i,\ h_i,\ \underset{\sim}{B},\ \underset{\sim}{D};\ \text{under given operating conditions})\}_j \tag{22}$$

where subscript j indicates the jth solute, g(---) and h(---) are some functional forms established in the surface force-pore flow model (3,5,10a). Using the above equations numerical parameters involved in the interfacial interaction forces and the pore size distributions can be determined as follows. Let us first choose seven reference solutes. Setting Equations 21 and 22 for each reference solute we have seven Equation 21 and seven Equation 22 corresponding to j = 1,2,3...7. Then, setting $(\underset{\sim}{D})_j$ is equal to the Stokes law radius of jth reference solute, $(\underset{\sim}{B})_j$ can be calculated from Equation 21 set for the jth solute, so that experimental $(\Gamma_A/c_{A,b})_j$ can be satisfied. Further, five pore size distribution parameters ($\bar{R}_{b,1}$, σ_1, $\bar{R}_{b,2}$, σ_2 and h_2) can be calculated by nonlinear regression analysis of seven Equation 22 using $\underset{\sim}{B}$ and $\underset{\sim}{D}$ values obtained above for each reference solute.

The foregoing method was applied to a membrane which was fabricated from the PA-10,500 polymer under study. As reference solutes, those listed in Table I as numbers 1-7 were used. The results are $\bar{R}_{b,1} = 6.65 \times 10^{-10}$ m, $\sigma_1/\bar{R}_{b,1} = 0.01$, $\bar{R}_{b,2} = 50.0 \times 10^{-10}$ m, $\sigma_2/\bar{R}_{b,2} = 0.48$, and $h_2 = 0.04$, thus confirming the existence of the bimodal pore size distribution on the membrane surface made of the particular polymer. The $\bar{R}_{b,2}$ value obtained is significantly smaller than 90×10^{-10} m (corresponding to the radius of space unfilled by polymer aggregate), which result is due to the partial evaporation (9 min) of the solvent involved in the membrane formation process. Table II shows other examples of bimodal pore size distribution with respect to cellulose acetate membranes, which were also determined using the set of reference solutes (numbers 3-7) given in Table I. Table II shows that the shrinkage process applied to cellulose acetate membranes transforms progressively more aggregate pores into the network pores (indicated by decrease in h_2) as the shrinkage temperature is increased for each membrane batch produced from different casting compositions and different casting conditions (for details see 6).

Control of the Pore Size Distribution

As mentioned earlier, aggregate pores are the ones generated in the process of membrane formation and can be regarded as a transient structure, while network pores can be considered as the structure in the polymer aggregate which is ultimately acquired in equilibrium with water environment. Then, there must be a natural tendency for a transient state (aggregate pores) to be transformed to the stable state (network pores), and such a tendency may offer a means of controlling pore size distributions. The above transformation can be achieved in various ways. Two of the methods are shrinkage of as-cast membranes in hot shrinkage media in the case of cellulose

Table I. Some Surface Parameters Pertinent to Reference Solutes Used for the Determination of Pore Size Distribution

			CA-398 [a]			PA [b]		
No.	Solute	Molecular weight	$\Gamma_A/c_{A,b}\times10^{10}$ m	$D\times10^{10}$ m	$B\times10^{30}$ m^3	$\Gamma_A/c_{A,b}\times10^{10}$ m	$D\times10^{10}$ [c] m	$B\times10^{30}$ m^3
1	Ethyl alcohol	46.1				5.88	1.94	3.60
2	Trimethylene oxide	58.1				16.78	2.30	50.41
3	1,3-Dioxolane	74.1	45.98	1.96	42.43	15.32	2.41	54.53
4	p-Dioxane	88.1	28.22	2.23	53.87	-1.54	2.80	38.63
5	12-Crown-4	176.2	-0.608	3.19	28.68	-6.71	4.00	-13.26
6	15-Crown-5	220.3	-7.98	3.77	-321.8		4.65	25.18
7	18-Crown-6	264.3	-7.08	4.29	-202.6		5.30	102.0

a Cellulose acetate

b Aromatic polyamide

c D values are not set equal to Stokes' law radii - see literature (7).

Table II. Pore Size Distributions of Membranes

Membranes [a]	$\bar{R}_{b,1}\times 10^{10}$ m	$\sigma_1/\bar{R}_{b,1}$	$\bar{R}_{b,2}\times 10^{10}$ m	$\sigma_2/\bar{R}_{b,2}$	h_2
Batch-316-unshrunk	10.37	0.002	50.9	0.196	0.050
Batch-316-67	9.17	0.002	41.9	0.196	0.005
Batch-316-77	8.37	0.002	39.9	0.195	0.003
Batch-316-82	7.47	0.002	35.9	0.195	0.001
Batch-602-unshrunk	10.37	0.002	54.2	0.185	0.060
Batch-602-62.5	9.42	0.005	48.4	0.177	0.019
Batch-602-67.5	9.07	0.004	45.9	0.137	0.010
Batch-602-70.0	7.87	0.003	42.9	0.108	0.006
Batch-602-80.0	6.87	0.002	37.9	0.049	0.002
Batch-18-unshrunk	10.37	0.002	60.8	0.160	0.080
Batch-18-77.5	8.37	0.004	47.9	0.147	0.030
Batch-18-85.0	7.29	0.003	44.9	0.098	0.010
Batch-18-87.5	7.20	0.003	44.2	0.098	0.005
Batch-18-90.0	7.20	0.002	41.9	0.098	0.001

[a] For example, the membrane code Batch-316-67 indicates a membrane from Batch-316 shrunk at 67°C.

acetate membranes (7), and a prolonged period of solvent evaporation in the case of aromatic polyamide hydrazide membranes (8).

Considering the contributions from the steric hindrance and the van der Waals force working between two polymer walls, the distance of which is designated as $\hat{d}$, and that from the entropy decrease induced by the progressive approach of above polymer walls (due to the improved ordering of polymer segments) to the total free energy, ΔG, of the membrane system, a free energy curve as a function of $\hat{d}$ can be obtained as illustrated in Figure 2 by a solid line. Assuming that $\hat{d}$ represents the pore diameter, it was found that the small energy barrier of < 7 kJ/mol, which is depicted in Figure 2 as a shoulder of the ΔG versus $\hat{d}$ curve, separates the pores on the right side of the energy barrier (corresponding to aggregate pores) from those on the left side of the energy barrier (corresponding to network pores). Therefore, the transformation of aggregate pores into network pores require the supply of energy or work from outside in order to overcome the energy barrier. Such an energy is supplied usually in a form of thermal energy. Shrinkage in hot shrinkage media and solvent evaporation are typical examples. However, it can be supplied also in a form of mechanical energy. Thus, stretching membranes sometimes has the same effect as shrinkage, since improved order of polymer segments can be achieved during the stretching procedure.

Experimental Evidences of Bimodal Pore Size Distribution and Its Significance for Membrane Separation and Fractionation Processes

The bimodal pore size distribution can explain and predict different permselectivities demonstrated by membranes from the same polymer material. For example, the separation of lactose was calculated for the corresponding ethanol separation for two series of membranes (Batch-316 and Batch-18) whose pore size distributions are presented in Table II. The difference in these two series of membranes is that $\bar{R}_{b,2}/\bar{R}_{b,1}$ ratio is greater for Batch-18 series than that in Batch-316 series. While h_2 of membranes in both series is progressively decreased, the separations of both solutes under consideration increase. As a result we obtain two distinctive lines of lactose separation versus ethanol separation as shown in Figure 3. Obviously, for a given ethanol separation, membranes of Batch-316 series show much greater lactose separation. This was confirmed by experimental results which are also included in Figure 3 (23). Though the agreement of calculated and experimental data is only fair, probably due to the difference in membrane samples used for the generation of pore size distributions and for RO experiment of lactose and ethanol separations, the tendency of calculated values was very well reproduced by experimental values, viz. the fractionation of lactose and ethanol can be accomplished far better by Batch-316 membranes than by Batch-18 membranes. Thus, the transport equations based on the surface force-pore flow model predict a considerably wide range of separation of a solute from membranes of the same polymer material, while the separation of another solute is fixed at a single value. The above prediction is possible only when bimodal pore size distributions are assumed. A

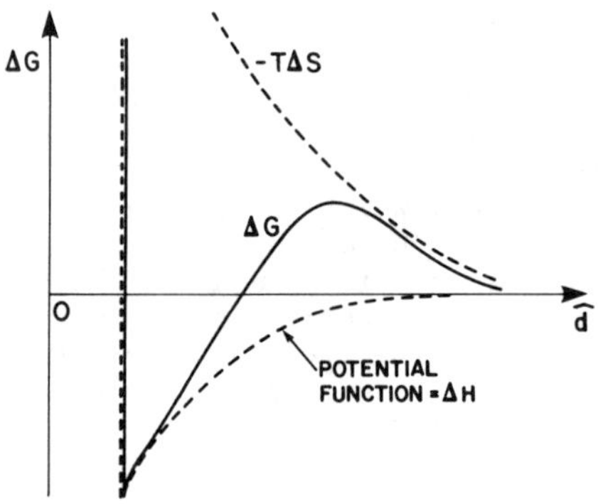

Figure 2. Change in Gibbs'free energy as the function of the distance between two polymer walls $\hat{d}$. (Adapted from Refs. 7 and 8.)

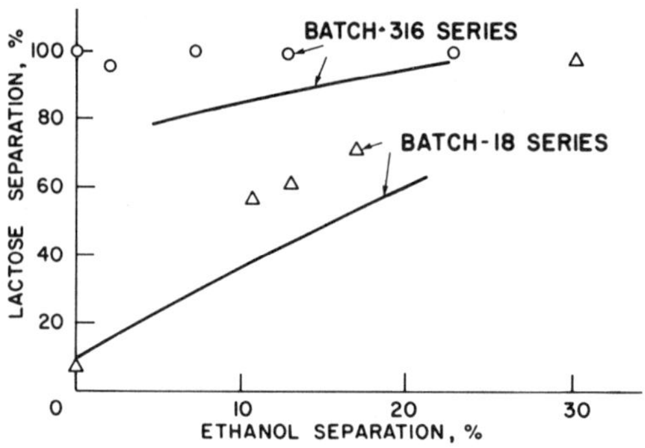

Figure 3. Separation of lactose versus separation of ethanol membranes, cellulose acetate Batch-316 series and Batch-18 series; operating pressure, 1724 kPag (= 250 psig); feed flow rate, 400 cm^3/min; solute concentration in feed 100 ppm. Key: ○ and △, experimental; and ——, calculated.

uniform pore size distribution and a single normal distribution do not enable such predictions (2,24).

Conclusions

There have been indications of the bimodal pore size distributions in the literature. Glückauf and Sammon have given a concept of two types of pores on the basis of their NaCℓ separation data in an amazingly similar fashion as our description (25). Lonsdale suggested the existence of "imperfection" in the membrane on the basis of difference in salt permeability coefficients observed in immersion and in reverse- and direct-osmosis experiments, though he assigned an order of microns to such "imperfect" pores (26). Paner et al. (11) showed the possibility of salt passage through the gap caused by "the imperfect ordering of polymer-micelles (same as our polymer aggregates)" at the membrane surface. In our approach we do not regard the pores of larger sizes on the membrane surface as "imperfections" or as "defects"; instead, we consider them "natural". Thus, our attempt to discuss the origin of bimodal distributions, to relate the pore size distribution to the membrane casting conditions, and to predict the membrane performance data such as separation, fractionation and product permeation rate by applying appropriate transport equations to such pore size distributions, is a new approach to the materials science of RO/UF membranes. A variety of pore size distributions, as described in this work, when combined with a variety of membrane materials which affect the interfacial interaction force in the membrane material-solvent-solute systems enables us to obtain an enormous variety of membrane performance data, and more importantly, such possibility is predictable by the approach presented.

Nomenclature

$\underset{\sim}{B}$ = constant characterizing the van der Waals attraction force, m^3

$c_{A,b}$ = bulk solute concentration, mol/m^3

$\underset{\sim}{D}$ = constant characterizing the steric repulsion at the interface, m

d_s = the distance between two polymer segments, m

$\underline{d}$ = the distance from polymer material surface and the center of solute molecule, m

$\widehat{d}$ = the distance between polymer walls, equivalent to pore diameter, m

E = Young's modulus, N/m^2

f_j = fraction solute separation based on the feed concentation with respect to the jth solute

ΔG = change in Gibbs free energy, energy/mass

ΔG_{cube} = ΔG associated with the splitting of a large vacant cube, J/kg

ΔG_{square} = ΔG associated with the splitting of a large vacant square, J/kg

ΔG_{disp} = ΔG associated with the dispersion of water molecules in a polymer aggregate, J/weight of a polymer aggregate

ΔH = change in enthalpy, energy/mass

ΔH_{cube} = ΔH associated with the splitting of a large vacant cube, J/kg

$\Delta H_{interface}$ = ΔH associated with the generation of interface at the water-polymer boundary, J/weight of a polymer aggregate

$\Delta H_{stretch}$ = ΔH associated with the stretching of polymer segments, J/weight of a polymer aggregate

h_i = ratio defined by Equation 19

$\underset{\sim}{k}$ = the Boltzmann constant

L = the length of the edge of a unit cube, m

L_o = the initial length of a polymer segment involved in the noncrystalline region, m

ΔL = amount of the stretching of a polymer segment, m

ℓ^3 = number of vacant unit cubes

m^3 = number of total unit cubes

n_i = number of pores belonging to the ith normal distribution

R_b = pore radius, m

$\bar{R}_b$ = average pore radius, m

$\bar{R}_{b,i}$ = average pore radius of the ith distribution, m

$\underset{\sim}{R}$ = gas constant

r_{water} = Stokes' law radius of water molecule (0.87×10^{-10} m)

S = the cross-sectional area of the polymer segment, m^2

ΔS = change in entropy, energy/mass•temperature

ΔS_{cube} = ΔS associated with the splitting of a large vacant cube, J/kg•K

ΔS_{disp} = ΔS associated with the dispersion of water molecule in a polymer aggregate, J/weight of a polymer aggregate•K

T = absolute temperature, K

w_1, w_2 = quantities defined by Equations 1 and 2, respectively

$Y_i(R_b)$ = normal pore size distribution function, 1/m

Greek Letters

Γ_A = surface excess of the solute A, mol/m^2

Λ = the length of the side of a unit square, m

λ^2 = number of vacant unit squares

$(2\mu)^2$ = number of total unit squares

N = number of water molecules in a polymer aggregate

ν = number of water molecules in a network pore

σ = surface tension, N/m

σ_i = standard deviation of the ith normal pore size distribution, m

τ = thickness of a flat polymer aggregate, m

ϕ = potential function of interaction force exerted on the solute from the pore wall, J/mol

Acknowledgments

This paper was issued as NRC No. 24033.

Literature Cited

1. Sourirajan, S. "Reverse Osmosis", Academic: New York, 1970; Chap. 1.

2. Matsuura, T.; Sourirajan, S. Ind. Eng. Chem. Process Des. Dev. 1981, 20, 273.

3. Matsuura, T.; Taketani, Y.; Sourirajan, S. "Synthetic Membranes" Vol. II, Turbak, A.F., Ed.; ACS Symp. Ser. 154, 1981; p. 315.

4. Matsuura, T.; Sourirajan, S. J. Colloid Interface Sci. 1978, 66, 589.

5. Chan, K.; Matsuura, T.; Sourirajan, S. Ind. Eng. Chem. Prod. Res. Dev. 1982, 21, 605.

6. Liu, T.; Chan, K.; Matsuura, T.; Sourirajan, S. Ind. Eng. Chem. Prod. Res. Dev. 1984, 23, 116.

7. Chan, K.; Liu, T.; Matsuura, T.; Sourirajan, S. Ind. Eng. Chem. Prod. Res. Dev. 1984, 23, 124.

8. Chan, K.; Matsuura, T.; Sourirajan, S. Ind. Eng. Chem. Prod. Res. Dev. 1984, 23, 492.

9. Nguyen, T.D.; Chan, K.; Matsuura, T.; Sourirajan, S. Ind. Eng. Chem. Prod. Res. Dev. 1984, 23, 501.

10. Sourirajan, S. "Lectures on Reverse Osmosis" National Research Council of Canada: Ottawa, 1983; (a) Lecture 4, (b) Lecture 6.

11. Panar, M.; Hoehn, M.H.; Hebert, R.R. Macromolecules 1973, 6, 777.

12. Kunst, B.; Sourirajan, S. J. Appl. Polym. Sci. 1970, 14, 1983.

13. Kunst, B.; Sourirajan, S. Desalination 1970, 8, 139.

14. Nguyen, T.D.; Chan, K.; Matsuura, T.; Sourirajan, S. Unpublished data.

15. Matsuura, T.; Blais, P.; Dickson, J.M.; Sourirajan, S. J. Appl. Polym. Sci. 1974, 18, 3671.

16. Gan, L.H.; Blais, P.; Carlsson, D.J.; Sprunchuk, J.; Wiles, D.M. J. Appl. Polym. Sci. 1975, 19, 69.

17. Rudin, A.; Johnston, H.K. J. Paint Technol. 1971, 43, 39.

18. Kesting, R.E.; Barsh, M.K.; Vincent, A.L. J. Appl. Polym. Sci. 1965, 9, 1876.

19. Strathmann, H.; Scheible, P.; Baker, R.W. J. Appl. Polym. Sci. 1971, 15, 811.

20. Matsuura, T.; Taketani, Y.; Sourirajan, S. J. Colloid Interface Sci. 1983, 95, 10.

21. Sprague, B.S.; Singleton, R.W. Text. Res. J. 1965, 35, 999.

22. Herlinger, V.H.; Hörner, H.-P.; Druschke, F.; Knöll, H.; Haiber, F. Angew. Makromolek. Chem. 1973, 29/30, 229.

23. Matsuura, T.; Sourirajan, S. "On the Predictability of RO/UF Performance from Interfacial Interaction Forces and Membrane Pore Structure", paper presented at the Symposium on Membrane Processes for Water Reuse and Material Recovery in the AIChE Annual Meeting (Diamond Jubilee), Oct. 30-Nov. 4, 1983, Washington, D.C.

24. Matsuura, T.; Tweddle, T.A.; Sourirajan, S. Ind. Eng. Chem. Process Des. Dev. 1984, 23, 674.

25. Glückauf, E.; Sammon, D.C. "Transport of Ions and Water through Cellulose Acetate Membranes" Proceedings 3rd International Symposium on Fresh Water from the Sea, Delyannis, A. and Delyannis, E., Eds.; Vol. 2: Athens, 1970; p. 397.

26. Lonsdale, H.K. "Desalination by Reverse Osmosis" Merten, U., Ed.; The M.I.T. Press: Cambridge, Massachusetts, 1966, pp. 93-160.

RECEIVED February 22, 1985

2

Development of Polyether Sulfone Ultrafiltration Membranes

M. KAI, K. ISHII, H. TSUGAYA, and T. MIYANO

Daicel Chemical Industries, Ltd., 1 Teppo-cho, Sakai-shi 590, Japan

A parameter was introduced successfully to collectively describe the effect of casting conditions on the performance of polyethersulfone ultrafiltration membranes. This parameter is related to the amount of water to be absorbed by as cast solution during evaporation period. Another parameter, Coagulation Value which corresponds to the amount of water necessary to start the phase separation of casting solution clearly elucidated the role of the additives. Between these independent parameters was found a relationship which extends the applicable range of solvents, additives and casting conditions. On the basis of this study, commercial ultrafiltration membranes thermally more stable than conventional polysulfone membranes have been developed.

Increasing demands for mechanically strong and stable UF membranes which should stand strong chemicals and high temperature resulted in the development and commercial production of Udel polysulfone (PSF) membranes by several manufactures. In contrast to many studies on PSF membranes (1,2) which are not stable enough for heat sterilization at a temperature higher than 100°C, little work (3) has been reported on UF membranes made from Victrex polyethersulfone (PES) which should be more promising than PSF due to its higher thermal stability. Their chemical structures are as follows.
Polyethersulfone (PES)

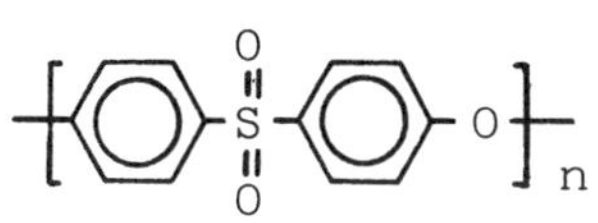

Polysulfone (PSF)

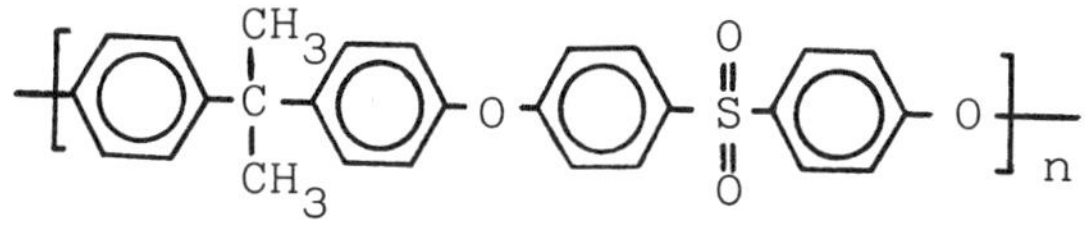

0097-6156/85/0281-0021$06.00/0

In this paper, the effect of parameters on membrane performance in membrane fabrication by phase inversion method, such as casting solution composition and casting conditions, is discussed with emphasis on the characteristic importance of water.

Experimental

Polymers. Commercially available polymers, Victrex 300P, one of PES polymers supplied by ICI and Udel P-1700, PSF by UCC were used without further purification after the purchase.

Glass Transition Temperature. The glass transition temperatures (Tg) of the polymers were measured by a torsional braid analyzer, Model 100-B1, Chemical Instruments Corp., under following conditions; pulse interval, 5 minutes and heating rate, 1 °C/min. The effect of glycerine absorbed by the polymers on Tg was measured with the samples soaked overnight in 50 % glycerine aqueous solution and then dried.

Preparation of Membranes. Membranes were prepared by the following process unless otherwise mentioned. The polymer was completely dissolved with a solvent or a mixture of solvent and non-solvent to obtain a homogeneous solution (dope). The dope was cast 180 microns thick on a glass plate. The cast dope was allowed to stand for 30 seconds. The casting atmosphere was controlled at 25 °C and 55 % RH, and the velocity of air stream over the cast dope, below 20 cm/min. The cast dope was immersed into a water bath kept at 17 °C. The membrane obtained was left in the water bath over 30 minutes. The residual solvent was leached out of the membrane completely in a hot water bath at 90°C for 15 minutes.

Evaluation of Membranes. Membranes were evaluated by use of thin channel flat cells with 200 microns channel height and 25 cm^2 effective membrane area and at the mean flow velocity of 1.3 X 10^2 cm/sec. Pure water and 0.2 % ovalbumin aqueous solution were fed under 0.5 Kg/cm^2 and at 25 °C. The concentration of the solute was determined by high performance GPC.

Coagulation Value. The coagulation value (C.V.) of a dope, a parameter which represents the resistance to yield polymer precipitate from the dope by the addition of water, was measured as follows. The test solution was prepared in the same way to make the corresponding dope, except only 2g of polymer was dissolved in 98g of the same liquid composition. Water was added dropwise like titration into the test solution kept stirred at 30 °C. The amount of water necessary to yield the first precipitate refers to the C.V. of the corresponding dope.

Results and Discussion

Thermal Stability of Polymers. It is well known that the thermal stability of amorphous polymer can be estimated by Tg in greneral. This criterion alone, however, is not enough for membrane, due to its highly fine structure and strong interaction with water. For instance, hydrophobic PSF UF membranes are thermally more stable than relatively hydrophilic cellulose acetate UF membranes, in spite of their almost the same Tgs observed of solid polymers.

The experimental results given in Table I show that the Tgs of cellulose acetates were lowered 40-50 °C by the glycerine treatment, while that of PSF decreased only 15°C. This observation gives a likely account to the above mentioned difference. Judging from its high Tg and insensitivity to the glycerine treatment, PES should be more promising than PSF to obtain thermally stable UF membranes.

Table I. Thermal Stability of Some Polymers (4-6)

Polymer		Water *1 Absorption (%)	Observed Tg (°C)	
			non-treatment	glycerine treatment
Hydrophilic Polymers	CDA *2	1.7-6.5	195	145-155
	CTA *3		195-205	150-160
Hydrophobic Polymers	PSF	0.3	190	175
	PES	0.43	225	225

*1 ASTM, D570 *2 Cellulose diacetate *3 Cellulose triacetate

Effect of Dope Composition. In order to determine an optimal solvent for PES and PSF, several high boiling and water miscible solvents were examined in binary dopes.

As shown in Table II, no water flux was measured of the membranes prepared with other solvents than 2-pyrrolidone (2-PN). Observation by scanning electron microscopy (SEM) revealed that these unproductive membranes had porous structure but with dense skin on the bottom surface in addition to top skin.

Table II. The Performance of the Membranes Prepared from Binary Dopes (polymer concentration, 20 wt%)

Solvents	PES		PSF	
	Flux (m^3/m^2 day)	Rejection (%)	Flux (m^3/m^2 day)	Rejection (%)
N-methyl-2-pyrrolidone	0	-	0	-
2-pyrrolidone	0.92	100	1.15	79
N,N-dimethyl-formamide	0	-	0	-
N,N-dimethyl-acetamide	0	-	0	-

In order to see whether the bottom skin prevented the water permeation, this layer was scraped off the membrane prepared with N-methyl-2-pyrrolidone (NMP). This pealed membrane, the cross-section of which is shown in Figure 1-A, exhibited water flux and solute rejection. On the other hand, the removal of the top skin showed no effect at all. The mechanism of the formation of these non-permeable bottom skin is not clear, but specific interaction between the dope and the glass plate (7), as well as the gelation mechanism, should be taken into consideration.

According to well-known procedure to improve the membrane performance, ternary systems with a solvent and an additive were examined. As a solvent, NMP and 2-PN were used; as an additive, all the compounds listed in Table III were tried and the concentration ranged from 5 to 20 wt% according to the nature of the additive.

Table III. Water Miscible Organic Additives

Acids	Acetic Acid, Propionic Acid, Lactic Acid
Esters	Ethyl Lactate, Triethyl Phosphate
Amide	N-Methylformamide
Amine	Pyridine
Ketone	Cyclohexanone

An important aspect of the additives' role is the effect to increase the sensitivity of the dope to imbibed water, that is, to decrease C.V.. Among many parameters proposed to elucidate the effect of additives, C.V. was chosen, because this parameter well represents the sensitivity of each dope to intruded water.

The effect of the additives to change the unproductive NMP binary dopes into productive ones is clearly shown in Figure 2, wherein pure water flux under 1 kg/cm^2 (PWF/P) is plotted against C.V.. PES and PSF membranes were obtained from NMP ternary dopes of which polymer concentration was 17 wt%. Common to both polymers, PWF/P takes off from zero at critical C.V. somewhat below that of the original binary dope and increases with decrease in C.V. regardless of the kind of additives. Both curves show that the more sensitive to the intruded water the dope is, the higher the PWF/P of the membrane is.

In Figure 3 similar relationship can be seen for the PES membranes obtained from the ternary dopes composed of PES (20 wt%), an additive and 2-PN, which gave productive membrane from the binary dope among the solvents examined (Table II).

Effect of Casting Conditions. In order to improve the membrane performance, the effect of casting conditions were studied with PES/2-PN binary dope. As 2-PN is not volatile, the standing time (evaporation time) was not expected to have significant influence on the membrane performance. The outcome, however, was more than a simple effect as seen in Figure 4, where the performance was shown as the function of the standing time.

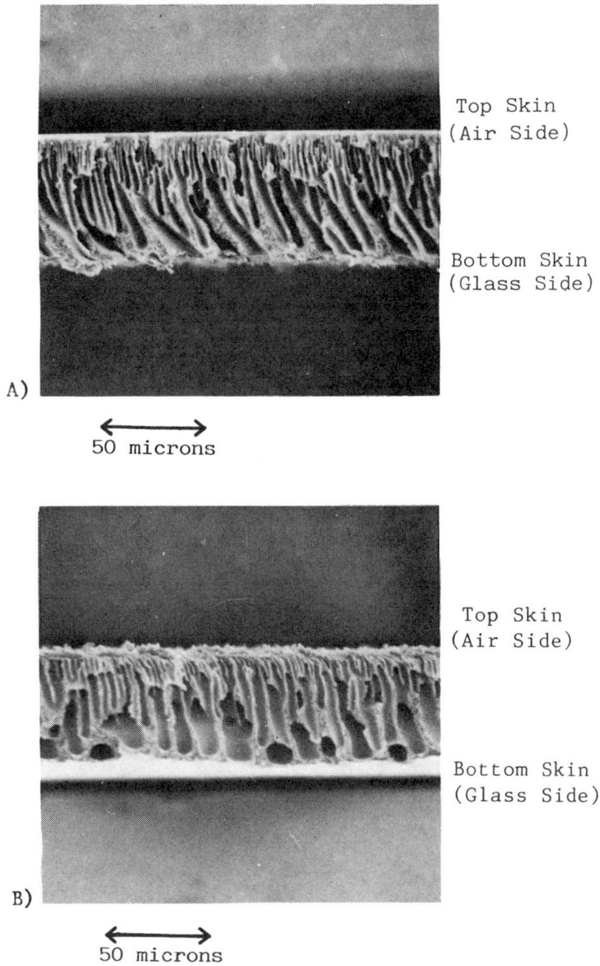

Figure 1. The SEM pictures of the cross-sections after the scraping treatment. A) The bottom skin was removed. B) The top skin was removed.

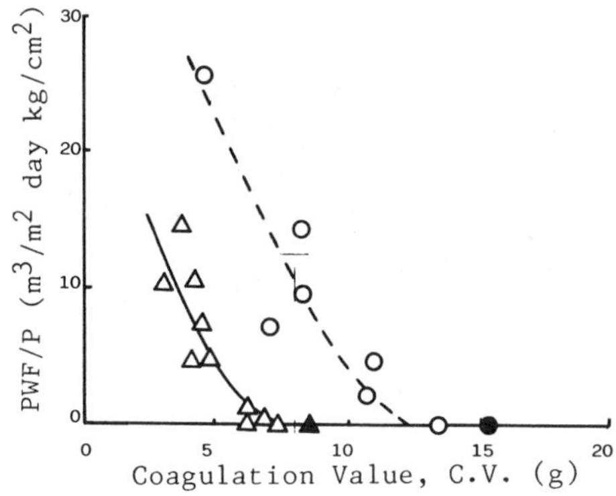

Figure 2. The performance of PES and PSF membranes obtained from NMP ternary dopes (polymer concentration, 17 wt%): ○, PES; △, PSF. Dark points represent binary dopes.

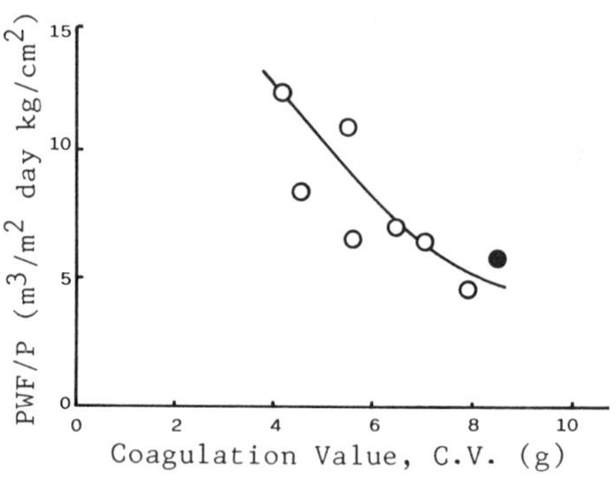

Figure 3. The performance of PES membranes obtained from 2-PN ternary dopes (polymer concentration, 20 wt%). Dark point represents binary dope.

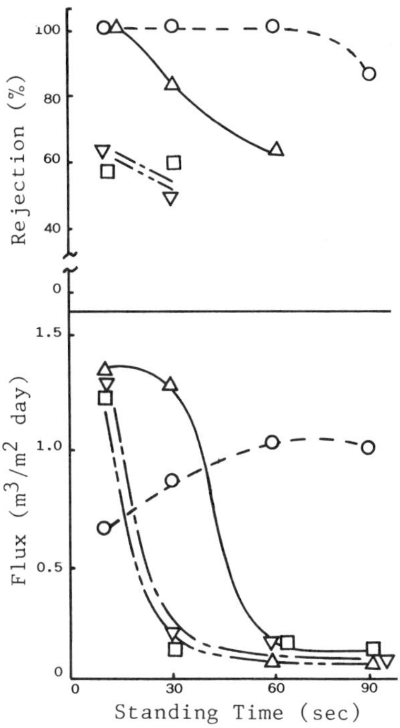

Figure 4. The effect of casting conditions. The binary dope composed of 20 wt% PES and 2-PN was cast on PET paper. The velocities of air stream over cast dopes were, ○ =20, △ =4.1X10^3, ▽=1.26X10^4 and □=1.76X10^4 cm/min. The humidity was 75 % RH.

In order to elucidate these complicate results, the effect of the water absorbed from air by the cast dope should be taken into consideration, because the dopes were very sensitive to water (8). The amount of the water to be imbibed during the standing time was calculated as follows.

The absorption rate of water from the air to the unit area of the cast dope (dW/dt, in g/cm^2 min) is proportional to the concentration of water in the air stream (h, in g/cm^3)

$$dW/dt = k \cdot h \tag{1}$$

where k (cm/min) is the overall coefficient of mass transfer. Equation 1 is integrated to obtain

$$W = k \cdot h \cdot t + Wo \tag{2}$$

where Wo (g/cm^2) is the amount of water present in the dope before casting. The effect of the velocity of the air stream over the cast dope (U, in cm/min) can be written as follows.

$$k = F \cdot U^a \tag{3}$$

where exponent a is 0.5 for laminar flow, and 0.8 for turbulent flow and F is a constant which mainly depends on the apparatus. Replacing h by the relative humidity (hr) and the saturated vapor pressure of water (Ps, in mmHg) and combining Equations 2 and 3, we obtain

$$W = F \cdot Ps \cdot hr \cdot U^a \cdot t + Wo$$

$$= F \cdot G + Wo \tag{4}$$

where G (Gelation Factor) $= Ps \cdot hr \cdot U^a \cdot t$ (5)

Equation 4 represents the amount of the water contained in the unit surface layer of the cast dope at the moment of the immersion as the sum of the water contained before casting and that imbibed during the standing time. The latter is given by the product of two parameters F and G. Although their physical meaning can not be clearly defined, G is the proper product of all the principal measurable variables. So, if the dope and the apparatus are given, the effect of different casting conditions should be able to be estimated by G.

As shown in Figure 5, four curves in Figure 4 converged into one curve by use of G. This curve covered the data taken under the casting conditions varied over wide range of 40-80 % RH for hr, 0.17-5.0 minutes for t, and 20-1.26 x 10^4 cm/min for U, as well as the data shown in Figure 4.

The curve in Figure 5 may be divided in three regions. In region A membrane performances do not change with G; in region B water flux increases but rejection decreases; in region C water flux decreases steeply to zero with increase in G. This nature of the curve might be explained as follows.

In region A, the amount of the absorbed water is too little to influence the membrane performance. Further imbibition of

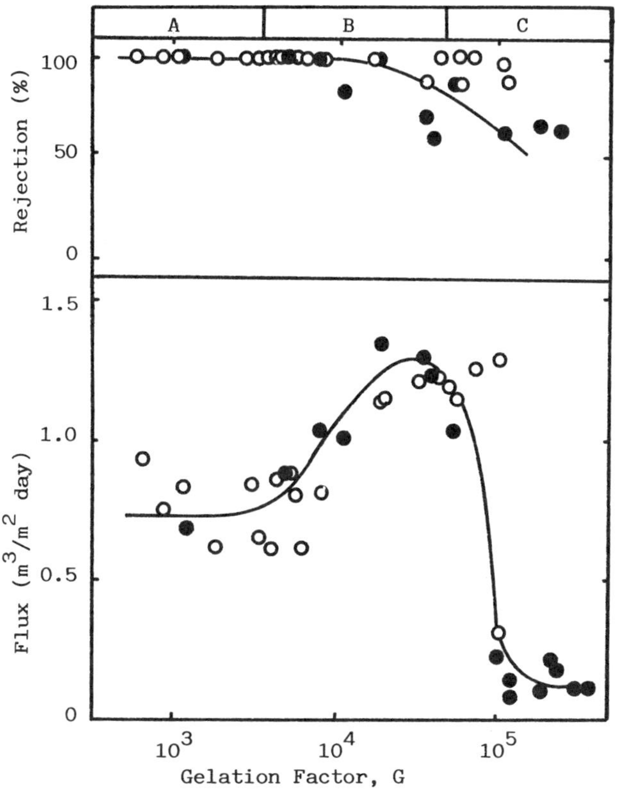

Figure 5. The relationship between G and membrane performance. Dark points represent the data shown in Figure 4.

water beyond region A advances the phase separation in top surface and grows the pores greater. This results in the increase of water flux and the decrease of solute rejection. In region C, further advancement in the phase separation breaks the finely pored structure and forms impermeable layer leaving much less but large-sized pores.

The introduction of G, through which the effects of independent variables of casting conditions can be converted each other, and the curve, which represents the relationship between membrane performance and G for a given dope, made it easier to set up optimal casting conditions.

Relationship between C.V. and G. A ternary dope was prepared from PES, 2-PN and lactic acid. From this dope, C.V. of which was smaller than that of the binary dope, membranes were made under various casting conditions, that is, various G values. The plotting of water flux vs. G for these membranes formed a curve which had similar shape to that for the binary dope membranes and was shifted in the direction to decrease G as shown in Figure 6.

This suggests that the effect of the decrease in C.V. on membrane performance can be obtained by the increase in G controlled independently of C.V.. In order to see the relationship between the effects of C.V. and G, G values at specific points of the curves, the border of region B and region C for instance, were plotted against C.V.s of the corresponding curves. As shown in Figure 7, a linear relationship was found between C.V. and log G. If one wants to obtain the membranes of the same performance, that at the border of region B and C for instance, from different dopes, the effect of the difference in C.V. can be compensated by the corresponding change in G given by the line in this figure.

The utilization of the relationship described above extends the range of choice of the additives and the casting conditions to achieve superior performance.

Thermal Stability of PES UF membrane

On the basis of the study described above, commercial PES UF membranes, Molsep DUS tubular and flat membranes have been developed successfully.

The separation performance of DUS-40 is characterized by the sharp cutoff at 4×10^4 Dalton as shown in Figure 8.

The thermal stability of DUS-40 is shown in Figure 9, in comparison with that of PSF membrane which has nearly the same cutoff point as DUS-40. DUS-40 kept its performance uninfluenced up to 130 °C, in contrast with a PSF membrane of which performance changed remarkably after thermal treatment over 100°C.

Conclusion

Commercial polyethersulfone UF membranes thermally more stable than conventional polysulfone membranes were developed through:

1. The application of 2-PN as the solvent.
2. The control of the sensitivity of the casting dope to water by the proper choise of additives.
3. The control of the casting conditions by use of a newly

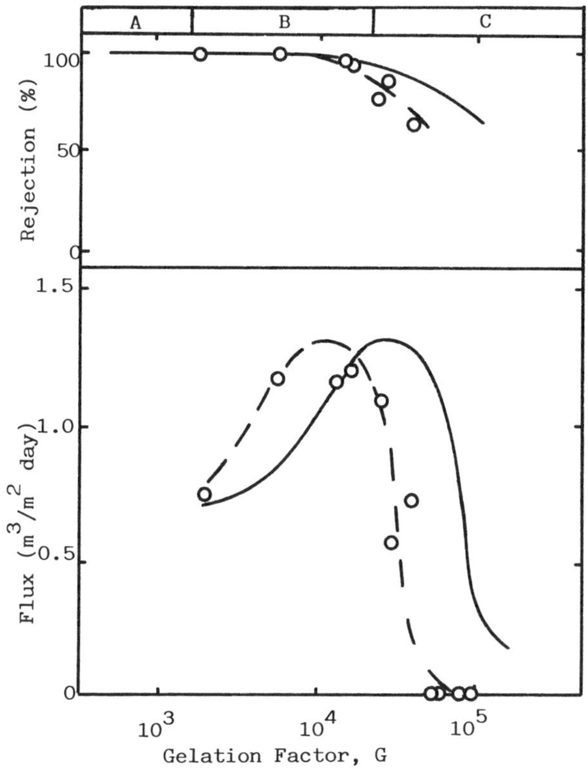

Figure 6. The relationship between G and the performance of membranes made from a ternary dope (20 wt% PES, 6 wt% lactic acid and 74 wt% 2-PN) in comparision with that for the binary dope membranes. Points and broken line indicate the ternary dope membranes. Solid line indicates the binary dope shown in Figure 5.

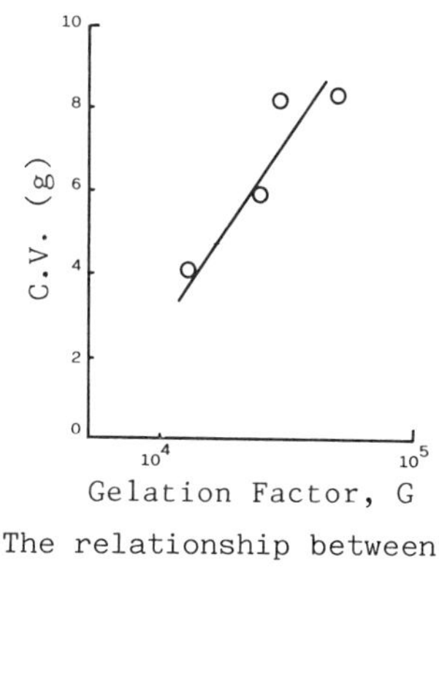

Figure 7. The relationship between C.V. and G.

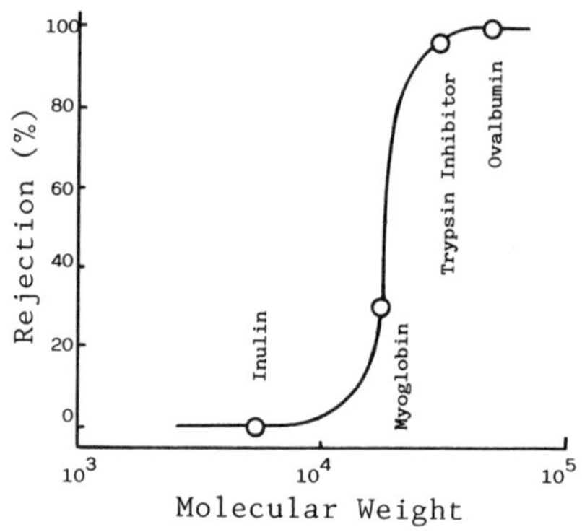

Figure 8. The separation performance of DUS-40.

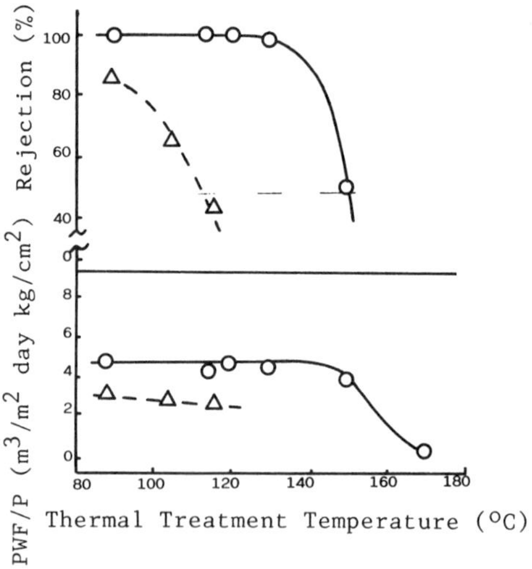

Figure 9. The thermal stability of DUS-40: ○, DUS-40; △, a UF membrane made from PSF. Membranes were immersed into a hot water bath in an autoclave for 30 min.

introduced parameter which corresponds to the amount of the water absorbed by the cast membrane during the evaporation time.

Literature Cited

1. Nishimura, M.; Muro, T.; Tsujisaka, Y. Kobunshi Ronbunshu 1977, 34, 713-8. (an example for flat membranes)
2. Cabasso, I.; Klein, E.; Smith, J. K. J. Appl. Polym. Sci. 1977, 21, 165-180. (an example for hollow fibers)
3. Tweddle, T. A.; Kutowy, O.; Thayer, W. L.; Sourirajan, S. Ind. Eng. Chem. Prod. Res. Dev. 1983, 22, 320-6.
4. Union Carbide Corp., Product Data, 1981.
5. Sumitomo Chemical Industries, Ltd., Technical Service Information Note, 1979.
6. Daicel Chemical Industries, Ltd., Technical Bulletin.
7. Tsugaya, H.; Miyano, T. Japanese Patent Kokai 57-207505.
8. Kamizawa, C.; Matsuda, M.; Masuda, H.; Tanaka, H. Koubunshi Ronbunshu 1983, 40, 401-4.

RECEIVED February 22, 1985

Nature of Dynamically Formed Ultrafiltration Membranes

SHOJI KIMURA[1], TOSHIRO OHTANI[2], and ATSUO WATANABE[2]

[1] Institute of Industrial Science, University of Tokyo, 7-22-1 Roppongi, Minatoku, Tokyo 106, Japan

[2] Ministry of Agriculture, Fishery, and Forestry, National Food Research Institute, 2-1-2 Kan-nondai, Yatabe, Tsukuba Ibaragi 305, Japan

Ultrafiltration membranes were made by filtering various high molecular weight solutes and suspended solids on a porous ceramic tube surface, which was originally used for dynamic reverse osmosis membranes. Nature of the dynamically formed ovalbumin membrane depends on concentration and pH of the solution, but generally the rejection of albumin itself is high and the flux is significantly larger than commercial membranes. Membranes formed can easily be removed by washing with NaClO. Since the ovalbumin membrane can also reject other solutes, its molecular weight cut off value was determined. Membranes made of dextran do not show such good results. Modules made of porous ceramic tubes were developed and tested using various solutions, whose components formed self-rejecting membranes. Results of secondary sewage and pulp factory effluents are shown.

A dynamically formed membrane has been applied for various effluent treatments since it was discovered by ORNL (1), because the formation and the cleaning procedures are rather simple compared to polymeric membranes. Materials used for dynamic membranes were Zr colloid which was later supplemented by polyacrylic acid (2). The latter membrane is now called ZOPA membrane. The development of the modular form of these membranes and results of various applications have been reported by D.G. Thomas (3), by Carre Inc. and Clemson University (4), by University of Natal (5), and by authors (6,7).

Applications of dynamic membranes to various ultrafiltration processes are very interesting, because they can be used at high temperatures, formed and cleaned very easily. But in the actual ultrafiltration process the gel layer formed on membranes controls the flux and rejection primarily and natures of original membranes become obscure. In this regard the gel layer can be considered as a kind of a dynamically formed membrane, which may be applicable for practical applications.

0097-6156/85/0281-0035$06.00/0

In this investigation natures of various gel layers formed on a bare porous ceramic tube, which was originally used as a support of ZOPA membrane, are studied as those of a dynamic membrane. The first part of this report deals with various natures of dynamically formed ovalbumin and dextran membranes. It is shown that the former membrane is self-rejecting and can also reject other solutes, but at the same time the concentration polarization effect is significant. The latter membrane is not self-rejecting and the flux decrease is due to the osmotic pressure effect. The second part deals with practical applications of Dynaceram module, which has been scaled up from a single ceramic tube such as the one used above to a module which contains 151 tubes. Results of treating an activated sludge and kraft pulp effluents are shown.

Formation of ovalbumin membranes

The experimental apparatus used is shown in Figure 1. A ceramic tube, whose O.D. is 10 mm and length 30 cm, is used as a supporting substrate of dynamically formed membranes. This ceramic tube is made of alumina and its pore sizes are from 0.5 to 1.5 microns. The surface layer is made of extremely fine particles with an average size of 0.05 microns produced by a special treatment. When aqueous ovalbumin solutions of various concentrations were fed over the ceramic tube under pressure, flux started to decrease and rejection of ovalbumin itself was increased by the gel layer formation. Both flux and rejection reached steady values after about an hour. These results are shown in Figure 2, where it is found that steady fluxes are dependent on the albumin concentration and rejections are almost over 90%. Experimental conditions in this case are as follows: temperature: 25°C, pressure: 5 atm, and flow rate: 5L/min.

The same procedure was applied using dextran as a solute to form membranes. But as it is shown in Figure 3, rejection did not increase, although flux declined considerably. This is considered to show that a gel layer of dextran was not formed and the flux decline was due to the osmotic pressure of dextran (8). A clear distinction between a gel layer formation mechanism and a osmotic pressure mechanism is difficult to define at this stage, but these two figures clearly show the difference of these two mechanisms.

Rejections of various solutes

Ovalbumin membranes, made from 800 and 5,000 ppm albumin solutions, were used to measure rejections of various molecular weight solutes, such as PEG and dextran, which are listed in Table I. Solute concentrations were kept as low as 200 ppm, so that no gel layer formation occurred. A membrane formed from 800 ppm solution had a large flux value of 30-35 10^{-3} kg/m^2 s, but observed rejections were as low as about 20% for solutes whose molecular weights were less than 100,000. A membrane from 5,000 ppm solution had about a half of the above flux but rejections were about 50-60%. These observed rejections were corrected to give true rejections by eliminating the effect of concentration polarization. The correction procedure is exlained later. Results are plotted in Figure 4, where it is found that true rejections of both membranes are almost the same. So it

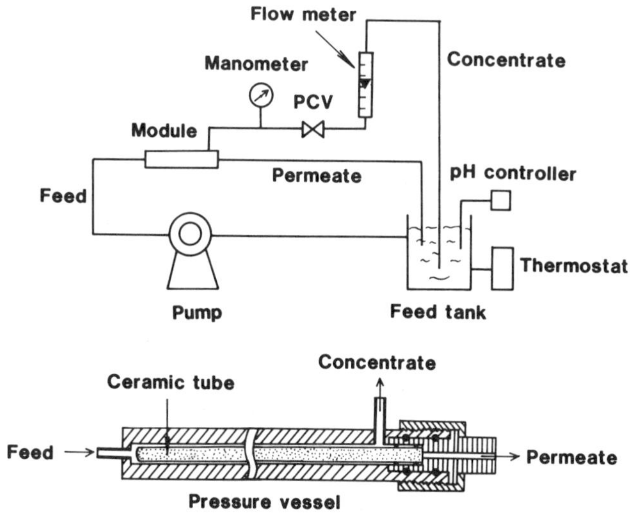

Figure 1. Experimental apparatus and a test cell.

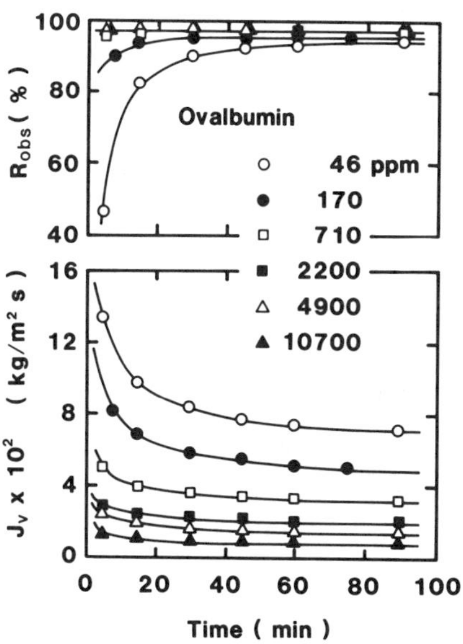

Figure 2. Dynamically formed ovalbumin membranes; gel layers rejecting ovalbumin.

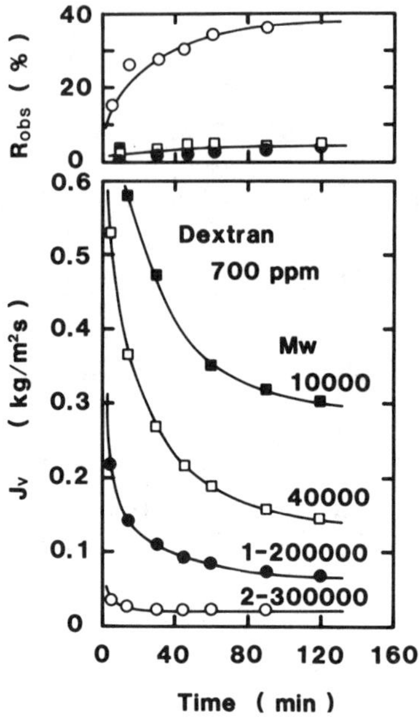

Figure 3. Dynamically formed dextran membranes; gel layers are not formed and dextran is not rejected.

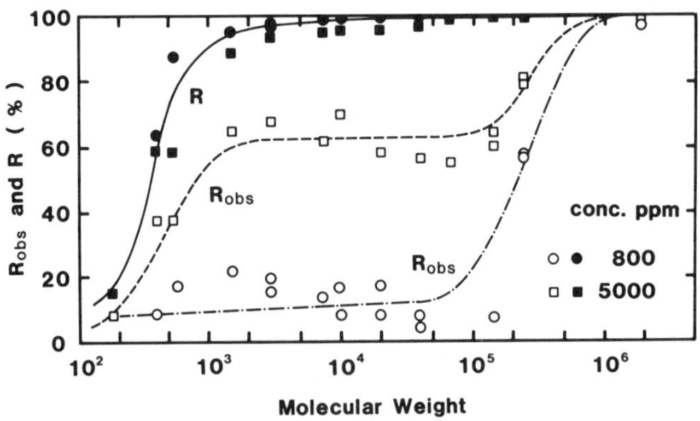

Figure 4. Rejections of various solutes by dynamically formed ovalbumin membranes.

becomes clear that low observed rejections are due to the concentration polarization effect.

Correction of concentration polarization effect

The experimentally observed rejection, R_{obs}, and the true rejection, R, are defined as

$$R_{obs} = (C_1 - C_3)/C_1 \tag{1}$$

$$R = (C_2 - C_3)/C_2 \tag{2}$$

where C_1, C_2, and C_3 are, respectively, the concentration in the bulk, at the membrane surface and in the permeate. The relation between C_1 and C_2 is given by Equation 3 as

$$(C_2 - C_3)/(C_1 - C_3) = \exp(J_v/k) \tag{3}$$

where J_v is a volume flux.

To correct the effect of concentration polarization and to obtain the true rejection from data of observed rejections using Equation 3, mass transfer coefficients, k, are needed and they can be obtained by the velocity variation method (9). But results first obtained in this experiment showed a wide scattering of data, which is considered to be both due to the uncertainty of diffusion coefficients reported in literatures, and to the wide molecular weight distribution of solutes. Since the mass transfer coefficient obtained from the reverse osmosis experiment can be applicable for the ultrafiltration case also, the velocity variation method was performed by measuring NaCl rejection using a Zr(IV)-polyacrylic acid dynamic membrane formed on the same ceramic tube. This result is shown in Figure 5, where it is seen that Leveque equation can be used. The latter equation is given as

$$N_{Sh} = 1.62(N_{Re} \cdot N_{Sc} \cdot d/l)^{1/3} \tag{4}$$

and is a typical mass transfer correlation for a laminar flow inside a tube.

In the case of ultrafiltration N_{Sh} becomes larger than the values obtained from Equation 4 for dextran. This led us to measure diffusion coefficients by a ultracentrifuge method. Results are given in Table I, where it is seen that measured values are larger than literature values. Using measured values it is proved that Leveque equation is also valid for ultrafiltration, as is shown in Figure 6. This correlation and diffusion coefficients were used for the correction of data shown in Figure 4.

Effect of pH on natures of ovalbumin membranes

As an isoelectric point of ovalbumin is at pH 4.5, the nature of membranes may be changed by changing pH of a solution before and after the dynamic membrane formation. Results are shown in Figure 7, where it is seen that the flux reaches minimum near the isoelectric point in both cases, although there are some

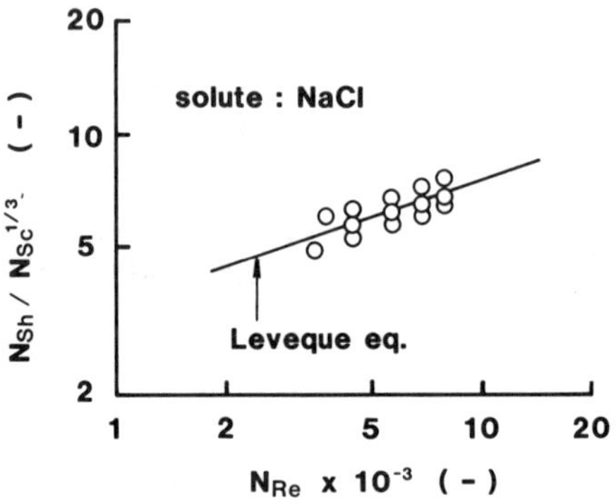

Figure 5. Correlation of mass transfer coefficients obtained by NaCl rejection using Zr(IV)-PAA dynamic membranes.

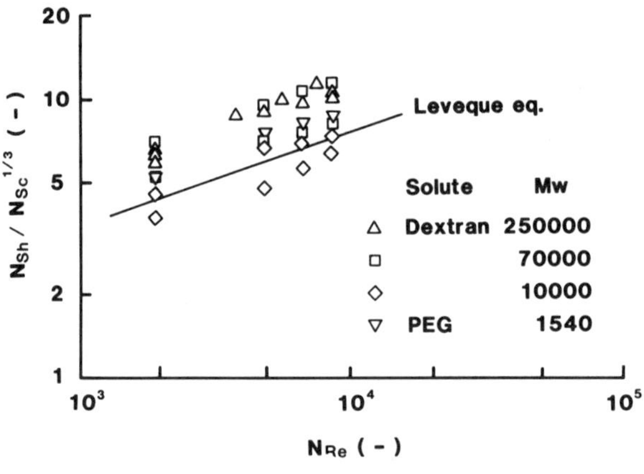

Figure 6. Correlation of mass transfer coefficients obtained by dextran and PEG using ovalbumin membranes.

discrepancies of data at large pH. This shows the membrane formed at pH 5-6 is the most dense one.

Table I. Molecular Weight and Diffusion Coefficients of Solutes

Solute	Molecular weight	D × 10^7 (cm^2/s)	Reference
Glucose	180	69	(10)
PEG #400	400	45	(10)
Raffinose	504	42	(10)
PEG #1540	1,500	23	(10)
PEG #4000	3,000	19	(11)
PEG #6000	7,500	14	(11)
Dextran	10,000	13	(11)
PEG #20000	20,000	10	(11)
Dextran	40,000	8.0	(11)
Dextran	70,000	5.8*	(3.9)(12)
Dextran	150,000	5.4*	(2.6)(12)
Dextran	250,000	6.2*	(2.0)(12)

*) measured in this work by ultracentrifuge.

Next, membranes formed at various pH values were used to measure rejections of various solutes listed above. Results are shown in Figure 8, where the general trend is similar to Figure 4. Though the membrane having a larger flux gives lower rejections and vice versa, true rejections of membranes remain the same at a very high level, and the molecular weight cut-off value is about 1000.

Development of Dynaceram module

Modules, which were scaled up from a single tube to multitubular ones, have been developed by TDK Electronic Co. Ltd., Japan, and named Dynaceram module. These modules were originally developed for the reverse osmosis process. But based on the above experimental data these modules were tested directly, without covering by particular dynamic membranes, for various actual effluents. Some of these results are reported here.

The specifications of modules are listed in Table II, and its

Table II. Specifications of Dynaceram Modules

	DC 0005	DC 0305	DC 0505	DC 0610
Flux (m^3/day kg/cm^2)	0.022	0.46	3.16	6.32
No. of tubes	1	22	151	302
Area (m^2)	0.0077	0.16	1.10	2.20
Dimensions (mm)	10ϕ×550	64ϕ×540	120ϕ×631	120ϕ×1100
Max. temperature (°C)	100	100	100	100
pH range	1-12	1-12	1-12	1-12
Max. pressure (kg/cm^2)	80	80	80	80

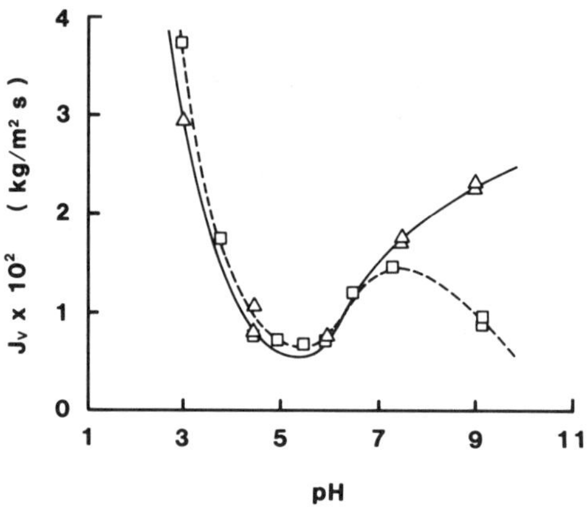

Figure 7. Effect of pH before and after membrane formation. pH was changed during (△) and after (□) membrane formation.

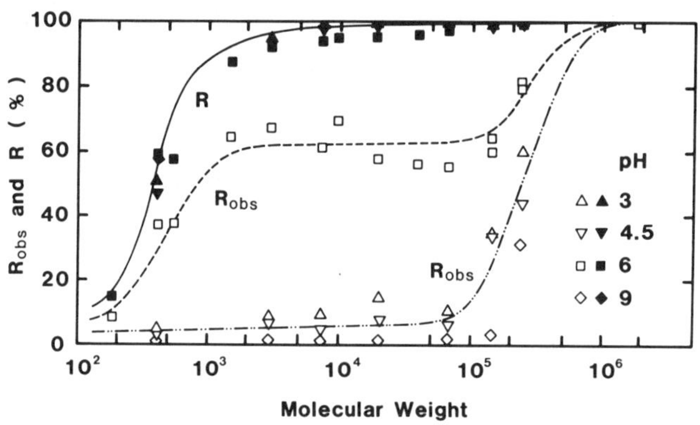

Figure 8. Nature of membranes formed at different pH values.

structure is shown in Figure 9. Dimensions of one ceramic tube is 5 mm O.D. and 550 mm long. DC-0305 has 22 tubes and DC-0605 has 151 tubes. DC-0610 is a combined form of two DC-0605.

Direct combination with activated sludge reactor

The activated sludge process is a most important water processing technology of sewage effluents. But the key point of this process is to make microbiological floc, which can easily be flocculated. Sometimes bulking phenomena occurs and flocculation becomes difficult. To avoid such a instability of activated sludge process, application of the membrane process, instead of sedimentation, is becoming a great concern of water processing engineers recently in Japan.

Following results have been obtained by combining bare Dynaceram modules with a 100 L/day activated sludge unit, which is treating effluent from a research laboratory of a food factory. Figure 10 shows a flux-time behavior of membranes, which depends on the concentration of sludge. Fluxes were very large, considering the low operating pressure of 1 kg/cm^2, and reached steady values after several hours. Recovery of the flux was possible by a back washing by air, but the most effective washing was performed by NaClO treatment. An example of flux recovery is shown in Figure 11, where it is seen that about 30 min washing by 500 ppm NaClO is sufficient. Permeate obtained contained 1300 mg/L TDS, but suspended solids (S.S.) of the feed, 6 mg/L, was completely removed. Amount of bacteria in the feed was 21,000, in which E.coli was 2200, while in the permeate bacteria count was reduced to 300, in which E.coli was 0.

Treatment of effluent from bleached kraft pulp process

Recently two major pulp factories in Japan installed large ultrafiltration units to process effluents from bleached kraft pulp washing processes, the pH and the temperature of which are 10-11 and over 50°C respectively. The main purpose of this process is to concentrate and recover lignin, which is further concentrated and used as fuel, while the permeate can be discarded without biological treatment. This results in the reduction of the amount of sludge. Membranes presently used are made of polysulfone.

Ceramic membranes have been tested for the comparison and results are as follows. Pressure dependence of the flux and the rejection of colors are shown in Figure 12. The particular feature in this case is that flux can be recovered easily by backwashing. Flux did not depend on the concentration ratio up to the value of 5. COD rejection was about 80%, color rejection was about 90%, over 5-fold concentration.

Conclusion

Natures of dynamically formed ultrafiltration membranes were investigated using a ceramic support tube and ovalbumin as a solute. This membrane has intrinsically high rejections, but due to the concentration polarization the observed rejection is low.

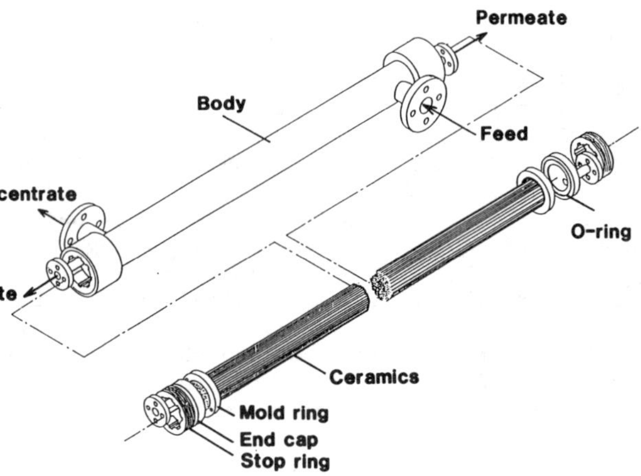

Fig. 9

Figure 9. Dynaceram module DC-0610.

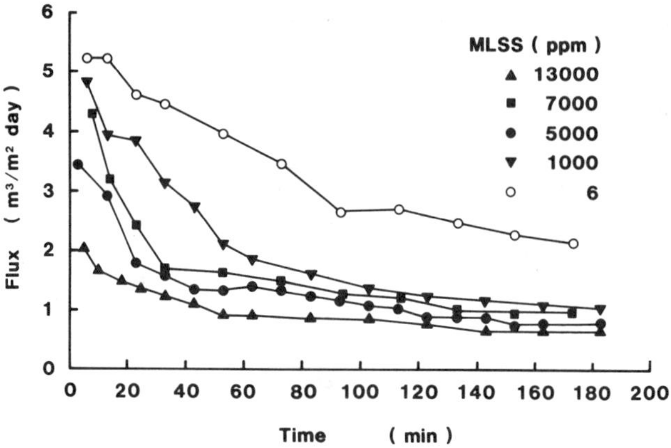

Figure 10. Direct ultrafiltration of activated sludge effluents of different MLSS concentrations. Pressure, 1 atm; and flow rate, 1 l/min.

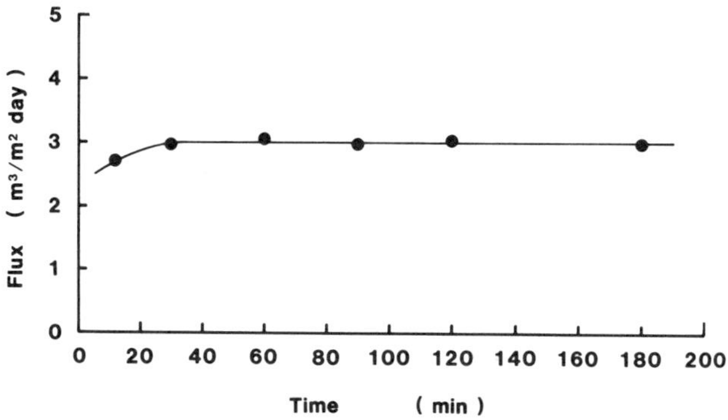

Figure 11. Recovery of flux by washing with NaClO. Washing conditions: NaClO, 490 ppm; pressure, 1 atm; flow rate, 4 l/min; and temperature, 16-17 °C.

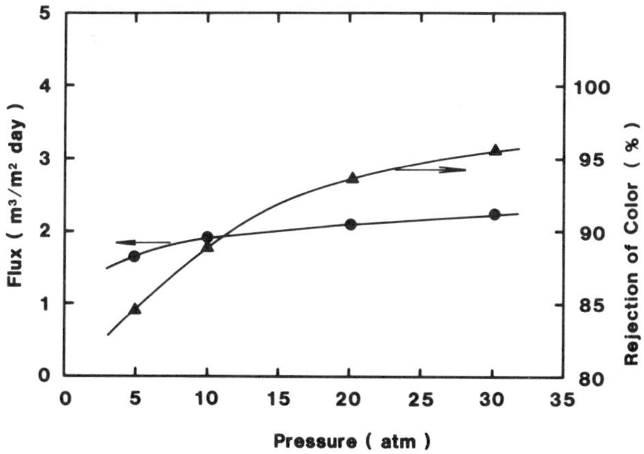

Figure 12. Flux and color removal of effluent from bleached kraft pulp plant. A membrane was formed at 5 atm. Feed velocity, 1 m/s; and temperature, 40 °C.

Examples of direct applications of bare Dynaceram modules, developed by TDK based on the above results, were also reported.

Nomenclature

C = solute concentration (g/cm^3)
D = diffusion coefficient (cm^2/s)
d = equivalent diameter (cm)
J_v = volume flux (cm^3/cm^2 s)
k = mass transfer coefficient (cm/s)
l = membrane length (cm)
N_{Re} = ud/ν
N_{Sc} = ν/D
N_{Sh} = kd/ν
u = flow velocity (cm/s)
ν = kinetic viscosity (cm^2/s)

Acknowledgment

Authors are grateful to Mr. Shoichi Wakabayashi, TDK Electronic Co. Ltd. for supplying us their modules and permitting us to publish some of his data.

Literature Cited

1. Marcinkowsky, A.E.; Kraus, K.A.; Phillips, O.H.; Johnson, J.S.; Shor, A.J. J. Am. Chem. Soc. 1966, 88, 5744.
2. Johnson, J.S.; Minturn, R.E.; Wadia, P. J. Electroanaly. Chem. 1972, 37, 267.
3. Thomas, D.G. In "Reverse Osmosis and Synthetic Membranes"; Sourirajan, S., Ed.; National Research Council of Canada: Ottawa, 1977; Chap. 14.
4. Brandon, C.A.; Gaddis, J.L.; Spencer, H.G. In "Synthetic Membranes vol. II"; Turbak, A.F., Ed.; ACS SYMPOSIUM SERIES No. 154, American Chemical Society: Washington, D.C., 1981; pp. 435-453.
5. Groves, G.R.; Buckley, C.A.; Cox, J.M.; Kirk, A.; MacMillan, C.D.; Simpson, M.J. Desalination 1983, 47, 305-12.
6. Nomura, T.; Kimura, S. Desalination 1980, 32, 57-63.
7. Kimura, S., Nomura, T. Desalination 1981, 38, 373-382.
8. Jonsson, G.; Christensen, P.M. Paper presented at Europe-Japan Joint Congress on Membranes and Membrane Processes, June 18-22, 1984, Stresa, Italy.
9. Nakao, S., Kimura, S. J. Chem. Eng. Japan 1981, 14, 32-37.
10. Brandrup, J.; Immergut, E.H., Ed. "Polymer Handbook" 1978.
11. Sherwood, T.K., Pigford, R.L.; Wilke, C.R. "Mass Transfer" McGraw-Hill, 1975.
12. Ohta, K.; Yamamoto, H.; Kawahara, K. Poly. Preprints, Japan 1976, 25, 1449.

RECEIVED February 22, 1985

4

Properties of Dynamic Polyblend Membranes in the Hyperfiltration of Electrolyte Solutions

H. G. SPENCER, H. C. RYCHLICKI, and K. S. MENON

Department of Chemistry, Clemson University, Clemson, SC 29631

Dynamic polyblend membranes prepared by deposition of a weak acid and a weak base polyelectrolyte pair on a zirconium hydrous oxide ultrafilter were used in hyperfiltration experiments with feed solutions containing simple electrolytes. Ion exclusion model parameters were determined. Effects of ionic surfactants on membrane permeability and electrolyte rejection were also evaluated. The membranes exhibited hyperfiltration properties characteristic of charged gel membranes possessing fixed-charge concentrations and signs dependent on pH.

Dynamic polyblend membranes on stainless steel were introduced to expand the family of dynamic membranes to include a membrane providing a high rejection of simple sugars at high temperatures (1). Fructose rejection by zirconium hydrous oxide-polyacrylate membranes on stainless steel rarely exceeds 0.4. However, fructose rejection greater tha 0.95 is readily obtained with polyblend membranes. These two types of dynamnic membranes also differ significantly in their rejection of simple electrolytes, especially in the dependence of rejection on pH, type of electrolyte, and concentration of electrolyte. This paper presents the hyperfiltration characteristics of a representative polyblend membrane and uses the results to obtain membrane parameters.

Polyblend Membranes

Dynamic polyblend membranes have been prepared by the sequential deposition of pairs of miscible polymers in a mutual solvent on zirconium hydrous oxide membranes supported on porous stainless steel tubes (1). The membranes described in this paper were prepared by deposition of a weak acid polyelectrolyte and a weak base polyelectrolyte to form the blend. The dependence of the concentration and sign of the fixed charge in the layer of the weak acid-weak base polyelectrolyte blend is expected to vary in a predictable manner as the pH of the feed solution is changed. At low pH the polybase will have a full positive charge while the polyacid will be only partially ionized and the membrane is expected to exhibit the electrolyte rejection characteristics of a porous positively charged membrane. At high pH the polybase will not be fully charged and the

0097-6156/85/0281-0047$06.00/0

polyacid will be completely ionized to produce a porous negatively charged membrane. The isoelectric point (iep) will occur at an intermediate pH. This sensitivity to pH, when weak acid and base groups occur in the membrane, provides an opportunity to investigate the dependence of hyperfiltration properties of hydrophilic membranes on the net fixed charge concentration and sign.

Ion Exclusion Models

Glueckauf (2) has modified the Donnon equilibrium expression for a single electrolyte to account for the micro-homogeneity of the distribution of free fixed charges. For electrolyte A_aY_y, where A is the counter ion and Y the co-ion with respect to the net free fixed charge in the rejection layer of the membrane containing singly charged fixed groups, the expression is (2 - 4)

$$(ay\bar{m})^y(ay\bar{m} + \bar{M})^a = (aym/G)^{nb} \quad (1)$$

where $\bar{m}$ and m are the molalites of the electrolyte in the rejection layer and adjacent feed solution respectively; $\bar{M}$ is the molality of the free fixed charges; b is an empirical constant of the system interpreted to be related to the micro-homogeneity of the distribution of fixed charges; n is the sum of a and y; and G is the ratio of the mean ionic activity coefficient in the rejection layer, $\bar{\gamma}_\pm$, to the mean ionic activity coefficient in the feed, $\gamma_\pm$, i.e.,

$$G = \bar{\gamma}_\pm / \gamma_\pm \quad (2)$$

By relating the apparent passage of the electrolyte, s, to the electrolyte concentrations in hyperfiltration (5),

$$s = \bar{m}/m \quad (3)$$

and assuming the coupling constant for the transport of the solvent and solute, B, is unity, Equation 1 becomes

$$(aysm)^y(aysm + \bar{M})^a = (aym/G)^{nb}. \quad (4)$$

In a more common, and more theoretically satisfying, treatment of the electrolyte coupling constant, B, the coupling constant for each of the ions of the electrolyte is assumed to be unity (5, 6). Although the values of $\bar{M}$ are affected by the choice of estimating B, the order of the values with respect to measurements for series of electrolytes with one membrane or an electrolyte with a series of membranes is not altered (6). The simpler estimation of B=1 is used in this discussion.

Three models can be represented by Equation 4 and limiting cases of it. The ideal model is obtained by assuming G = 1, b = 1. A non-ideal model is obtained by assuming b = 1 and evaluating G in Equation 2 by assuming the activity coefficient dependence on ionic strength, I, is the same in the feed solution and the rejection layer; where $I = \frac{1}{2} \sum m_i z_i^2$ in the feed solution and $\bar{I} = S\bar{m} + \bar{M}$ in the rejection layer, with $S^1 = 1$ for $NaNO_3$ and S = 3 for Na_2SO_4, etc. (7). The inhomogeneity model for ion exclusion, Equation 1, was introduced to retain G = 1 and yet account for the enhanced penetration of electrolyte at low concentration (2 - 4). The inhomogeneity model for hyperfiltration preserves this assignment. The resulting operational equations are presented for three types of electrolytes in Table I, assuming $\bar{M} > aysm$, i.e., at low electrolyte concentrations.

Table I. Ion Exclusion Model Equations for Electrolyte Hyperfiltration by Charges Gel Membranes in the Limit of Low Concentration

	Electrolyte Type		
Model	a=y=1	a=2, y=1	a=1, y=2
Ideal	$s\bar{M} = m$	$s\bar{M}^2 = (2m)^2$	$s\bar{M}^{\frac{1}{2}} = (2m)^{\frac{1}{2}}$
Non-ideal	$s\bar{M}G^2 = m$	$s\bar{M}^2G^3 = (2m)^2$	$s(\bar{M}G^3)^{\frac{1}{2}} = (2m)^{\frac{1}{2}}$
Inhomogeneity	$s\bar{M} = m^{2b-1}$	$s\bar{M}^2 = (2m)^{3b-1}$	$s\bar{M}^{\frac{1}{2}} = (2M)^{(3b-2)/2}$

Using these equations for low concentrations the dependence of s on m for a single electrolyte and the dependence of s on the type of electrolyte at fixed m can be used to determine membrane parameters. The ideal model indicates that slopes of log s vs. log m should be 1, 2 and $\frac{1}{2}$ for a=y=1; a=2, y=1; a=1, y=2 electrolytes respectively. Although dynamic zirconium hydrous oxide membranes exhibit slopes near the ideal values, the zirconium hydrous oxide-polyacrylate membranes and other dynamic hyperfiltration membranes exhibit slopes that are significantly less than ideal (8, 9).

The non-ideal model has been applied to a zirconium hydrous oxide membrane to account for the concentration dependence of s using different types of electrolytes (7). The value of $\bar{M}$ obtained using the non-ideal model was about six times as large as $\bar{M}$ determined by the ideal model and was constant for the various electrolyte types. However, application of this non-ideal model when the slope of log s vs. log m is significantly different from the ideal slope does not give constant values of $\bar{M}$ (6).

The inhomogeneity model can account for large deviations in the slope of log s vs. log m and will be applied to the hyperfiltration results obtained with the polyblend membranes. It provides two parameters from hyperfiltration experiments, $\bar{M}$ and b, that are useful in characterizing the membranes.

The values of the fixed charge concentration index, $\bar{M}$, and the fixed charge distribution index, b, change as intuitively expected (6). The fixed charge concentration in a zirconium oxide-polyacrylate membrane at pH near 7 should be much higher than in a zirconium oxide membrane at the same pH, which is near its iep. $\bar{M}$ is larger and b is smaller for the polymer-containing membrane. The presence of an organic polyelectrolyte on a zirconium oxide substrate generally reduce b, making the electrolyte rejection less sensitive to concentration. A change in pH from 4 to 7 increases the ionization of the poly(acrylic acid) and should result in a higher fixed charge concentration unless the swelling is too great. Hyperfiltration of $NaNO_3$ solutions using dynamic (formed-in-place) zirconium oxide-poly(acrylic acid) membranes usually gives a small increase in $\bar{M}$ and a significant decrease in b when the pH is changed from 4 to 7 which accounts for its greater rejection of simple electrolytes at concentrations used in the normal test evaluation; 0.01-0.03*m*.

Experimental

Membranes. Three materials were deposited in sequence under hyperfiltration pressure and cross flow conditions; zirconium hydrous oxide, poly(acrylic acid) (Acrysol from Rohm and Haas), and a weak base polymer containing primary amine groups. The porous stainless steel support tube was 0.203 cm in diameter by 30.5 cm in length, providing a membrane area of 0.00195 m^2. The details of forming zirconium hydrous oxide-polyacrylate have been reported by Thomas (10). The formation of polyblend membranes and the effects of each step on the membrane performance has been described (1).

Hyperfiltration Procedures. A hyperfiltration system provided by CARRE, Inc. was used as a test apparatus (11). The temperature, pressure, and cross flow velocity could be varied and measured. Experiments were carried out over a broad pH range, at temperatures between 30 and 70°C, under gauge pressures up to 5.5 MPa, and at cross flow velocities up to 18 m/s. Unless otherwise stated the test solution concentrations were 2 g/L for the electrolytes. The electrolyte rejections were determined by measuring the conductivity of the feed and permeate solutions. When an ionic surfactant was present conductivities of the simple electrolytes in the feed were corrected to remove the contribution of the surfactant. Rejection of surfactant by the membranes was effectively unity.

Materials. The simple electrolytes used in membrane formation and in the hyperfiltration tests were reagent grade. Distilled water was used in all formations and experiments. The surfactants were used as obtained; sodium dodecyl sulfate (SDS) and polyoxyethylene (E_{23}) dodecyl ether (BRIJ-35) from Fisher Scientific Co. and dodecyl trimethyl ammonium chloride (DTAC) from Pfaltz and Bauer, Inc. The CMC of these surfactants are (12): 2.4 g/L at 50°C for SDS, 0.056 g/L at 50°C for BRIJ-35, and 0.9 g/L at 30°C for DTAC.

Results and Discussion

Hyperfiltration Parameters. The effects of applied pressure, cross flow velocity, and temperature on the rejection, r, of $NaNO_3$ and volume flux, J, were investigated using one of the polyblend membranes, membrane A. The gauge pressure was varied from 2.0 to 4.2 MPa at pH 2.6 and 6.3 maintaining the temperature, T, at 325 ± 1 K and the cross flow velocity, F, in the range 13 to 17 m/s. The flux increased linearly with gauge pressure, but the intercept on the gauge pressure axis was offset from zero. In all calculations the applied pressure, p, is the gauge pressure corrected for this offset. The offset and slope were identical at both pH 2.6 and 6.3. The rejection was independent of p at pH 6.3 and increased only slightly with increasing p at pH 2.6. The rejections are not corrected for pressure differences.

The cross flow velocity was investigated at pH 6.3, T = 330K, and p = 2.7 MPa over the range 2 to 14 m/s. Both the flux and rejection of $NaNO_3$ were essentially constant; J = 4.4 ± 0.2 m/s and r = 0.27 ± 0.01. For a cross flow velocity of 2 m/s the Reynolds number is approximately 6000, indicating turbulent flow in all measurements.

The effect of temperature was investigated over the range 305 to 330K at pH 6.3, F = 16 ± 2 m/s, and p = 2.7 MPa. The rejection of $NaNO_3$ was independent of the temperature in this range and the dependence of the membrane permeability, J/p, on T for all electrolyte solutions was described by (11)

$$(\frac{J}{p})_T = (\frac{J}{p})_{323}\exp[-1,980(\frac{1}{T} - \frac{1}{323})] \quad (5)$$

The membrane permeabilities are adjusted to 323K using Equation 5. No adjustment of rejection for temperature is necessary.

Effect of pH. The effects of pH on the passage, s, where

$$s = 1 - r \quad (6)$$

and on membrane permeability are indicated for four test polyblend membranes in Figures 1 and 2. The passages exhibit maxima for the three membranes at different values of pH; 5 for A, 8 for B, and 7 for C. The membrane permeabilities of membranes A and B also exhibit maxima at pH 5 for A and 7 for B. The membrane permeability appears to increase at low pH for membrane C, similar to zirconium hydrous oxide-polyacrylate membranes.

The effect of pH on the rejection of Na_2SO_4 and membrane permeability was also determined using membrane A. The results with Na_2SO_4 are compared with results with $NaNO_3$ in Figure 3. The electrolyte exclusion models predict that for dilute solutions the passage for Na_2SO_4 will exceed $NaNO_3$ when the net free fixed charge in the rejection layer is positive and the order will be reversed when the membrane charge is negative. The model is consistent with a positive membrane charge at low pH and a negative charge at high pH with the crossover point indicative of the isoelectric point, iep, at pH 4 to 5. This ionic character of a weak acid-weak base polyelectrolyte pair is expected. The location of the iep at pH 4 to 5 and the smaller passage at pH 10 than at low pH are interpreted by the model to indicate an excess of the polyacid over the polybase in the rejection layer of the membrane. The pKa's of the polyelectrolytes are approximately 4 and 10. The maximum in s appears to occur at a higher pH than the iep defined above.

Effect of Electrolyte Concentration. Membrane A, with iep near pH 5, was also used to investigate the effect of $NaNO_3$ concentration on its passage and on the membrane permeability. The experiments were carried out at pH 7.0 and 10.2 with T = 329K, p = 2.7 MPa, and F between 12 and 16 m/s. The dependence of the passage on concentration is presented in Figure 4 as log s vs. log m. This dependence is linear and described by

$$\log s = 0.098 \log m \quad (7)$$

at pH 7.0 and

$$\log s = 0.310 \log m + 0.102 \quad (8)$$

at pH = 10.2. Applying the inhomogeneous ion exclusion model $\overline{M} = 1.0\underline{m}$ and b = 0.55 at pH 7.0 and $\overline{M} = 0.8\underline{m}$ and b = 0.66 at pH 10.2. The model indexes imply a higher fixed charge concentration and a lower fixed charge distribution homogeneity at the pH nearer the iep, i.e., at pH 7. This effect of pH on the fixed charge concentration is unexpected unless the swelling at the higher pH is much greater than at pH 7; probably the result of fewer

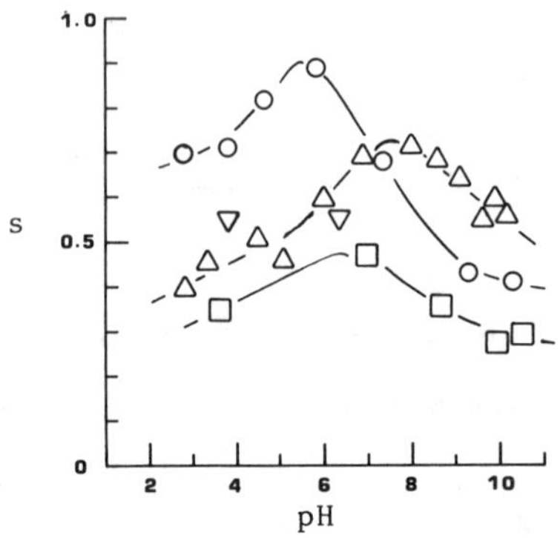

Figure 1. Passage of $NaNO_3$ vs. pH for polyblend membranes: ○, A; △, B; ▽, C; and □, D.

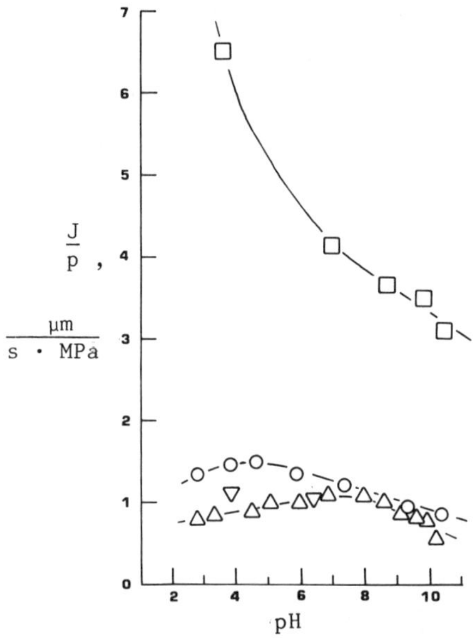

Figure 2. Membrane permeability vs. pH for polyblend membranes: ○, A; △, B; ▽, C; and □, D.

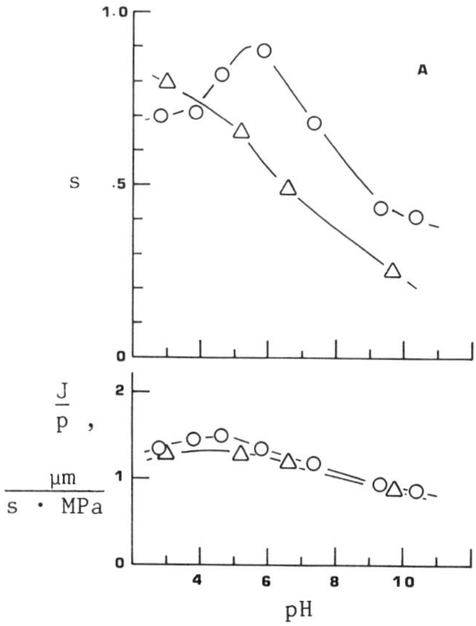

Figure 3. Electrolyte passage and membrane permeability vs. pH: O, $NaNO_3$ and △, Na_2SO_4.

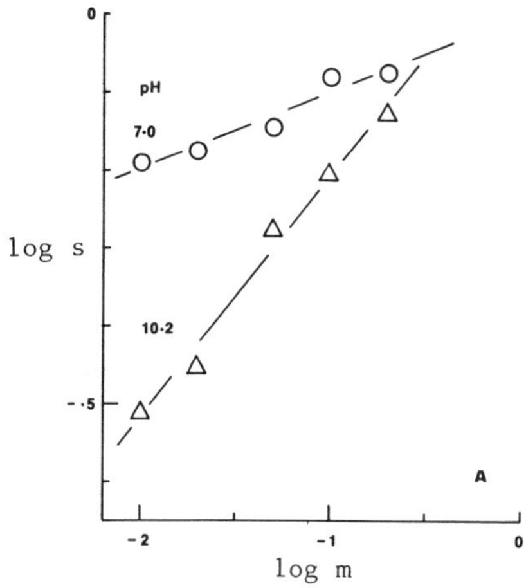

Figure 4. Logarithm of passage vs. logarithm of electrolyte concentration in the feed.

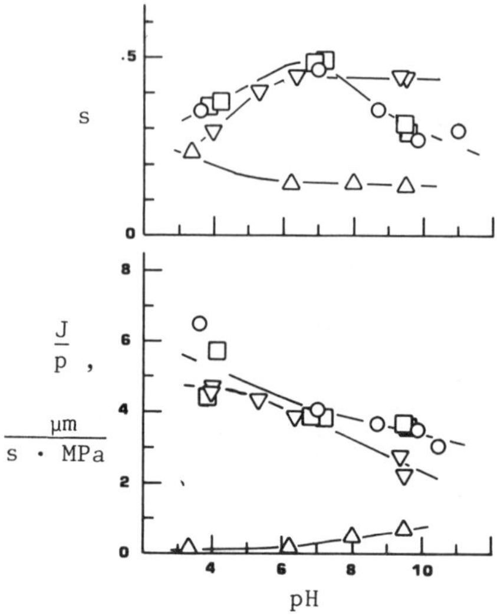

Figure 5. $NaNO_3$ passage and membrane permeability vs. pH for polyblend membrane D: O, $NaNO_3$ only; □, $NaNO_3$ + BRIJ-35; ▽, $NaNO_3$ + DTAC; and △, $NaNO_3$ + SDS.

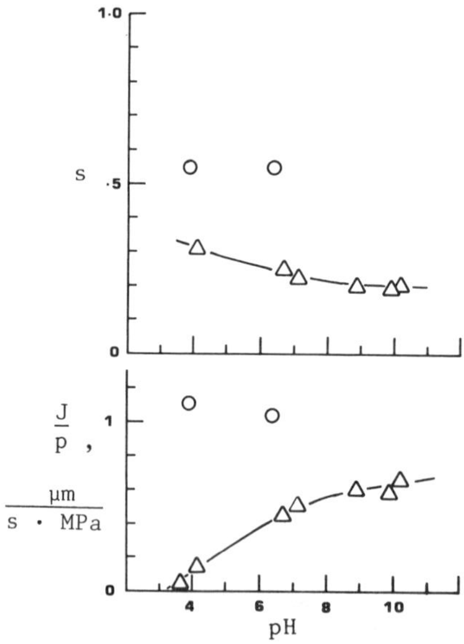

Figure 6. $NaNO_3$ passage and membrane permeability vs. pH for polyblend membrane C: O, $NaNO_3$ only and △, $NaNO_3$ + SDS.

ionic cross links at pH 10.2 where the membrane is less stoichiometric with respect to polyelectrolyte charges. Also, the analysis could be modified somewhat by the possibility of significant hydroxide ion transport at the higher pH, which is not accounted for in the present model (6). The membrane permeabilities were nearly independent of $NaNO_3$ concentration; with a slight increase observed at the lowest concentration.

Effects of Surfactants on Membrane Permeability and $NaNO_3$ Rejection. Inclusion of surfactants in the hyperfiltration test solutions containing $NaNO_3$ provides another means of determining the sign of the fixed charge of the membrane. For example, the effects of non-ionic and ionic surfactants on the membrane permeability and $NaNO_3$ rejection by a zirconium oxide-polyacrylate membrane were characteristic of a membrane with a fixed negative charge (12). The cationic surfactant caused a very large reduction in membrane permeability and was not readily removed by washing with acidic and basic solutions. The non-ionic and anionic surfactants had little effect on either the membrane permeability or the electrolyte rejection.

The membrane permeability and $NaNO_3$ passage of a polyblend membrane, membrane D, in the presence of surfactants are presented in Figure 5. The order for the experiments was: no surfactant; polyoxyethylene (E_{23}) dodecyl ether (BRIJ-35), a non-ionic surfactant; dodecyl trimethyl ammonium chloride(DTAC); a cationic surfactant; and sodium dodecyl sulfate(SDS), an anionic surfactant. The non-ionic surfactant produced no change from the test without a surfactant. The cationic surfactant appears to cause a measureable decrease in membrane permeability and a small increase in passage at the higher pH, 9 to 10, where the net charge of the membrane is expected to be negative. The anionic surfactant reduced the passage and the membrane permeability significantly. The effect on the membrane permeability was greatest at low pH where the net charge of the membrane is expected to be positive. The greatest effect on the passage was near the iep of the membrane.

A similar effect of the presence of the anionic surfactant was observed in a second experiment using another polyblend membrane, membrane C. See Figure 6. In this case no other surfactants were used in the experiments before testing the anionic surfactant.

The effects of the surfactants appear to be consistent with an electrolyte exclusion mechanism. The ionic surfactant behaves like a fixed charge when the fixed charge of the membrane is opposite in sign, but has little effect when the fixed charge of the membrane is of the same sign.

Literature Cited

1. Spencer, H. G.; Todd, D. K.; McLellan, D. B. Desalination 1984, 49, 193.
2. Glueckauf, E. Proc. Royal Soc. A 1962, 268, 339, 350.
3. Meares, P. In "Membrane Separation Processes;" Meares, P., Ed.; Scientific: Oxford, 1976; Ch. 1.
4. Meares, P. In "Ion Exchange Membranes;" Flett, D. S. F. Ed.; Ellis Harwood: Chichester, 1983; pp. 9 - 30.
5. Dresner, L.; Johnson, J. S., Jr. In "Principles of Desalination;" Spiegler, K. S., Ed.; Academic: New York, 1980; Part B, Second Ed., pp. 401 - 450.

6. Spencer, H. G. "Electrolyte Exclusion Model for Hyperfiltration of Electrolyte Solutions by Charged Gel Membranes;" Desalination, in press.
7. Shor, A. J.; Kraus, K. A.; Smith, W. T., Jr.,; Johnson, J. S., Jr. J. Phys. Chem. 1968, 72, 2200.
8. Kraus, K. A.; Shor, A. J.; Johnson, J. S., Jr. Desalination 1967, 2, 243.
9. Johnson, J. S., Jr.; Minturn, K. E.; Wadia, P. H. J Electroanal. Chem. 1972, 37, 267.
10. Thomas, D. G. In "Reverse Osmosis and Synthetic Membranes;" Sourirajan, S., Ed.; National Research Council Canada: Ottawa, 1977, pp. 295 - 312.
11. Brandon, C. A.; Gaddis, J. L.; Spencer, H. G. In "Synthetic Membranes;" Turbak, A. F., Ed.; ACS Symposium Series No. 154, America Chemical Society: Washington, DC, 1981; Ch. 24.
12. Luppino, R. J.; Spencer, H. G. Desalination 1982, 41, 33.

RECEIVED February 22, 1985

5

Structure, Permeability, and Separation Characteristics of Porous Alumina Membranes

A. F. M. LEENAARS[1], K. KEIZER, and A. J. BURGGRAAF

Laboratory of Inorganic Chemistry, Materials Science, and Catalysis, Department of Chemical Engineering, Twente University of Technology, P.O. Box 217, 7500 AE Enschede, The Netherlands

In this paper a survey is given on ceramic, porous γ-Al_2O_3 membranes. The information of the membranes can be described with a slibcasting model. This means that the layer thickness of the membranes increases linearly with the square root of the dipping time. The pores of these membranes are slit-shaped and the pore size depends on the temperature/time treatment. It amounts 2.7 nm and 4.8 nm at temperatures of 400°C and 800°C respectively while the porosity remains constant in this temperature range (> 50%).
The water permeability is proportional to the pressure drop till at least 50 bar and inversely proportional to the membrane thickness. In ultrafiltration experiments the 'cut off' value of the membrane with 2.7 nm pores is 2000 for polyethylene glycols. The membrane surface can be modified with several types of ions and therefore these types of membranes can be made suitable for gas separation and hyperfiltration applications.

In the literature the use of inorganic membranes is not frequently mentioned. The most-cited membranes in this class are the inorganic Vycor-type glass membranes. For instances these materials with a porosity near 30% and a pore size as small as 3 nm can be used for the desalination or the separation of urea from water (1,2). Rejection values between 35% and 95% were obtained depending on the pore size distribution and the feed concentrations. A major problem is the chemical instability of these types of membranes.

In gas separations Kameyama et al. (3) and Shindo et al. (4) used these glass membranes for the separation of hydrogen from hot gas mixtures. At temperatures near 800°C the thermal stability was insufficient.

Because of these chemical and thermal instabilities other types of membranes were prepared and tested, as for instance porous

[1] Current address: Professor Holstlaan, Philips National Laboratory, 5656 AA Eindhoven, The Netherlands

0097–6156/85/0281–0057$06.00/0

alumina or zirconia films on ceramic or coal supports (3,5). The pore diameters of these membranes amount at least 10 nm, which is large for gas separation purposes and the separation of low molecular weight liquid mixtures.

In our laboratory (6,7,8,15,16) porous γ-alumina membrane films on porous supports were developed with pore diameters smaller than 3 nm and porosities larger than 50%.

In this paper the essential parts of the preparation of these membranes are given. Furthermore the formation of the membrane on a porous support is described by a slibcasting model. The pore structure as a function of temperature and time is presented and the permeability for water and some other liquids is given. Polymer solute rejection data from ultrafiltration experiments are given and finally some remarks are made concerning the applications of these membranes for gas separation purposes. Details of these subjects are to be found elsewhere (6,7,15,16).

Membrane Preparation

The preparation of the γ-alumina membranes is based on a gel/sol preparation technique. Important chemical features are reported by Yoldas (9) for bulk ceramic bodies. In Figure 1 preparation procedures are shown. It is essential that the boehmite (γ-AlOOH) phase is obtained as a precursor material because other phases cannot be peptised easily with acids. Therefore the decomposition of aluminium-s-butoxide should take place at temperature above 80°C. The crystallite size of γ-AlOOH is determined by the temperature and time treatment and the acid type and concentration during peptising.

The preparation of supported membranes is performed by a dipping procedure (7). By bringing one side of a support into contact with the sol capillary forces are exerted by the pores of the support, according to the equation

$$\Delta P_C = 2\gamma\cos\Theta/r \tag{1}$$

where P_C is the capillary pressure drop (Nm^{-2})
γ is the surface tension of the dispersion medium (Nm^{-1})
r is the pore radius of the support (m)
Θ is the contact angle

The contact angle between the liquid and the solid surface is assumed zero degrees furtheron in this paper.

When the pores of the support are small enough, the boehmite particles cannot enter with the same velocity as the dispersion medium (water) and the concentration of boehmite at the entrance of the pores of the support will increase. At a certain concentration of sol particles, the sol will be converted into a gel. The main parameters determining whether or not a gel layer is formed during dipping are:

a) the sol concentration
b) the dipping
c) the pore size of the support
d) the type and amount of acid used to peptise the sol.

In Table I the effect of c) and d) is shown for a constant sol concentration of 1.2 mol.L^{-1}. At sol concentrations smaller than

0.1 $mol.L^{-1}$ no gellayer is formed under any of the conditions given in Table I. For the gel formation the gelling concentration and the boehmite particle size play an important role.

Table I. Influence of Type of Support (Pore Size) and Peptising Acid on Formation of Gel Layers During Dipping

Modal pore size support (μm)	Peptising acid used HCl	HNO_3	$HClO_4$
0.12	yes	yes	yes
0.34	yes	yes	no
0.80	yes	no	no

Note: Yes, gel layer formed; no, no gel layer formed. Sol concentration 1.2 mol/L^{-1}.

In the series HCl-, HNO_3-, $HClO_4$-boehmite sols, the gelling concentration increases and the boehmite particle size decreases. This is in agreement with the data presented in Table I.

The gellayer growth can be described by a slibcasting process (7,10) according to equation (2)

$$Lg = \left(\frac{Cg}{\eta}\,t\right)^{\frac{1}{2}} \quad (2)$$

where Lg is gellayer thickness (m)
Cg is a constant depending on the structural properties of the gellayer and the support (N)
η is the viscosity of water ($\sim 10^{-3} Nsm^{-2}$)
t is the dipping time (s).

In practice a constant K_s has to be added to equation (2) because of the adherence of a thin sol layer to the gellayer, when the support is taken from the sol directly after dipping, so equation (2) becomes

$$Lg = \left(\frac{Cg}{\eta}\,t\right)^{\frac{1}{2}} + K_s \quad (3)$$

which also is shown in Figure 2. Using concentrations of 1 $mol.L^{-1}$ of a HNO_3-boehmite sol the Cg-value is of the order of (5-7.5) * 10^{-15} N and K_s = (6-7) * 10^{-7} m for calcined γ-Al_2O_3 films with a porosity of 50-55%. This means that with a dipping time of 10s a layer thickness of about 8μm is achieved after calcination. In this way supported membrane films of some μm's thick can be made within a few seconds of dipping. The time necessary to achieve a certain gellayer thickness increases with an increasing sol concentration, with a decreasing gelling concentration and with a decreasing pore size of the support.

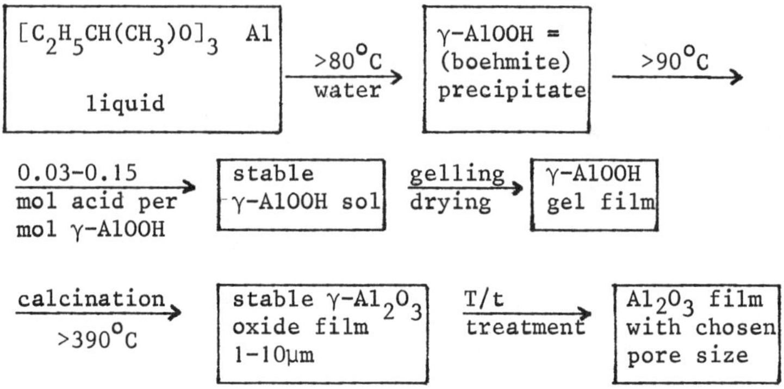

Figure 1. Diagram for the preparation of alumina membranes.

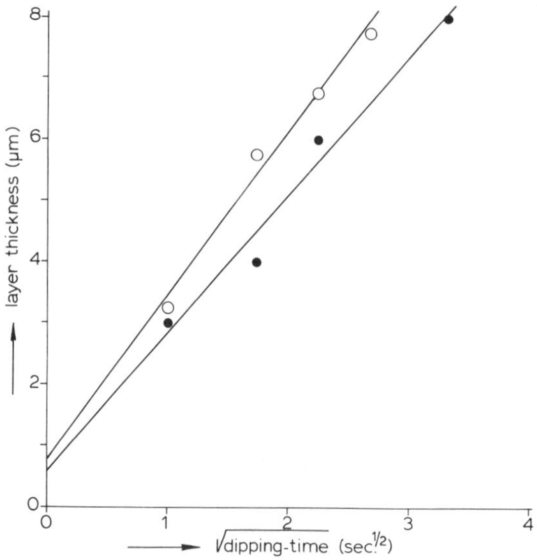

Figure 2. The membrane thickness after calcination plotted as the square root of the dipping time (closed circles: 0.72 mol boehmite, liter^{-1} and open circles: 1.22 mol litre^{-1}).

Structure of γ-Al_2O_3 Membranes

The microstructural characteristics of alumina membranes as a function of the temperature/time treatment are shown in Table II. Leenaars et al. (6) showed by means of gasadsorption measurements, X-ray diffraction and transmission electron microscopy (TEM) that the particles of boehmite are plate-shaped and are stacked in a card pack structure as shown in Figure 3. This structure has therefore slit-shaped pores in which the slit width is approximately as large as the thickness of the crystallites (Fig. 3). When the γ-AlOOH is heated to about 400°C, it is converted into γ-Al_2O_3. It has been shown that this γ-Al_2O_3 structure is closely related to that of the preceding γ-AlOOH (6).

Table II. Microstructural Characteristics of Alumina Films as a Function of the Sintering Temperature for Different Periods

Temperature °C	Time (hrs)	Phase	BET-surface (m^2g^{-1})	Modal pore size (nm)	Porosity (%)
200	34	γ-AlOOH	315(2)*	2.5	41
400 (A)	34	γ-Al_2O_3	301(7)	2.7	53
	170	γ-Al_2O_3	276(4)	2.9	53
	850	γ-Al_2O_3	249(2)	3.1	53
500	34	γ-Al_2O_3	240(1)	3.2	54
700	5	γ-Al_2O_3	207(2)	3.2	51
	120	γ-Al_2O_3	159(2)	3.8	51
	930	γ-Al_2O_3	149(2)	4.3	51
800 (B)	34	γ-Al_2O_3	154(2)	4.8	55
900	34	Θ-Al_2O_3	99(2)	5.4	48
1000 (C)	34	α-Al_2O_3	15(3)	78	41
550**	34	γ-Al_2O_3	147(4)	6.1	59

*The standard deviation is given in parentheses.
**prepared from a sol obtained by a autoclave treatment at 200°C.

The membranes were prepared from a 0.07 mol HNO_3/mol boehmite sol peptised at 90°C-100°C during 16 hrs. A part of this sol was hydrothermally treated in an autoclave at 200°C during 5 hrs. The plate thickness of the primary boehmite crystallites then increases from about 3 nm to 7 nm and the slit width of the γ-Al_2O_3 membranes is therefore also larger at about the same temperature/time treatment (see Table II). In the temperature region between 400°C and 900°C the same alumina phase is observed and the modal pore diameter varies between 2.7 and 4.8 nm. The porosity is independent of the temperature between 400°C and 900°C. At temperatures larger than 900°C the pore size increases strongly and the porosity decreases as a result of recrystallisation of the α-phase (corundum) and sintering effects.

In Figure 4 pore diameter distributions of the membranes A,B and C from Table II are shown. The pore diameters and porosities of membranes A and B are determined with an adsorption/desorption technique at liquid nitrogen temperatures (6). The other pore size

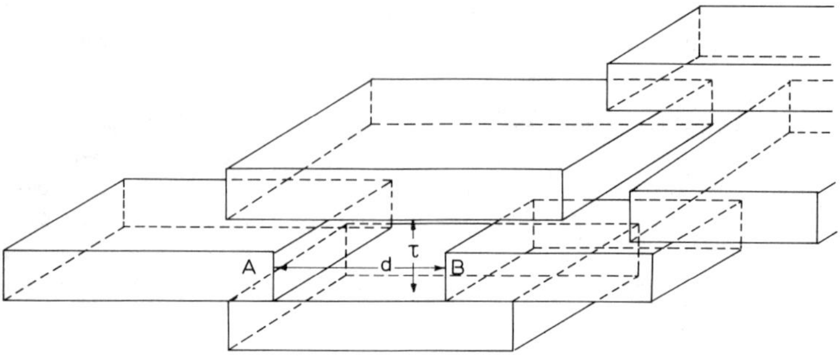

Figure 3. Idealized model of the boehmite membrane microstructure.

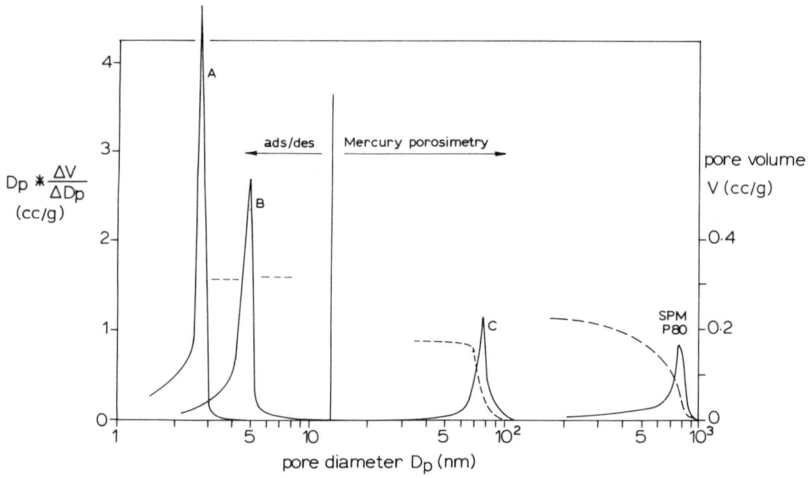

Figure 4. Pore diameter distributions of various samples as given in Table II. The broken line is the cumulative pore volume (cc/g) as a function of the pore diameter. SPM P80 is a commercially available material (Staatliche Porzellan Manifaktur, Berlin).

distributions are determined with mercury porosimetry (diameter > 15 nm).

In conclusion it can be stated that in the temperature range from 400°C upto and including 800°C the pore growth is rather small and the porosity is constant also after a long sintering period (1000 hrs).

The Permeability for Pure Liquids

The permeability of γ-Al_2O_3 membranes (Table II) is measured for the pure liquids water, ethanol, s-butanol and hexane at temperatures between 3-30°C. This permeability is measured for the membranes calcined at 400°C, 500°C and 800°C with pore diameters as given in Table II (2.7 nm - 4.8 nm). The values for the volume fluxes through these supported membranes were corrected for the support, assuming that the flow resistance of the membrane and support can be put in series. According to Strathmann (11) and using our own data this assumption is correct within 1°/oo for membranes thicker than 1μm.

For viscous flow through a porous system the Kozeny-Carman relation can be used (17):

$$\frac{1}{A}\frac{\Delta V}{\Delta t} = \frac{\Delta P \cdot \varepsilon^3}{K \cdot L \cdot \eta \cdot S_v^{\,2}(1-\varepsilon)^2} \tag{4}$$

where $\frac{\Delta V}{\Delta t}$ is the volume flux ($m^3 s^{-1}$)
A is the cross-sectional area of the membrane (m^2)
ΔP is the pressure drop over the membrane (Nm^{-2})
η is the viscosity of the liquid (Nsm^{-2})
ε is the porosity of the membrane (-)
S_v is the surface area of the membrane per unit volume non-porous material (m^{-1})
L is the thickness of the membrane system (m)
K is the Kozeny-Carman constant ($=K_oK_t$) which includes a shape (K_o) and tortuosity (K_t) factor (-).

According to this equation the volume flux multiplied with the viscosity of the liquid should be independent of the type of permeant for a given membrane. However, this value ($\frac{\Delta V \cdot \eta}{A \cdot \Delta t \cdot \Delta P}$; $cP \cdot cm \cdot bar^{-1} \cdot h^{-1}$) decreases at 20°C for a 3μm-thick membrane calcined at 500°C from 0.63 for water, 0.53 for ethanol, 0.50 for hexane to 0.38 for s-butanol. This means that these liquids do not act as 'free-bulk' liquids in the γ-Al_2O_3 membranes. A similar effect has been found for the permeability of hydrocarbons through Vycor glass (pore diameter 4 nm) (12).

For water it appears that the macroscopic viscosity change with temperature is in agreement with the change of the flux suggesting 'bulk-water' behaviour. It must be noted that this does not imply that all the water inside the pores of the membrane behaves like 'bulk-water'. Water, which is more or less immobile will not contribute to the temperature dependency of the mobile part but will nevertheless increase the apparent viscosity. Nevertheless, water was chosen as a permeant to characterize the porous structure, because its possible deviation from bulk behaviour is the least of all permeants investigated. When the reciprocal flux ($\frac{\Delta t}{\Delta V}$) is plotted as a function of the membrane thickness for membranes calcined at 400°C, 500°C and 800°C, the following equation is found:

$$\frac{\Delta t}{\Delta V} = \eta \cdot C \cdot t_m \quad (5)$$

where $\frac{\Delta t}{\Delta V}$ is the reciprocal flux ($cm^{-1} \cdot bar \cdot h$)
η is the viscosity of the liquid (cP)
C is a structural parameter of the membrane depending on porosity and pore size (at a cross-sectional area of 1)
t_m is the membrane layer thickness (μm).

This relation is independent of the pressure drop ΔP over the membrane until at least 50 bar. For membranes calcined at 400°C or 500°C the value of C is of the order of 0.7-0.8 and for membranes calcined at 800°C this value is 0.17. In principle the membrane thickness can be reduced to 1μm. This means that for ultrafiltration purposes, a water permeability of 1-6 cm h^{-1} bar^{-1} can be achieved.

With the water permeability values and the measured porosities and surface areas the Kozeny-Carman constant K (including tortuosity) can be calculated. When this calculation is carried out for the 800°C membrane data (because in that case the geometrical surface is equal to the BET-surface), then a value of 13 ± 2 is found for $K(=K_o K_t)$. This is larger than the value of 5 found for the systems with a packing of spherical particles. This differece can mainly be attributed to the slit-shape of the pores and with this K-value and some literature data a length to thickness ratio of about 15 to 1 for the crystallites can be calculated (15). This gives a lateral size of about 40 nm which roughly agrees with the observations obtained by TEM (6).

Ultrafiltration and Hyperfiltration Experiments

Ultrafiltration experiments are carried out on γ-Al_2O_3 membranes of about 6μm thickness calcined at temperatures of 400°C and 800°C during a few hrs. These filtration experiments were performed in a stirred dead-end permeation cell by using polyethylene glycol (PEG) and dextran with mol-weights (M_w) varying between 600 and 70,000. The experiments were performed at pressure drops of 1 bar (800°C membrane) and 3 bar (400°C membrane). The permeability for the PEG solutions are equal to the pure water permeabilities and are independent of time (24 hrs.). The permeabilities for dextran solutions decrease as a function of time and are after a few hours about 20% lower than those found for pure water. The rejection values (R) are shown in Fig. 5 and are defined as

$$R = 1 - \frac{c_p}{c_f} \quad (6)$$

where c_p is the concentration in the permeate (g,L^{-1}) and c_f is the concentration in the feed (g,L^{-1}).
Two rejection values for the Carbowax (= polyethylene oxide) measured with a 150 Å Nuclepore filter (Zeman, Wales (13)) are also shown. A cut-off value (R > 90%) of 2000 for membrane A is about the value to be expected from the structural parameters of the

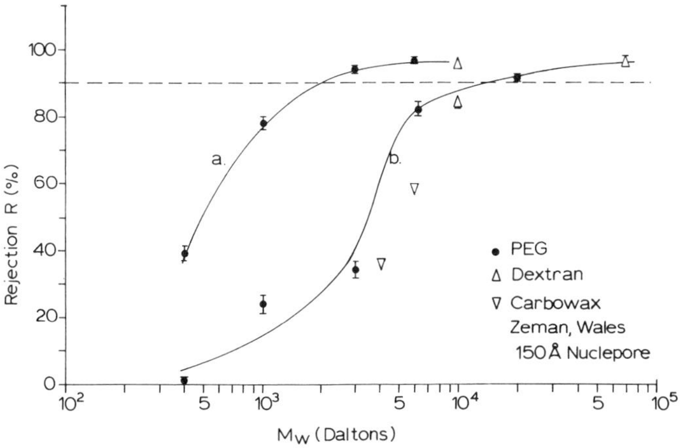

Figure 5. The rejection values of PEG and dextran ploymer solutes as a fuction of the mol.weight (M_w) for a membrane calcined at 400°C (a) and 800°C (b).

membrane, if no pore wall interactions occur (16). The same conclusion can be drawn for membrane B with a 'cut-off' value of 20,000. The rejections found for both membranes are larger than those for a Nuclepore filter with pores, which are cylindrically shaped. The difference between rejections with membrane B and with a 150 Å Nuclepore filter is somewhat smaller than expected from the given pore sizes of 4.8 nm and 15 nm, however the value of 4.8 nm is a modal pore size and for Nuclepore filters this is a maximum pore size. Furthermore for Nuclepore filters the method to determine the pore size is not given in the paper of Zeman, Wales (13) and a comparison of pore sizes is therefore difficult. Some hyperfiltration results are shown in Table III. In all cases the rejections are not very large and for phenol/water they are smaller than those obtained with Vycor-type glass membranes.

Table III. Rejection of Molecules Dissolved in Water for a Membrane A at a Pressure Drop ΔP = 40 bar

system	R (%)	Flux (cm $h^{-1} bar^{-1}$)
phenol/water	8	0.096
ethanol/water	12	0.091
s-butanol/water	16	0.086
raffinose/water	28	0.082

Note: All concentrations 10 $g.L^{-1}$ except phenol, 1 $g.L^{-1}$. See Table II for a membrane A.

There is a small interaction of alcohols with the pore wall surface of pure γ-Al_2O_3 membranes as shown before. In our opinion therefore the rejections of these alcohols are larger than of phenol but this interaction is too small to achieve large rejections. The larger rejection of raffinose is a molecule size effect. It is possible to change the chemical character of the membrane pore surface as is described in the next section, which can have a positive effect on the found rejections. It is not clear for the moment why Vycor-type exhibits the reported, larger, rejection values for phenol.

Application of Ceramic Membranes for Gas and Liquid Separation Purpose

γ-Alumina membranes with a pore diameter of 3-5 nm can be used for gasseparation applications according to a Knudsen diffusion mechanism (14). This mechanism is only useful for the separation of hydrogen or helium from other gases because the separation factor is about equal to the inverse square root of the molar mass ratio. For Knudsen diffusion (> 90%) our type of membranes can be used at temperatures of 500°C-800°C and pressures of 20-40 bar or lower.

For lower temperature gas separations (20°C-400°C) the mechanism of surface diffusion (in some cases including capillary condensation) has to be used. In this case the pores should be as small as possible and there has to be a specific interaction (preferential sorption) of one of the components with the membrane pore surface. Our membranes are suitable for these types of application because of their small pores. Moreover it is easy to modify the surface with several ions. Specific interactions with for instance CO_2 and H_2O have been shown already. An additional advantage of this type of modification is the further decrease in pore sizes which might favor hyperfiltration with larger rejections than sofar.

Conclusions

γ-Al_2O_3 porous membranes with a thickness of 1-10 μm are prepared by dipping porous ceramic supports in a boehmite sol and calcining at at least 400°C. The formation of gel layers on a porous support (during dipping) can be well described as a slibcasting process. The modal slit width of these membranes calcined at 400°C and 800°C are 2.7 nm and 4.8 nm respectively and these values increase only slightly as a function of time. Larger pore sizes can be achieved by appropriate heat treatment at higher temperatures. The porosity of 50-55% and the crystal structure do not change in this temperature range of 400°C-800°C.

The waterpermeability is proportional to the applied pressure drop and is inversely proportional to the membrane thickness. With a membrane thickness of 3 μm these fluxes are 0.3 cm $h^{-1}bar^{-1}$ and 2 cm $h^{-1}bar^{-1}$ for membranes calcined at 400°C and 800°C respectively. For other liquids the permeability (multiplied with the viscosity) is lower.

In ultrafiltration experiments the 'cut-off' value (M_w) is 2000 for a membrane calcined at 400°C and 20,000 for a membrane calcined at 800°C. Solute rejections in hyperfiltration experiments are not larger than 28% (saccharose/water). However it is possible to change the chemical features of the membrane surface by modifying with other types of cations and thereby changing the preferential sorption characteristics and the hyperfiltration behaviour.

Acknowledgments

We would like to thank Prof.dr. C.A. Smolders and Dr. D. Bargeman for the stimulating discussions, Mrs. S.M. van Hengstum-Nijhuis, Mr. G.B.J. Borggreve and Mr. D. Wesseling for their assistance and Ing. L. Dewael, Dr.ir. de Lau and Ir. C.A.M. v.d. Broek of the ceramic laboratories of Philips N.V. (Elcoma) for supplying part of the supports. The financial assistance from the ministries of science affairs and economical affairs is gratefully appreciated.

Literature Cited

1. Ballou, E.V.; Wijdeven, Th.; Leban, M.I. Envir. Sci. Technol. 1971, 98, 1032-8.
2. Schnabel, R.; Valont, W. Desalination 1978, 24, 249-72.
3. Kameyama, T.; Fukuda, K.; Fushishige, M.; Yokokawa, H.; Dokiya, M. Proc. 3rd world energy conference 1980, 2, 569-79.
4. Shindo, T.; Obata, K.; Hakuta, T.; Yoshitone, H.; Todo, N.; Kato, J. Adv. Hydr. Energy 1981, 2, 325-33.
5. Cacciola, A.; Leung, P.S. Europ. Patent 40 282, 1980.

6. Leenaars, A.F.M.; Keizer, K.; Burggraaf, A.J. J. Mater. Science 1984, 19, 1077-88.
7. Leenaars, A.F.M.; Burggraaf, A.J. accepted for publication in J. Colloid Interface Science 1985.
8. Keizer, K.; Leenaars, A.F.M.; Burggraaf, A.J. Science of Ceramics 1983, 12, 101-6.
9. Yoldas, B.E. J. Mater. Sci. 1975, 10, 1856-60.
10. Dal, P.H.; Berden, W.J.H. Science of Ceramics 1968, 4, 113-31.
11. Strathmann, H. Chem. Ztg. 1979, 103, 211-9.
12. Debije, P.; Cleland, R.L. J. Appl. Phys. 1959, 36, 843-9.
13. Zeman, L; Wales, M. In "Synthetic Membranes II: Hyper - and Ultrafiltration Uses", Turbak, A.F. ed. ACS symposium Series No 154, Amer. Chem. Soc., Washington, DC 1981, 411-34.
14. Keizer, K.; Leenaars, A.F.M.; Burggraaf, A.J. In "Ceramics in Advanced Energy Technologies", Kröckel, H.; Merz, M.; van der Biest, O. eds., D. Reidel Publishing Company: Dordrecht 1984, 367-85.
15. Leenaars, A.F.M.; Burggraaf, A.J. part 3, submitted to J. Membr. Science 1985.
16. Leenaars, A.F.M.; Burggraaf, A.J. part 4, submitted to J. Membr. Science 1985.
17. Carman, P.C. Trans. Inst. Chem. Eng. 1937, 15, 150-166.

RECEIVED March 4, 1985

6

Plasma-Polymerized Membranes of 4-Vinylpyridine in Reverse Osmosis

S. SUZUKI, T. OGAWA, and N. HITOTSUYANAGI

Department of Chemistry, Science University of Tokyo, Shinjuku-ku, Tokyo 162, Japan

The reverse osmosis membranes, prepared by plasma-polymerization of 4-vinylpyridine using both 50 Hz low-frequency (Lf) and 13.5 MHz (Rf) generator were studied in terms of their membrane characteristics, morphologies, properties of polymerized substances and life. The membrane characteristics were evaluated for both inorganic and organic solutes and compared with those of cellulose acetate membranes. The surface and the cross-section of membranes were observed by scanning electron microscopy and the effects of pH-values and temperature of feed solution on the membrane life were investigated. The formation of nitril group by glow discharge was observed.

There are two methods of preparing reverse osmosis membranes by plasma polymerization. In the first method the plasma-polymerized substance precipitates on the substrate surface to form a layer, while in the second method the plasma polymerization takes place directly on the top of the substrate which is placed on one of two electrodes (1-5). Not many studies have been reported so far on the choice of the method.

In the present study the membrane preparation using 4-vinylpyridine and the characterization of the membrane properties were studied applying the second method for the membrane formation. Further attempt was made to use a luminous tube transformer to generate low frequency (50 Hz) instead of the high frequency generator (13.5 MHz). The properties of both low frequency (50 Hz) and high frequency (13.5 MHz) membranes were compared in terms of their transport properties, morphologies and infrared spectroscopies. Furthermore, the effects of glow discharge on millipore substrate were investigated.

Experimental

Reactor. The reactor used is shown in Figure 1. It consists of two 47-mm-diameter stainless steel electrodes enclosed in a 90-mm-

0097-6156/85/0281-0069$06.00/0

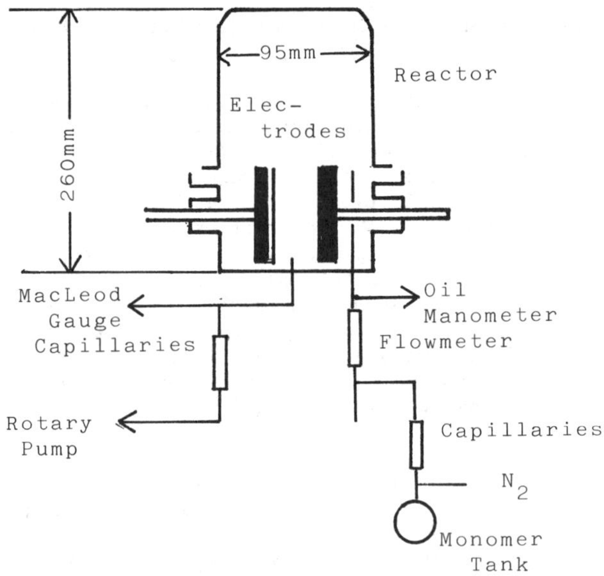

Figure 1. Schematic of experimental apparatus.

diameter reactor, the volume of which is 1500 ml. The distance between electrodes was chosen to 15 mm and a substrate was fixed to one of the electrodes with a set of teflon frames. The plasma-polymerized membranes (PPM) were made by using both 50 Hz and 13.5 MHz generators. The 50 Hz low frequency (Lf) generator was a luminous tube transformer (Matsushita Electric Co. Ltd.) and its primary voltage was adjusted by a slide rheostat. The 13.5 MHz radio frequency (Rf) generator was a radio frequency transmitter (Nippon Kohshuha Co. Ltd. Type SKN-05P, power input 20 W). These were connected to the electrodes through an impedance matching network.

Chemicals. Reagent grade 4-vinylpyridine was purified for plasma polymerization by means of distillation under a reduced pressure. As a substrate Millipore filter (0.025) was used.

Operating Conditions. All PPM membranes were pressurized at 80 kg/cm^2 for 2 h prior to reverse osmosis experiment. The operating conditions were as follow: operating pressure; 80, 70, 60 and 40 kg/cm^2, flow rate; 400 ml/min, feed solution temperature; 25 ± 1°C, effective membrane area; 9.26 cm, feed concentration; 0.01 mol/l.

Data Processing. The solute rejection R, the volume flux J_v and the solute flux Js were measured. The hydraulic permeability coefficients Lp was obtained from the pure water permeation rate. The reflexion coefficient was determined by extrapolating the rejection R to the infinite pressure. The solute permeability coefficients P was caluculated by the equation derived from the Spiegler-Kedem equation (6).

To evaluate the polymer deposition rate, an aluminum foil was used as substrate, on which the increase of weight was measured.

The effect of pH of the feed solution on the membrane performance was studied in the following way: the membrane performances were measured before and after the immersion of membranes in various pH solutions for 100 h. The pH value of the solution was adjusted by Britton-Robinson buffer solution and NaOH solution.

Scanning Electron Microscopy. The morphological observation of the surface, the cross-section of membranes and the effects of glow discharge on Millipore filters were carried out by an ALPH-10 scanning electron microscope (Akashi Seisakusho Co. Ltd.). The samples were coated under high vacuum with carbon and gold. In order to observe the cross-section the membranes were fractured after hardening in liquid nitrogen.

Infrared Spectroscopy. The infrared spectra were obtained using a Japan Spectroscopic Model DS-403G, in which the samples were prepared by plasma polymerization on KBr disk. The observation of the surface of membrane by an attenuated reflection (ATR) technique was performed using an IR435-ATR-2A (Shimazu Co.).

Results and Discussion

Effect of the Deposition Pressure on Membrane Performances. The effect of the deposition pressure on membrane performances is shown in Figure 2. Up to the deposition pressure of 1.0 torr, both rejection and water flux increase. Then the former decreases while the latter increases. It is noted that rejection exhibits a maximum value. The membrane corresponding to the maximum shows properties such as rejection of 83 - 96% and water flux of 0.4 - 0.5 m^3/m^2.day (0.1% NaCl 0.1 in feed solution and at 40 kg/cm^2). The effect of the deposition pressure on the deposited quantity is shown in Figure 3. The deposited quantity also passes through a maximum at a pressure of about 0.4 torr. However, the pressures, at which maxima are exhibited, are not identical, therefore it is considered that there is no connection between the rejection and the deposited quantity. It is worth noticing that the rejection has a maximum value and after the maximum the membrane loses its function gradually and accordingly the selectivity decreases.

Comparison of the Membrane Performances of CAM and PPM. In Table I the selectivity of the plasma polymerized membrane by using low frequency (50 Hz) generator (Lf-PPM) is shown in comarison with that of the Manjikian type cellulose acetate membrane (CAM). The reflexion coefficients of Lf-PPM membrane for Na_2SO_4, glucose and NaCl, are almost the same values as those of CAM, however in case of organic solutes the selectivity of Lf-PPM is better than that of CAM. The same conclusion can be derived from the solute permeability coefficient P, which is also shown in Table I. The reason why the PPM membrane shows better selectivity for organic solutes than CAM membrane, might be attributed to the weak hydrophilicity of PPM membrane.

Comparison of the Membrane Performances of Lf- and Rf-PPM Membranes. As the transport properties of both Lf- and Rf-PPM membranes with respect to inorganic and organic solutes are shown in Table II and III. In case of inorganic solutes the values of reflexion coefficient of Rf-PPM membrane are larger and those of the solute permeability coefficient P, smaller than Lf-PPM membrane. While in

Table I. Comparison of Transport Coefficients in PPM-Lf and CAM

Solute	P P M(Lf) σ (——)	P P M(Lf) P (cm/sec) ×10^{-5}	C A M σ (——)	C A M P (cm/sec) ×10^{-5}
Na_2SO_4	0.99	0.12	0.99	0.44
Glucose	0.97	0.92	0.99	
NaCl	0.95	2.54	0.99	1.56
n-C_4H_9NH	0.81	5.14	0.66	22.6
n-C_3H_7CHO	0.79	6.33	0.45	26.7
n-$CH_3COC_2H_5$	0.73	8.90	0.46	26.2
n-C_3H_7COOH	0.71	8.07	0.47	22.16
n-C_4H_9OH	0.57	8.56	0.42	61.4
C_6H_5OH	0.34	29.91	-0.27	

$L_{P,PPM(Lf)}$: 4.9×10^{-6} (cm^3/kg·sec)

$L_{P,CAM}$: 5.5×10^{-6} (cm^3kg·sec)

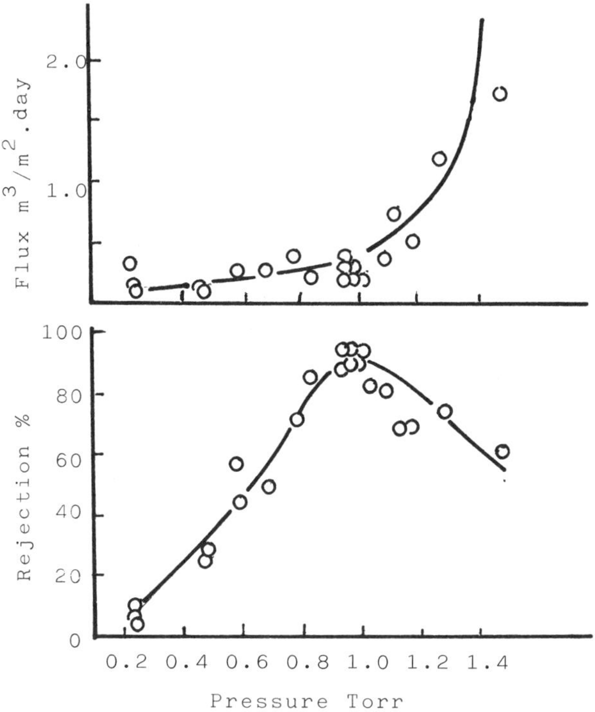

Figure 2. Effect of gas pressure on reverse osmosis performance: discharge power, 13 - 25 W; discharge time, 1 min; flow rate, 0.2 - 0.45 ml/min, STP.

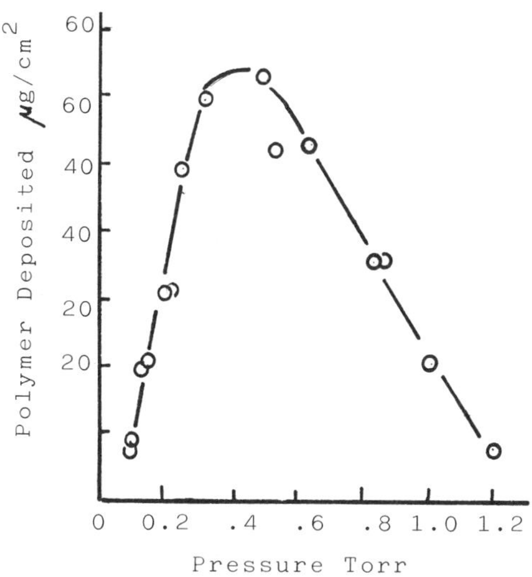

Figure 3. Effect of gas pressure on quantity deposited: discharge power, 13 - 25 W; discharge time 1 min; flow rate, 0.2 - 0.45 ml/min, STP.

case of organic solutes, the situations differ and the values of reflexion coefficients of Rf-PPM membrane are smaller. There are some exception in both cases, for instance, with respect to $NaNO_3$ and n-C_3H_7CHO. Though the plasma-polymerized substances on both Lf- and Rf-membrane are identical as shown later, the latter is superior in inorganic solute separation, while for organic solutes the former is superior. The difference in transport properties may be due to the difference in the adhesive state between the surface of substrate and the plasma-polymerized layer or due to the difference in the density of the layer deposited by plasma discharge.

Effects of Operating Conditions on Membrane Performances. The variation of membrane performances over 300 h operation time is shown in Figure 4. In case of the Lf-PPM the salt rejection decreases about 10% from 97 to 85% at 300 h operation, however the water flux increases about 5 times. In case of the Rf-PPM membrane, after a slight increase at the beginning of operation the rejection levels off to about 78% and the water flux increases during first 80 h and then decreases gradually to a constant value (about 24 gfd) at 300 h operation. These results show that the Rf-PPM membrane has a higher durability.

Effect of pH. The effect of pH of the feed solution on the membrane performances is shown in Figure 5. The rejection shows maxima at certain pH values for both membranes i.e. at pH 5 in case of the Lf-PPM membrane, at pH 9 in case of the Rf-PPM membrane. Similarly the data on water flux show minima at certain pH values, where the membrane shows the most stable behavior. In case of the Lf-PPM membrane, the water flux shows a minimum at pH 7 and in case of the Rf-PPM at pH 8.

Effect of Temperature. Figure 6 shows the variation of the membrane performance with the operating temperature in case of the Rf-PPM membrane. The rejection decreases slightly between 20° - 60°C, but

Table II. Transport Coefficients for Inorganic Solutes

	P P M(Lf)		P P M(Rf)	
	σ	P	σ	P
	(——)	(cm/sec)	(——)	(cm/sec)
Solute		$\times 10^{-4}$		$\times 10^{-4}$
LiCl	0.71	1.03	0.89	0.82
NaCl	0.70	0.88	0.95	0.63
KCl	0.66	2.28	0.84	1.64
RbCl	0.65	1.90	0.86	1.70
CsCl	0.65	2.04	0.84	1.87
Na_2SO_4	0.87	0.61	0.99	0.08
$CaCl_2$	0.75	1.23	0.95	0.55
$NaNO_3$	0.57	2.71	0.69	3.16

$L_{p,PPM(Lf)}$: 1.59×10^{-5} (cm^3/kg·sec)
$L_{P,PPM(Rf)}$: 2.92×10^{-5} (cm^3/kg·sec)

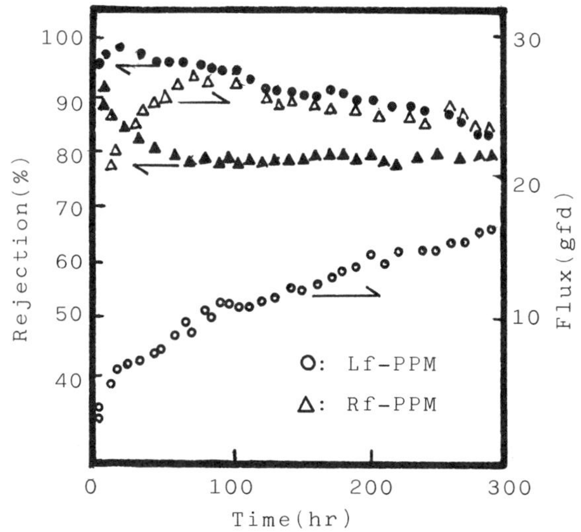

Figure 4. Variation of membrane performance with operation time: O, Lf-PPM; and Δ, Rf-PPM.

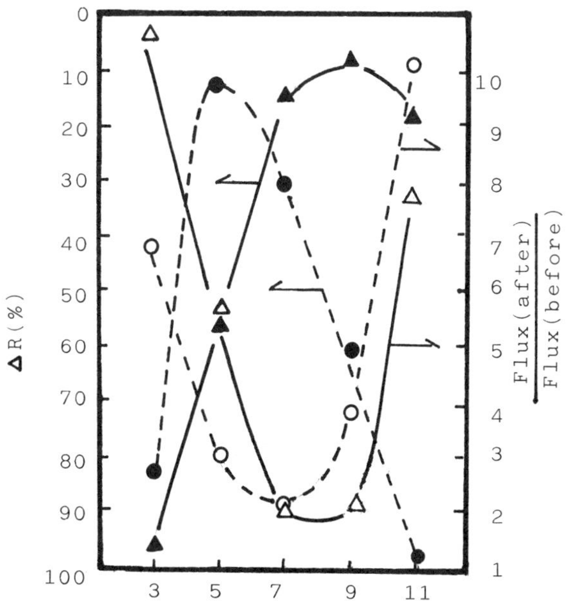

Figure 5. Effect of feed's pH on membrane performance. Symbols are the same as in Figure 4. R means increase of rejection during the operation.

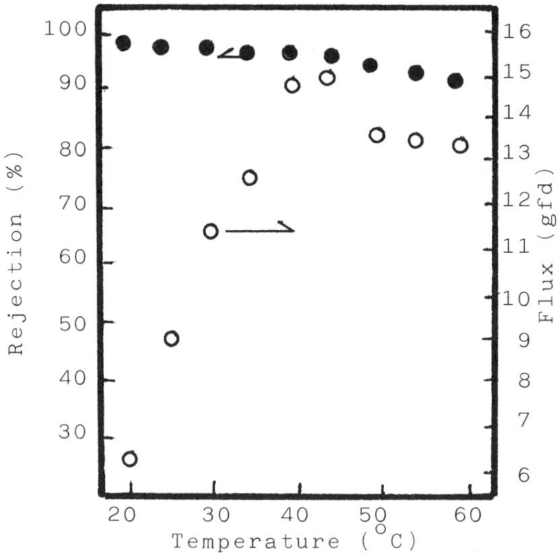

Figure 6. Effect of operating temperature on membrane performance in case of Rf-PPM.

Table III. Transport Coefficients for Organic Solutes

Solute	PPM(Lf) σ (—)	PPM(Lf) P (cm/sec) $\times 10^{-5}$	PPM(Rf) σ (—)	PPM(Rf) P (cm/sec) $\times 10^{-5}$
Glucose	0.97	0.92	0.99	1.96
$n\text{-}C_4H_9NH_2$	0.81	4.97	0.59	29.1
$n\text{-}C_3H_7CHO$	0.79	6.13	0.87	10.3
$CH_3COC_2H_5$	0.73	8.61	0.46	65.2
$n\text{-}C_3H_7COOH$	0.71	7.81	0.47	58.0
$n\text{-}C_4H_9OH$	0.57	8.28	0.47	52.8
Urea	0.65	2.67	0.46	76.7
C_6H_5OH	0.34	28.2	0.22	49.8

$L_{P,PPM(Lf)}$: 4.90×10^{-6} (cm^3/kg·sec)
$L_{P,PPM(Rf)}$: 1.98×10^{-5} (cm^3/kg·sec)

the water flux is subjected to a large variation. From these results it is concluded that both Lf- and Rf-PPM membranes do not have good durability for a long term operation and the membranes are susceptible to the change in pH and temperature of feed solutions.

Morphology of Membranes. Scanning electron micrographs of the surface of Lf- and Rf-PPM membranes are shown in Figure 7 and 8, where the spherical polymers are observed. In both cases the size of spheres ranges from 0.07 to 0.4 μm in diameter. In Figure 9 and 10, the cross-sections of both membranes are shown. The plasma-polymerized layer of Lf-PPM membrane is about 0.6 μm and of Rf-PPM, 0.4 μm respectively. In the latter case it is noted that the interface between plasma-polymerized layer and the surface of the substrate is hardly discernible and the adhesive state between them is tighter.

Infrared Spectroscopy. Figure 11 shows the infrared spectra of both 4-vinylpyridine and the plasma-polymerized layer. the absorption bands at 2200 cm^{-1} and in range of 1470 - 1600 cm^{-1} could be assigned to the nitrile group and the pyridine ring respectively (7,8). The absorption bands of the pyridine ring were observed in all cases, but that of the nitrile group could be observed only for the plasma-polymerized polymer. Consequently it seems reasonable to assume that some of pyridine rings were decomposed by a glow discharge and the nitrile group was formed.

By means of ATR technique the infrared spectra of both Lf- and Rf-PPM membranes were identical and exhibited the characteristics peak at 2200 cm^{-1} assigned to the nitrile group.

Effects of Glow Discharge on Millipore Substrate. The surface of Millipore substrate was damaged by glow discharge and the pore size on the surface became larger. The variation of the pore size, as effected by increase of the power supply, the discharge time and the pressure of the reactor in nitrogen atmosphere, was studied by means

Figure 7. Scanning electron micrograph of the surface of Lf-PPM. x10000

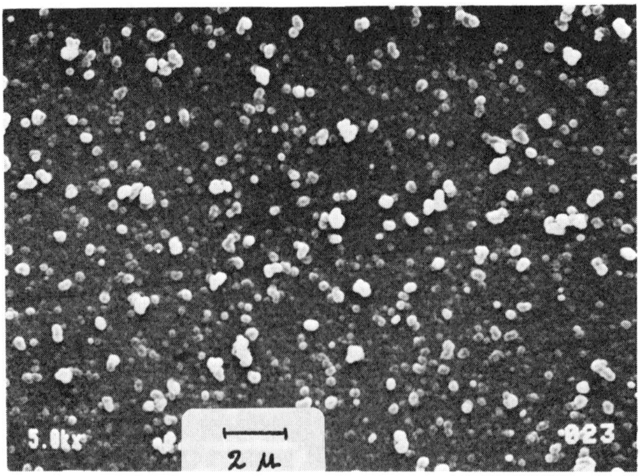

Figure 8. Scanning electron micrograph of the surface of Rf-PPM. x5000

Figure 9. Scanning electron micrograph of the cross-section of Lf-PPM. x5000

Figure 10. Scanning electron micrograph of the cross-section of Rf-PPM. x10000

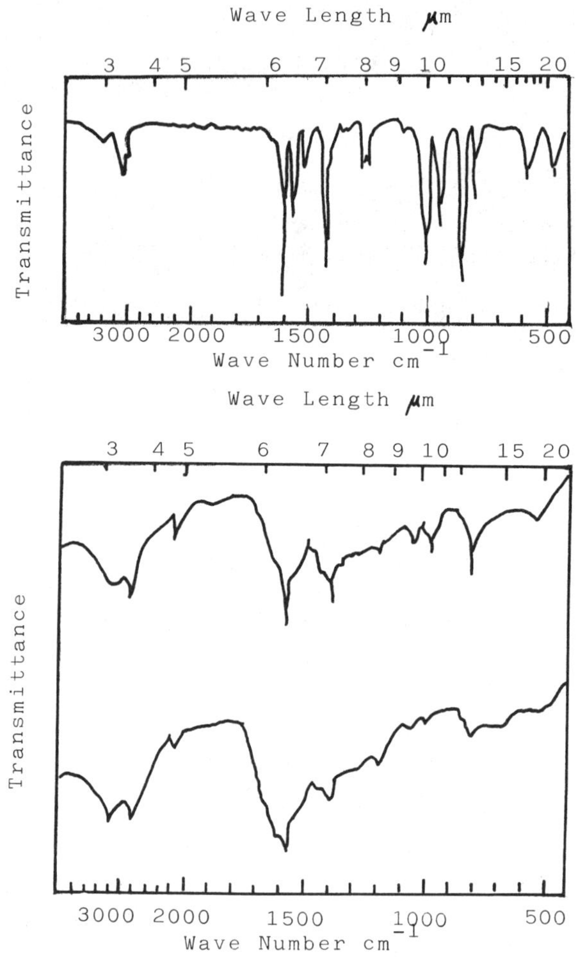

Figure 11. Infrared spectra of monomer and plasma-polymerized 4-vinylpyridine deposited on KBr.

of scanning electron microscopy. In all cases, the pore size grows larger and reached certain values between 0.15 and 0.25 μm, which are close to the size of the pores in the interior of membrane. Therefore it may be concluded that the surface layer was destroyed at the beginning of discharge and that the interior of the membrane could be exposed.

Conclusion

By using a low frequency 50 Hz generator, an attempt was made to prepare plasma-polymerized membranes. The rejection by the membrane so produced was in the range of 83 - 96% and the water flux was in the range of 0.4 - 0.5 $m^3/m^2 \cdot day$(0.1% NaCl in feed solution and at 40 kg/cm^2). The transport property of the Lf-PPM membrane was discussed in terms of reflexion coefficients, the hydraulic permeability coefficients and the solute permeability coefficients obtainable for those membranes and compared with that of the Manjikian type cellulose acetate membrane.

The performances of both Lf- and Rf-PPM membranes were compared. The selectivity of the Rf-PPM membrane for inorganic solutes was superior, on the other hand that of the Lf-PPM membrane was superior for organic solutes.

The durability of both Lf- and Rf-PPM membrane in the long term operation and at different pH values and temperatures of feed solution was studied. They did not show an excellent durability and therefore it seems necessary to find an appropriate way of membrane reinforcement.

The spherical polymers between 0.07 - 0.4 μm were observed on the surface of both plasma-polymerized membranes. The thickness of plasma polymerized layer ranged from 0.4 to 0.6 μm.

By infrared spectroscopy it was concluded that the pyridine ring decomposed by glow dicharge and the nitrile group formed.

The surface layer of millipore substrate was destroyed by glow discharge and the interior of the membranes was exposed.

Acknowledgments

The authors wish to acknowledge Prof. Drs. Matsuura of NRC, Ohya of Yokohama University and Mrs. E. Tsukuda of the Institute of Physical and Chemical Research, for their valuable advices.

Literature Cited

1. Wydeven, Th; Hollahan R., in "Techniques and Applications of Plasma Chemistry"; Hollahan, J. R. and Bell, A. T., Eds.; John Willey & Sons; New York, 1974; p. 215.
2. Buck, K. R.; Davar, V. K. Br. Polym. J. 1970, 2, 238.
3. Yasuda, H.; Lamaze, C. E. J. Appl. Polym. Sci. 1973, 17, 201.
4. Bell, A. T.; Wydeven, Th.; Johnson, C. C. J. Appl. Polym. Sci. 1975, 19, 1911.
5. Peric, D.; Bell, A. T.; Shen, M. J. Appl. Polym. Sci. 1977, 21, 2661.
6. Spiegler, K. S.; K edem, O. Desalination, 1966, 1, 311.
7. Tazuke, S.; Okamura, S. J. Polym. Sci. A-1. 1967, 5, 1083.

8. "Atlas of Spectral Data and Physical Constants for Organic Compounds"; Grasselli, J. G.; Ritchey, W. M., Eds.; CRC Press; Ohio, 1975, 2nd Ed.; Vol. I, p. 342 and 346.

RECEIVED February 22, 1985

7

γ-Ray Induced Enhancement Effect on Salt Rejection Properties of Irradiated Membranes

R. Y. M. HUANG and J. J. KIM

Department of Chemical Engineering, University of Waterloo, Waterloo, Ontario, Canada N2L 3G1

The effect of gamma-rays on the thin film composite membranes of sulfonated poly(phenylene oxide) (SPPO) was investigated. Composite thin film membranes of sulfonated poly(phenylene oxide) on polysulfone (PS) substrate (SPPO-PS) were prepared by coating the hydrogen or sodium form of SPPO onto the porous polysulfone (PS) substrate in various thicknesses. The SPPO-PS membranes were then irradiated in air, oxygen or nitrogen atmospheres to doses of 0.5 megarads from a Cobalt-60 gamma-ray source at a dose rate of 0.11 megarads/hour. The effect of the gamma-ray irradiation was found to be beneficial in enhancing the separation characteristics while monitoring constant flux rates. The improvement in salt rejections in reverse osmosis of samples of Alberta tar sands waste waters containing Na^+, Ca^{++}, Mg^{++}, Cl^-, SO_4^{--}, HCO_3^-, CO_3^{--}, SiO_2^{--}, and heavy oil residue was found to be quite significant and are presented in detail. The best results were obtained at radiation doses of 2-5 megarads and resulted in increases of salt rejections from 88% to over 98% of the irradiated SPPO-PS membranes, at flux rates of 10-30 gfd.

Although success has been achieved in the past decade in reclaiming non-potable waters by RO through various types of thin cellulosic ester membrane films, there exists the need for more stable longer life materials. One of the most promising noncellulosic membrane systems for reverse osmosis (RO) applications appears to be sulfonated poly (2,6-dimethyl-1,4-phenylene oxide) (SPPO) (1).

During the past few years considerable progress has been made in the development of thin film composite membranes of various polymers and their development has been described in detail in an article by Cadotte and Petersen (2) in 1981. The synthesis of poly-(phenylene oxide) and its sulfonation has been described in a report to the office of Saline Water Research and Development in 1971 by Chludzinski et al (3) of the General Electric Company. Huang et al have recently reported work on ionically crosslinked

0097-6156/85/0281-0083$06.00/0

poly(acrylic acid)composite thin film membrane cast onto a porous polysulfone substrate (4).

The present series of papers (5-7) is concerned with the synthesis and transport properties of sulfonated poly(phenylene oxide) thin film composite membranes for reverse osmosis applications. Part I (5) dealt with the synthesis of sulfonated poly(phenylene oxide) (SPPO) polymer and Part II (6) has described the preparation of thin film composite membranes of sulfonated poly(phenylene oxide) (SPPO)-(PS), (SPPO-PS) membranes and its application for the purification of Alberta tar sands waste waters using reverse osmosis. This study (Part III) will focus on the enhancement effect of gamma-ray irradiation of the SPPO-PS thin film composite membranes on the separation characteristics of the reverse osmosis separation process.

Experimental Procedures

Synthesis of Poly(phenylene oxide) PPO and its sulfonation to sulfonated poly(phenylene oxide) SPPO. The synthesis of PPO and its subsequent sulfonation to SPPO has been described in detail in Part I of this series.

Materials. 2,6-Dimethylphenol (Gold Label, 99.8% purity), Copper-(I)chloride (Purified Grade, 99% purity) and O-Dichlorobenzene (99% purity) were obtained from Aldrich Chemicals. Chlorosulfonic acid (Practical Grade), Anhydrous magnesium sulfate (Baker Analytical Reagent Grade, 99.4% purity) and Chloroform (Baker Analytical Reagent Grade, 99.2% purity) were obtained from J.T.Baker Chemicals. Pyridine (Fisher Certified Reagent, 99% purity) was from Fisher Scientific. The oxygen used was high purity, 99.6% supplied by Union Carbide Ltd., while the Polysulfone (MW = 30,000) was obtained from Polyscience Inc. The Alberta tar sands waste water samples were obtained from the Water Sample Bank, Alberta Research Council, Edmonton, Alberta.

Preparation of Porous Polysulfone Substrate. A solution containing 12.5 wt. % PS and 12.5 wt. % methyl cellosolve in dimethylformamide was cast onto a clean glass plate using a glass bar in the thickness of 0.3 mm. After the casting, the coated liquid film was immersed into 15 wt. % NaCl quenching bath immediately. The film gelled very quickly, it was then washed with water, and cut into the required size with a membrane die. Finally, it was put into deionized water for at least 24 hours and thoroughly dried before SPPO polymer solution casting.

Preparation of Sulfonated Poly(phenylene oxide) Composite Membrane. The SPPO polymer was exhaustively dried for 200 hours under vacuum at room temperature (hydrogen form) or 70°C (sodium form), and the dry SPPO polymer was dissolved in 2:1 chloroform/methanol or pure methanol solvents, to form a 4 wt. % casting solution and cast onto 1.0 mil microporous polypropylene or polysulfone substrate stuck onto a glass plate to form a coating of SPPO polymer of 0.2 mil dried thickness, and the composite membrane formed was dried for a minimum of 2 hours under cover and overnight without cover at ambient conditions. The dried membranes were removed from the glass plate immersed into 10 wt. % NaCl aqueous solution to convert the hydrogen form into sodium form and stored wet in a 10 wt. % NaCl aqueous solution container. Since polysulfone substrate was soluble in chloroform, the casting solvent was pure methanol (without chloroform).

Gamma-ray Irradiation of SPPO-PS Thin Film Composite Membrane.

To effect crosslinking, the SPPO-PS thin film composite membrane was Cobalt-60 gamma-ray irradiated in a Gammacell-220-Co-60 unit, supplied by Atomic Energy of Canada Ltd., to various irradiation doses up to 5 megarads at a dose rate of 0.11 megarad/hour in air, nitrogen or oxygen atmospheres at room temperature.

Membrane Durability to Chlorine, Acid and Base (8,9). Immersion tests were performed to evaluate the resistance of membranes to oxidative chlorine, acid and base. All membrane exposures were carried out under equilibrium conditions at fixed concentration and constant pH. Chlorine was added as sodium hypochlorite to buffer solution at pH 7. The selected chlorine level was 30 ppm, representing a tenfold higher level than 3 ppm, the average chlorine residual applied in water disinfection practice. Aqueous sulfuric acid solution (1N, pH=0.3) was used as acid and aqueous sodium hydroxide solution (1N, pH=14) as base. The solutions in one liter poly-ethylene jars were stirred magnetically and the jars tightly stoppered to prevent concentration change by volatility. Chlorine level and pH were checked periodically and adjusted, as needed, by the addition of a small volume of concentrated stock solution or deionized water. Solutions of these chemicals were quite stable when tighly stoppered. After immersion of membranes to various exposure times, these were washed with deionized water and membrane performance characteristics were then measured.

Measurement of Tensile Strength. The Universal testing instrument (Instron Model 1122) was used for tensile strength measurements of the membranes. The thin film membrane specimen, saturated with water, was about 8 mm wide and 20 mm long. The crosshead speed was 1 mm/min.

Analytical Procedures. Sodium chloride or other inorganic compound concentrations for single component solutions were determined using a Water Associates differetial refractometer Model R403. Quantitative analyses of the mixed component solution in aqueous phases were carried out by standard methods (10). Atomic absorption spectroscopy (Perkin-Elmer Model 303) was used for determination of dissolved sodium, calcium and magnesium. Sulfate was determined by the turbidimetric method with $BaCl_2$ using a spectrophotometer (Bausch and Lomb Spectronic 20) at 420 nm. Choride was determined by potentiometric titration with $AgNO_2$ solution using glass and silver-silver chloride electrodes (Potentiograph E576, Metrohm Herisaw Co., Switzerland). Heavy oil was determined by ultraviolet absorbance at 253.7 nm (Varian Techtron UV-VIS Spectrophotometer Model 635) (10).

The intrinsic viscosities of SPPO and gamma-ray irradiated SPPO polymers were determined in an Uhbelohde viscosimeter in methanol solutions at 25°C.

Purification of Waste Water from Heavy Oil Fields. Reverse osmosis tests were carried out on synthetic and natural waste waters from the heavy oil fields. The waste waters were obtained from the Esso facility at Cold Lake, Alberta, the Shell facility at Peace River, Alberta, and the Texaco facility at Athabasca, Alberta. The aqueous phase remaining from the bitumen-water separation contained residual minerals, usually about 5000 to about 15000 ppm total dissolved solids (TDS), and a certain amount of heavy oil (10). These waste waters were prefiltered before reverse osmosis tests to remove suspended solid particles for membrane protection. The reverse osmosis tests were conducted in 6 high pressure cells, which have been described in detail in the previous paper (6). The effective membrane area was 18.1 cm^2 (4.3 cm in diamter). The experiments were conducted at pressures of 300 to 700 psig, at a room temperature of approximately 20°C, and with feed solution circulated during the experiment. The recovery for each membrane sample was about 0.1%. With this low recovery, concentration polarization effects were minimized and a true measure of the intrinsic performance of the membrane could be approached.

Results and Discussion

Reverse Osmosis Results of Irradiated Membrane. Tables I, II and III show the results of reverse osmosis for the purification of waste waters from Esso Cold Lake, Alberta, Shell Peace River, Alberta, and Texaco Athabasca, Alberta heavy oil fields respectively, for a SPPO-PS composite membrane. These tables compare the separation characteristics of non-irradiated and gamma-ray irradiated (5M Rads) SPPO-PS composite membranes and show the significant improvement in % rejection of the various salts caused by the gamma-ray irradiation process with very little change in the production rate. These results indicate that the membrane rejection characteristics are excellent and can be considered adequate for application for the purification of waste waters from the heavy oil fields.

Both the hydrogen form and sodium form of sulfonated poly(phenylene oxide) in SPPO-PS composite membranes were exposed to irradiation doses ranging from 0 to 5 M Rads, in air, nitrogen and oxygen atmospheres and the results are shown in Figures 1 and 2. As can be seen, their salt rejection characteristics were increased considerably from 88 to 98% without a change in production rates. The best results were obtained at radiation doses ranging from 2-5 M Rads while differences in the irradiation atmospheres were found to be negligible. Figure 3 shows the effects of radiation dose and IEC (ion exchange capacity) of the SPPO-PS membranes on their salt rejection and production rates. The best results were obtained with an IEC of 2.10 at a radiation dose of 5 M Rads. The effect of membrane ion exchange capacity, membrane water content, coated polymer thickness as well as the dependence of feed composition on the production rate and salt rejection for non-irradiated membranes have been discussed in detail in Part II of this series (6).

Radiation Chemistry of Sulfonated Poly(phenylene oxide) Polymer The radiation chemistry of sulfonated poly(phenylene oxide) is not clearly understood and is currently still undergoing investigation. Chapiro (11-12) has reviewed the basic mechanisms of radiation-

Table I. Reverse Osmosis[a] Results for the Purification of Waste Waters from Esso Cold Lake Alberta

Parameter	Feed (ppm)	Permeate (ppm)								Rejection(%)							
Membrane[(b)]		SPPO[(e)]				SPPO-γ[(f)]				SPPO				SPPO-γ			
Run #		1	2	3	Avg	4	5	6	Avg	1	2	3	Avg	4	5	6	Avg
Ca^{++}	33.4	0.09	0.09	0.04	0.07	0.05	0.03	0.04	0.04	99.7	99.7	99.9	99.8	99.9	99.9	99.9	99.9
Mg^{++}	3.2	0.03	0.06	0.05	0.05	0.02	0.03	0.04	0.03	99.1	98.1	98.1	98.4	99.4	99.1	98.8	99.1
Na^{+}	1312	339	268	91	233	96	81	112	96	74.2	79.6	93.1	82.2	92.7	93.8	91.5	92.7
Cl^{-}	3129	578	470	239	429	242	224	283	250	81.5	85.0	92.4	86.3	92.3	92.8	91.0	92.0
HCO_3^{-}	259	38.1	35.0	25.9	33.0	24.4	22.9	28.9	25.4	85.3	86.5	90.0	87.3	90.6	91.2	88.8	90.2
CO_3^{--}	0	0	0	0	0	0	0	0	0								
SO_4^{--}	98	7.6	0.1	0.1	2.6	0.1	0.1	2.0	0.7	92.2	99.9	99.9	97.3	99.9	99.9	98.0	99.3
Total Hardness[(c)]	97	0.35	0.47	0.35	0.39	0.25	0.20	0.22	0.22								
Alkalinity[(d)]	425	63	58	43	55	40	38	48	42								
pH	7.45	6.48	7.00	6.80	6.76	6.75	6.70	6.95	6.80								
Production rate (gfd)		8.88	15.01	6.94	10.28	8.33	9.86	7.53	8.57								

(a) 600 psi, 20°C
(b) SPPO-PS composite membrane 0113-12NM Series, IEC = 2.10 meq/g, thickness = 0.2 mil/3 mil = 5 μm/75 μm
(c) ppm $CaCO_3$
(d) ppm $CaCO_3$, to pH 4.50
(e) Non-irradiated membranes
(f) Gamma-ray irradiated (5 M Rads) membranes

Table II. Reverse Osmosis[a] Results for the Purification of Waste Waters from Shell Peace River Alberta

Parameter	Feed (ppm)	Permeate (ppm)								Rejection(%)							
Membrane[(b)]		SPPO[(e)]				SPPO-γ[(f)]				SPPO				SPPO-γ			
Run #		1	2	3	Avg	4	5	6	Avg	1	2	3	Avg	4	5	6	Avg
Ca^{++}	43.7	0.17	0.10	0.06	0.11	0.01	0.03	0.03	0.02	99.6	99.8	99.9	99.7	99.9	99.9	99.9	99.9
Mg^{++}	11.2	0.07	0.07	0.05	0.06	0.02	0.03	0.04	0.03	99.4	99.4	99.6	99.5	99.8	99.7	99.6	99.7
Na^{+}	519	86	56	23	55	23	21	26	23	83.4	89.2	95.6	89.4	95.6	96.0	95.0	95.6
Cl^{-}	833	152	102	45	100	43	38	52	44	81.8	87.8	94.6	88.0	94.8	95.4	93.8	94.7
HCO_3^{-}	296	56.9	39.0	24.4	40.1	25.9	25.3	26.8	26.0	80.8	86.8	91.8	86.5	91.3	91.5	90.9	91.2
CO_3^{--}	0	0	0	0	0	0	0	0	0								
SO_4^{--}	170	9.1	2.8	1.2	4.4	0.6	1.4	1.0	1.0	94.6	98.4	99.3	97.4	99.6	99.2	99.4	99.4
Total Hardness[(c)]	155	0.71	0.54	0.36	0.54	0.11	0.20	0.24	0.18								
Alkalinity[(d)]	485	94	64	40	66	43	42	44	43								
pH	7.65	6.86	6.65	6.25	6.59	6.40	6.50	6.60	6.50								
Production rate (gfd)		8.21	12.87	6.03	9.04	7.01	8.47	6.39	7.29								

(a) 600 psi, 20°C
(b) SPPO-PS composite membrane 0113-12NM Series, IEC = 2.10 meq/g, thickness = 0.2 mil/3 mil = 5 μm/75 μm
(c) ppm $CaCO_3$
(d) ppm $CaCO_3$, to pH 4.50
(e) Non-irradiated membranes
(f) Gamma-ray irradiated (5 M Rads) membranes

Table III. Reverse Osmosis[a] Results for the Purification of Waste Waters from Texaco Athabasca Alberta

Parameter	Feed (ppm)	Permeate (ppm)								Rejection(%)							
Membrane[(b)]		SPPO[(e)]				SPPO-γ[(f)]				SPPO				SPPO-γ			
Run #		1	2	3	Avg	4	5	6	Avg	1	2	3	Avg	4	5	6	Avg
Ca^{++}	13.5	0.11	0.10	0.02	0.08	0.01	0.02	0.02	0.02	99.2	99.3	99.9	99.4	99.9	99.9	99.9	99.9
Mg^{++}	1.5	0.01	0.01	0.01	0.01	0.01	0.01	0.01	0.01	99.3	99.3	99.3	99.4	99.3	99.3	99.3	99.3
Na^{+}	355	26.0	6.6	7.0	13.2	5.1	4.0	4.8	4.6	92.7	98.1	98.0	96.3	98.6	98.9	98.6	98.7
Cl^{-}	550	43	13	14	23	8	7	9	8	92.2	97.6	97.5	95.8	98.5	98.7	98.4	98.5
HCO_3^{-}	119	11.6	8.8	9.8	10.1	11.6	10.1	9.4	10.4	90.3	92.6	91.8	91.5	90.3	91.5	92.1	91.3
CO_3^{--}	0	0	0	0	0	0	0	0	0								
SO_4^{--}	233	8.1	0.1	0.6	2.9	0.8	0.2	0.1	0.4	96.5	99.9	99.7	98.8	99.7	99.9	99.9	99.8
Total Hardness[(c)]	40	0.31	0.29	0.09	0.23	0.07	0.09	0.09	0.08								
Alkalinity[(d)]	195	19	15	16	17	19	17	16	17								
pH	7.05	6.25	5.85	6.15	6.08	6.45	6.20	6.10	6.25								
Production rate (gfd)		9.15	19.49	8.51	12.38	10.36	11.80	9.30	10.49								

(a) 600 psi, 20°C
(b) SPPO-PS composite membrane 0113-12NM Series, IEC = 2.10 meq/g, thickness = 0.2 mil/3 mil = 5 μm/75 μm
(c) ppm $CaCO_3$
(d) ppm $CaCO_3$, to pH 4.50
(e) Non-irradiated membranes
(f) Gamma-ray irradiated (5 M Rads) membranes

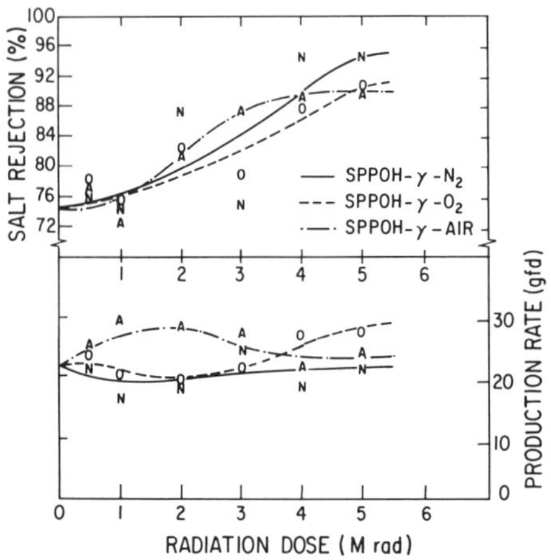

Figure 1. Effect of gamma-ray irradiation on sulfonated poly-(phenylene oxide) in SPPO-PS composite membrane (SPPO hydrogen form, IEC = 2.83 meq/g).

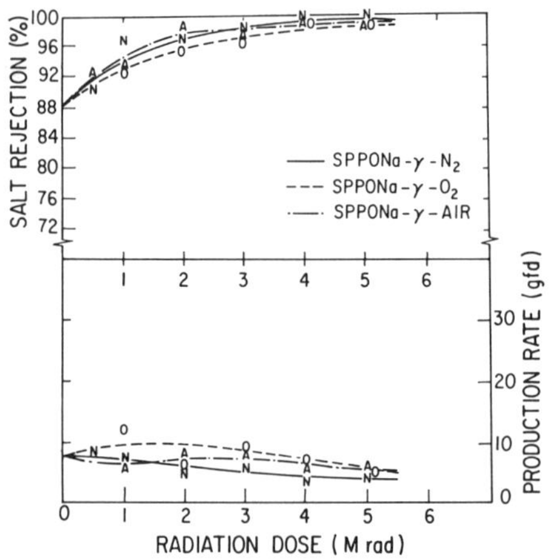

Figure 2. Effect of gamma-ray irradiation on sulfonated poly-(phenylene oxide) in SPPO-PS composite membrane (SPPO sodium form, IEC = 2.10 meq/g).

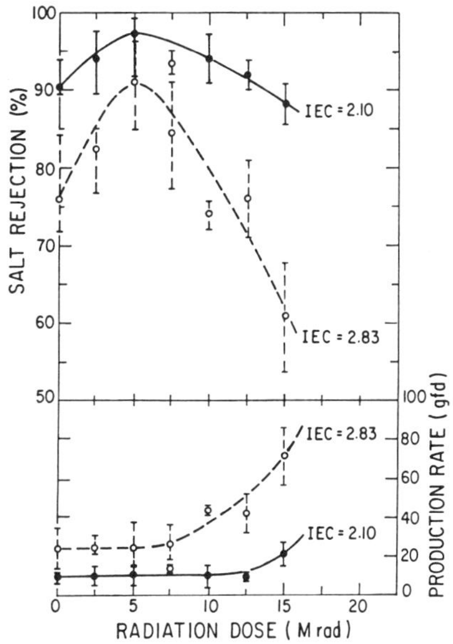

Figure 3. Reverse osmosis results of gamma-ray irradiated SPPO-PS composite membranes (sodium form SPPO, 600 psig, 1000 ppm NaCl, 25°C).

induced changes in polymers, most reactions are commonly interpreted on the basis of free radical processes, but other species, ions and other reactive intermediates may also play a significant role. The main end effects of gamma-rays on polymers are crosslinking and degradation of the polymer chains. There is no published literature on the radiation chemistry of sulfonated poly(phenylene oxide) but the main effects can be considered to be crosslinking coupled with some degradations. King (13) has reviewed the radiation chemistry of an analogous polymer poly(ethylene oxide) where the main-effects of gamma radiation have been found to be crosslinking at low doses after which chain degradation predominates. The final result was a combination of cross linking and oxidative degradation. No exact G-values for crosslinking and degradation have yet been reported. However, since the phenylene oxide unit in poly(phenylene oxide) contains a benzene ring structure which is more stable to radiation, it may be surmised that degradation is less prone to occur in favour of more crosslinking.

Figure 4 shows the intrinsic viscosity of SPPO and gamma-ray irradiated SPPO (2 M Rads), as can be seen there appears to be some degradation of the γ-ray irradiated SPPO. However, an indication of the crosslinking process can be seen in the tensile strength (psi) vs radiation dose plot as shown in Figure 5. The Δ(SPPO/PS-PS) psi plot shows an increase up to 5 M Rads after which it decreases. This is a strong indication that crosslinking occurs at low radiation doses followed by degradation at higher doses as observed for poly(ethylene oxide). The results of solubility tests for SPPO and gamma-ray irradiated SPPO polymers are shown in Table IV.

Durability and Stability of SPPO-PS Thin Film Composite Membranes to Chlorine, Acids and Bases. Figure 6 shows the changes in RO performance and tensile strength of irradiated or non-irradiated SPPO-PS memranes on continuous exposure to 30 ppm chlorine at pH 7.0 As can be seen, there was very little change in the % salt rejection and tensile strength even after exposure to 600 hrs, while there was a slight increase in the production rate. Figure 7, shows an identical plot on continuous exposure to 1N-NaOH (pH - 14.0). Similar trends can be noted, except for the significant increase in production rate as exposure time increases. Changes in RO performance and tensile strength on continuous exposure to 1N-H_2SO_4 (pH = 0.3) are shown in Figure 8, similar trends as for the case of chlorine exposure are observed. These durability and stability tests indicate the excellent stability of the irradiated and non-irradiated SPPO-PS thin film composite membranes to chlorine, acids and bases which make them very suitable for use in the purification of the Alberta tar sands waste waters. This compares favourably to conventional membranes such as assymetric cellulose acetate membranes which fail rapidly under these conditions.

Conclusions

The gamma-ray irradiation of SPPO-PS thin film composite membranes in air, oxygen or nitrogen atmospheres at radiation doses ranging from 2-5 M Rads has shown significant improvements in the % salt rejection and constant flux rates in the purification of Alberta Tar sands waste water for recycling use in steam generation. The

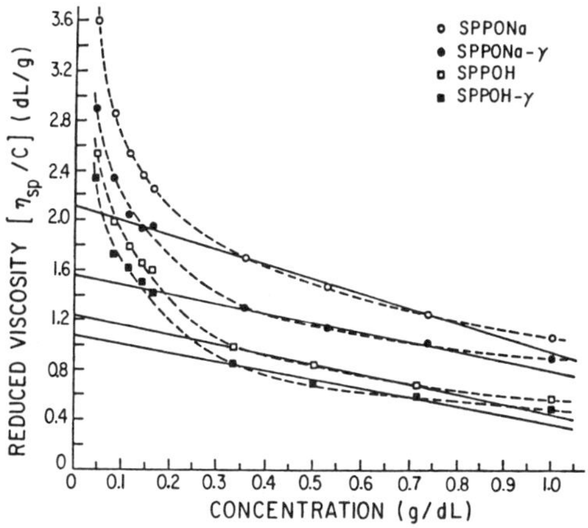

Figure 4. Measurement of intrinsic viscosity of SPPO and gamma-ray irradiated SPPO in methanol at 25°C. (a) SPPOH; sulfonated poly(phenylene oxide) hydrogen form, IEC = 2.07 meq/g. SPPONa; sulfonated poly(phenylene oxide) sodium form, IEC = 2.07 meq/g. (b) radiation dose; 2 M Rad

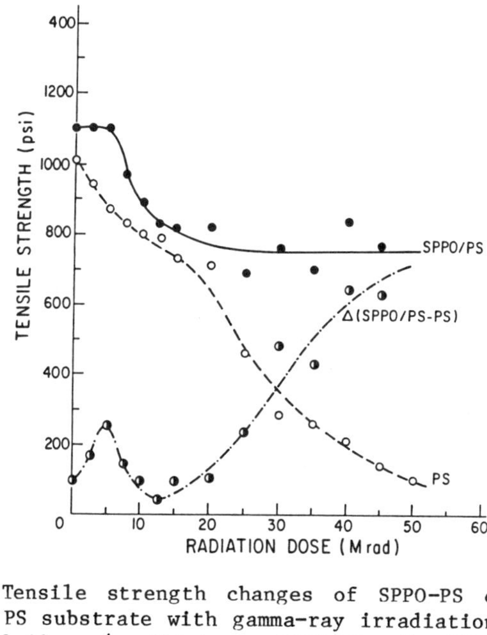

Figure 5. Tensile strength changes of SPPO-PS composite membrane and PS substrate with gamma-ray irradiation (sodium form SPPO, IEC=2.10 meq/g, thickness SPPO/PS = 5μm/75μm)

Table IV. Solubility of SPPO and Gamma Ray[a] Irradiated SPPO in Various Solvents

Solvent	Hydrogen[b] Bonding Group	Solubility Parameter $\delta (cal/cm^3)^{1/2}$	Solubility[c]			
			SPPOH	SPPOH-γ[d]	SPPONa	SPPONa-γ[d]
Water	S	23.4	–	–	–	–
Formamide	S	19.2	–	–	–	–
Glycerin	S	16.5	–	–	–	–
2-Pyrrolidone	S	14.7	+	+	+	+
Methanol	S	14.5	+	+	+	+
Formic acid	S	12.1	+	+	+	+
Acrylic acid	S	12.0	+	+	+	+
n-Propanol	S	11.9	+	+	–	–
Pyridine	S	10.7	+	+	–	–
Cresol	S	10.2	+	+	–	–
Dimethyl formamide	m	12.1	+	+	–	–
Acetone	m	9.9	+	+	–	–
Tetrahydrofuran	m	9.1	+	+	–	–
Methyl methacrylate	m	8.8	Δ	+	–	–
Ethyl methacrylate	m	8.3	Δ	Δ	–	–
2-Ethyl hexyl acrylate	m	7.8	–	–	–	–
Diethyl ether	m	7.4	–	–	–	–
Nitro methane	p	12.7	–	–	–	–
Chloroform	p	9.3	–	–	–	–
α-Methyl styrene	p	8.5	–	–	–	–
n-Pentane	p	7.0	–	–	–	–

[a] In air at atmospheric pressure; radiation dose 2 Mrad.

[b] S, strongly hydrogen bonded; m, moderately; and p, poorly.

[c] +, soluble; –, insoluble; and Δ, some part insoluble at 1 wt. % polymer in solvent.

[d] SPPOH-γ, gamma ray irradiated sulfonated poly(phenylene oxide) (IEC = 2.07 meq/g) hydrogen form.
SPPONa-γ, gamma ray irradiated sulfonated poly(phenylene oxide) (IEC = 2.07 meq/g) sodium form.

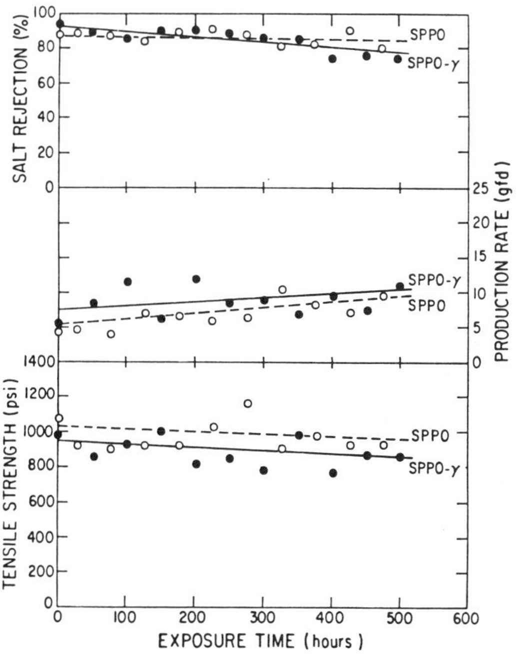

Figure 6. Changes in RO performance and tensile strength of SPPO-PS composite membranes on continuous exposure to 30 ppm chlorine at pH 7.0

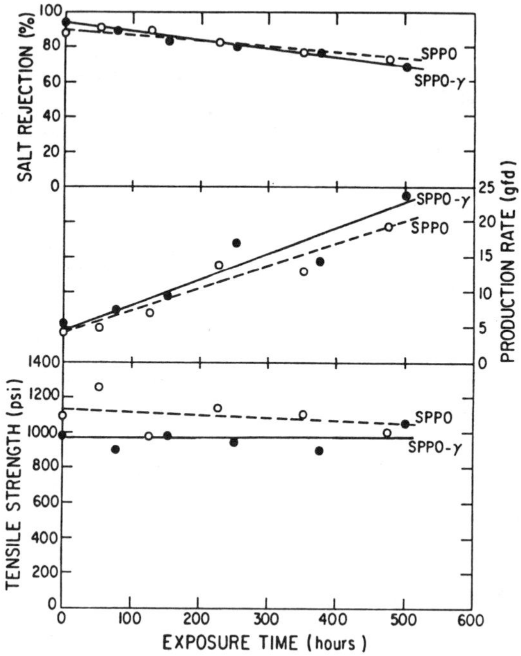

Figure 7. Changes in RO performance and tensile strength of SPPO-PS composite membranes on continuous exposure to 1N-NaOH (pH = 14.0)

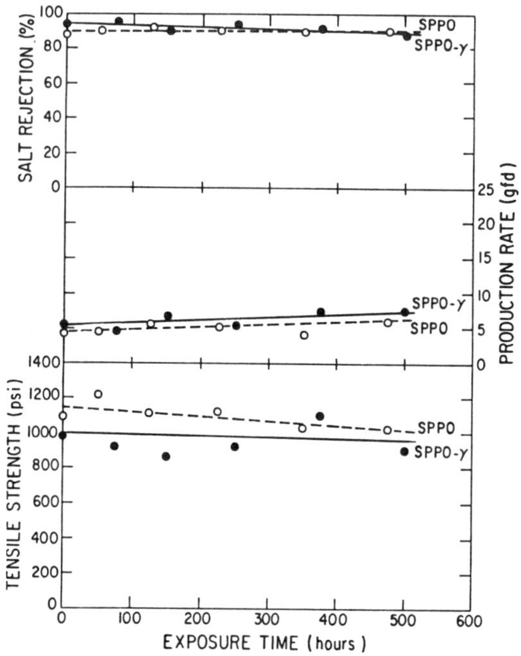

Figure 8. Changes in RO performance and tensile strength of SPPO-PS composite membranes on continuous exposure to 1N-H_2SO_4 (pH=0.3)

irradiated membranes also showed a considerable degree of stability to the effect of chlorine, acids and bases after long exposure to these media and is considered suitable for this application. The enhancement effect is believed to be caused by a crosslinking of the SPPO polymer in the thin film composite membranes caused by the gamma-ray irradiation.

Acknowledgments

We would like to thank the National Science and Engineering Research Council of Canada (NSERC) for its support of this research programme. Thanks are also due to Dean Wallace, Alberta Research Council, Edmonton, Alberta for providing the Alberta tar sands waste water samples.

Literature Cited

1. LaConti, A.B. "Advances in Development of Sulfonated PPO and Modified PPO Membrane Systems for Some Unique Reverse Osmosis Applications". In Reverse Osmosis and Synthetic Membranes, Edited by S. Sourirajan, Ottawa, 1977. pp.211-229,

2. Cadotte,J.E.; Petersen,R.J., ACS Symposium Series, No. 153, American Chemical Society, Washington, D.C., In "Synthetic Membranes:, Vol. 1, Desalination, (1981), pp305-325.
3. Chludzinski,P.J.; Austin,J.F.; Enos,J.F. "Development of Polyphenylene Oxide Membranes", Office of Saline Water Research and Development Progress Report No. 697, General Electric Co., Lynn, Mass., 1971.
4. Huang,R.Y.M.; Gao,C.J.; Kim,J.J., J. Appl. Polym. Sci., 28, 3063 (1983).
5. Huang,R.Y.M.; Kim,J.J. "Synthesis and Transport Properties of Thin Film Composite Membranes I. Synthesis of Poly(phenylene oxide) Polymer and its Sulfonation", J.Appl.Poly.Sci., accepted for publication, April 1984.
6. Huang,R.Y.M.; Kim,J.J. "Synthesis and Transport Properties of Thin Film Composite Membranes II. Preparation of Sulfonated Poly(phenylene oxide) Thin Film Composite Membranes for the Purification of Alberta Tar Sands Waste Waters", J. Appl.Poly. Sci., accepted for publication, April 1984.
7. Huang,R.Y.M.; Kim,J.J. "Treatment of Oil Recovery Process Waste Water", Canadian Patent Application filed July 18, 1984, Serial No. 459,169.
8. Slater, J.; Zacharian, M.R.; McCray, S.B.; McCutchon, J.W., Desalination, 48, 1(1983).
9. Kuwahara,H.; Yasuda,T.; Nakamura,M., Proc. of 7th Inter. Symp. on Fresh Water from the Sea, 2, 165 (1980).
10. Asano,B.H. et al., In "Review of Treatment and Recycling of Produced Water from Heavy Oil Fields", CH2M HILL Canada Ltd., Calgary, 1981.
11. Chapiro,A., In "Radiation Chemistry of Polymeric Systems", p.339, Interscience - John Wiley, 1962.
12. Chapiro,A., In "Advances in Chemistry Series", American Chemical Society, No.66, p.22, "Irradiation of Polymers", 1967.
13. King,R.A., In "Advances in Chemistry Series", American Chemical Society, No. 66, 113, "Irradiation of Polymers", 1967.

RECEIVED February 22, 1985

8

Design Features of the Polyether Composite (PEC)-1000 Spiral-Wound Membrane Element

T. ITOH, M. KURIHARA, N. KANAMARU, and T. TONOMURA

Technical Development Department, Toray Industries, Inc., 3-Chome, Sonoyama, Otsu, Shiga 520, Japan

The PEC-1000 membrane is a composite type of membrane consisting of (1) crosslinked polyether salt-rejecting barrier layer, (2) polysulfone supporting layer and (3) polyester fabric ((2) & (3) are both reinforced by crosslinked polyether resin). So it displays strong physical structure and zero m-value (no-compaction). As for the structural material of the RO element, extensive hydrodynamic study under high pressure exposure has been done specially for the selection of the permeate carrier. The above features, and the special modular design providing the spiral pattern for brine flow are the essential factors which assure the high temperature and pressure stability of the membrane, and uniform flow velocity over the membrane surface. The latter feature permits the use of feed waters with relatively high fouling index, and easy cleaning of the membrane surface even after heavy fouling. These advantages are actually shown in many commercial RO plants in the world.

The Special Feature of the PEC-1000 Membrane

Toray's composite membrane, designated as PEC-1000 (PEC is derived from Polyether Composite), has been developed from many years of research, development, testing and it represents the latest in membrane technology (1,2). PEC-1000 is specially designed for single-stage seawater desalination, and we took into consideration both its performance and mechanical structure.

Single-stage seawater desalination permits low TDS permeate from high temperature-high salinity seawater feed at a high recovery rate of feedwater. To meet the challenge of dependability and durability in basic seawater membrane performance, Toray has built into its design the highest salt rejection, permeate flow rate, and durability for high temperature and high pressure operation. This design feature is not found in other single-stage, partial single-stage or two-stage membranes.

0097-6156/85/0281-0099$06.00/0

Mechanical Structure and Performance of PEC-1000 Membrane

The PEC-1000 membrane shown in Figure 1 is made of a composite type structure (3) consisting of

1) a protective membrane layer (0.2μ)
2) a thin 300 Å salt barrier layer (being thicker than that in other composite membranes)
3) a porous supporting membrane made of polysulfone, and
4) polyester fabric as the base structure.

The ultrathin salt barrier layer, which determines membrane performance, is constructed from a crosslinked polyether, which yields

1) high salt rejection
2) high permeate flow rate
3) wide pH range operation and stability
4) high temperature durability, and
5) dependability

And it should be noticed that the resin of ultrathin layer also reinforces both the polysulfone supporting layer and the polyester fabric to give them more elasticity. As a result, the Young's Modulus of the polysulfone layer is increased for example, to 16.7 kg/mm^2 from 10.3 kg/mm^2 which is the common value for the general polysulfone substrate. However, as for the polyester fabric, the effect of the reinforcement is not so obvious. Such unique membrane substrate provides the Toray composite membrane with

6) high pressure durability
7) durability for handling and service

The PEC-1000 membrane exhibits excellent high salt rejection characteristic, an average of 99.85%, and high permeate flow rate of 0.4 m^3/m^2.day at 56 kg/cm^2 (800 psi), for 3.5% NaCℓ or seawater fed at 25°C (77°F) as shown in Figure 2 (1,3). The membrane also has the durability for high pressure operation (70 kg/cm^2-1000 psi) with high feed-water temperature. Many experimental data indicate there is no effect of compaction on membrane performance and therefore the m-value (flux decline slope) for the PEC-1000 membrane is zero.

Element Configuration and Its Flow Pattern in PEC-1000 Spiral Wound Membrane Element

PEC-1000 membrane elements are different in configuration (4,5) from other spiral-wound elements. One main design advantage is that the feedwater flows tangentially into the element along the cylindrical surface, flows spirally around towards the centre (permeate gathering) pipe in the element and exits the element from a circular opening located near the centre pipe at the downstream end of the element. PEC-1000 membrane element configuration is shown in Figure 3. In the case of other spiral-wound elements, water flows into the element from the feed end of the element, flows parallel with the center pipe of the element, and exits the element at the downstream end. The improved configuration of the PEC-1000 membrane element posseses the following advantages over other type of elements (4).

1) The uniform distribution and high flow velocity of the feedwater over the membrane surface minimizes concentration polarization. Hence, the membrane performance does not

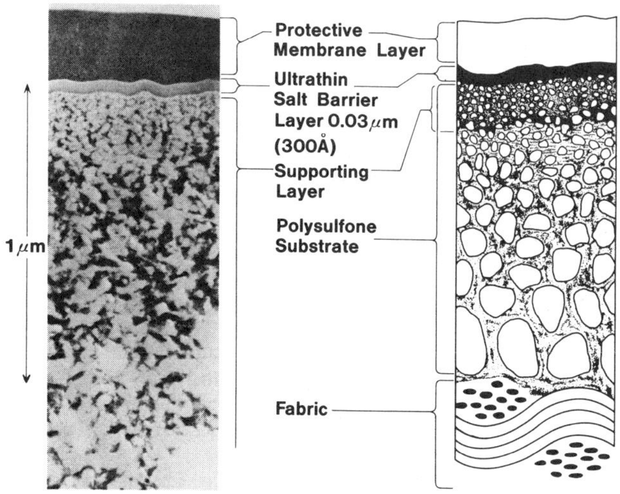

Figure 1. Electronmicrograph of the cross-section and sectional drawing of a PEC-1000 composite membrane. (Reproduced with permission from Ref. 3. Copyright 1983 Elsevier Science.)

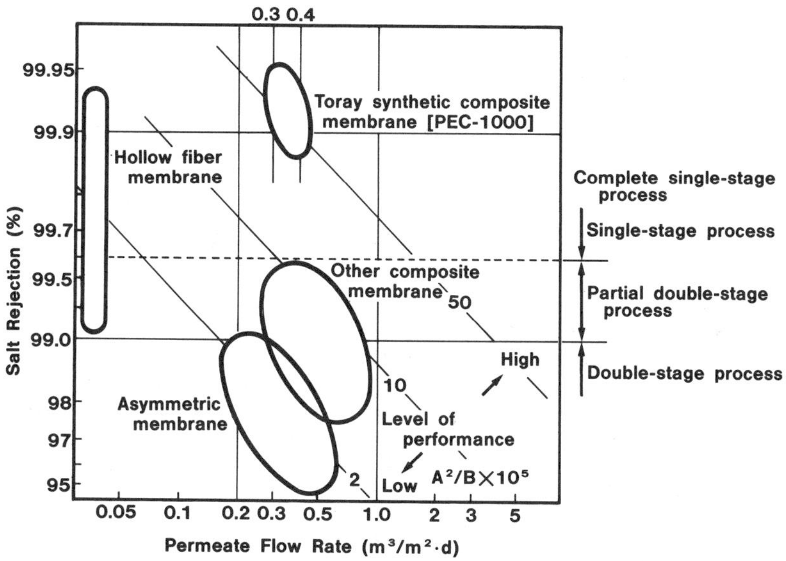

Figure 2. Diagrammatic comparison of desalination performance of different commercially available membranes for seawater (3.5% seawater, 56 kg/cm^2, 25 $^{\circ}$C. (Reproduced with permission from Ref. 3. Copyright 1983 Elsevier Science.)

decline even at a high recovery rate which results in low flow rate of brine, as illustrated in Figure 4. In the spiral type elements, the axial deformation of the flow path sometimes occurs and forms a few small ridges in rolling-up the membrane leaves. This results in non-uniform velocity distribution and decrease in element performance in other type of elements. Therefore the PEC-1000 membrane element could maintain the uniform flow velocity distribution and the element performance even if such deformation occurred.

2) There is no dead space when the element is inserted in the pressure vessel. As a result, the following advantages are obtained.
   a. Difficulty in bacteria growth. This is because the feedwater does not stagnate in the space between the outside surface of the element and the inside surface of the vessel.
   b. Rapid replacement of the feedwater after cleaning and disinfection of the module.
   c. Rapid approach to steady permeate flow rate after start-up of the plant.

3) There is no trouble based on telescopic deformation. When a large amount of the feedwater flows into a spiral-wound element, or when pressure drop is increased by fouling, other types of elements deform easily, and a telescopic deformation occurs. On the other hand, ends-sealed PEC-1000 membrane element resists greatly to the telescopic deformation; the telescopic deformation of the element does not occur even at a pressure drop of 3 kg/cm^2 per one element.

From the advantages mentioned above, Toray PEC-1000 composite membrane module makes it possible to produce pure water at lower cost than other types of modules. Data of long-term field tests and demonstration plants suggest that PEC-1000 RO process is now well established for the single-stage seawater desalination (6,7,8,9).

Structural Material of the PEC-1000 Membrane Element

Basic Quality of the Structural Material. PEC-1000 membrane element has been developed for single-stage desalination of seawater, especially for high salinity and high temperature seawater in the Middle East. This means that PEC-1000 membrane element would be used at very severe operating conditions such as 50-70 kg/cm^2 operating pressure, and 30-40°C feedwater temperature. Therefore the structural material must have the required durability against pressure and temperature. So we had paid much attention to select the material of each part of the element and examined its toughness and durability.

In the following section, we describe how we made the selection of permeate spacer, which has the most important influence on element performance.

Permeate Spacer of PEC-1000 Membrane Element. Permeate spacer is installed to maintain the permeate flow by supporting the membrane against the operating pressure. We can

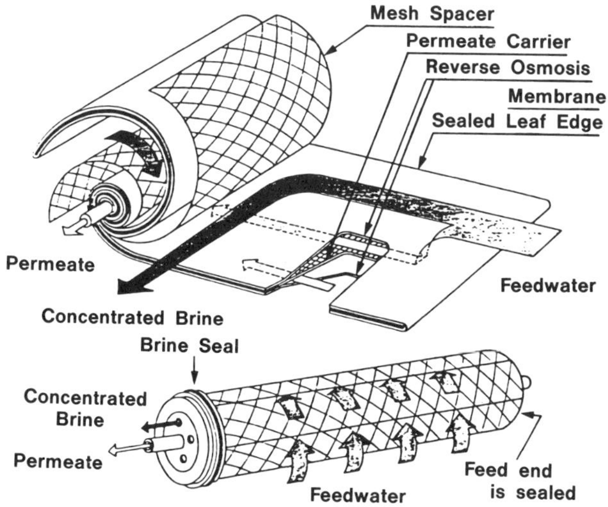

Figure 3. PEC-1000 element configuration. (Reproduced with permission from Ref. 4. Copyright 1983 Elsevier Science.)

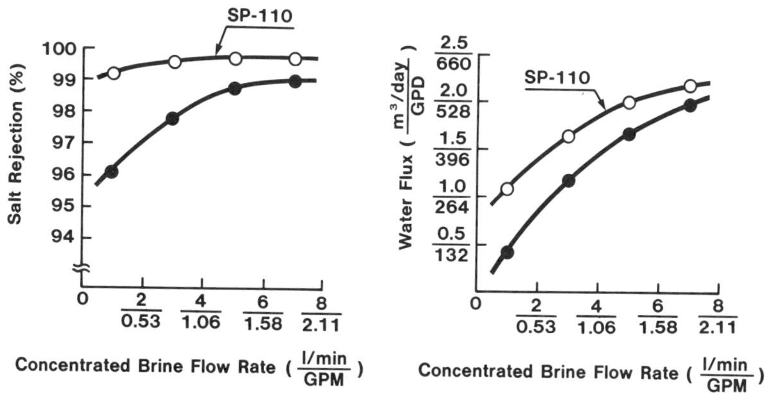

	Spiral Configuration	Membrane
—○—	Toray Type	PEC-1000
—●—	Another Type (Test sample)	PEC-1000

Test Condition NaCl 35000 ppm
56 Kg/cm^2, 25 °C
(800 psi, 77 °F)

Figure 4. Comparison of the performance between PEC-1000 and another type element.

express the above function in more specific terms as shown in Table I. These characteristics are naturally needed in any RO element. But in the case of the PEC-1000 membrane element, they are needed even more.

Table I. Basic Characteristics Needed for Permeate Spacer

Characteristic		Requirement
1. High pressure durability	→	rigid material
2. Low fluid resistance	→	proper configuration to prevent the membrane from being deformed
3. No dissolution of material into the permeate water		
4. Thinner material in thickness to gain more membrane area per unit volume		

As the permeate spacer of RO element, some kinds of tricot knit fabrics made rigid in some way, are commonly used. Figure 5 shows an example of tricot knit fabric. The specification of fabric is given in terms of wale No., the number of ribs per inch, and the course No., the number of cross points per inch. The groove between ribs is used as permeate water channel, so wale No. has much relevance to flow resistance as shown later.

As for the way to make tricot rigid, resin treatment commonly using melamine resin is applied in many cases. However, conventional melamine resin type tricot has the following two problems. One is less durability against pressure, particularly in the case of high operating pressures over 50 kg/cm^2, and the other is the dissolution of materials, mainly melamine resin, into the permeate water during long-term operation, by oxidative degradation.

We had started the study of the selection of the permeate spacer from this point of view. We discussed this problem with fabric and textile researchers in our company. Toray Ind. Inc. is a maker of synthetic fibers and textiles originally, so we have much potential and experience in this field. We studied about

1) the material and the size of tricot yarn,
2) the number of filament of each yarn,
3) fabric structure,
4) temperature of heat treatment,

and so on.

We resolved the problem of the melamine type tricot by applying an improved tricot knit fabric, the so called mixed or conjugate yarn type, as shown in Figure 6. It is composed of at least one composite yarn, each yarn having the first and the second components of high polymers, different in a specific characteristic, such as melting point, solubility, and degree of swelling (10). The first yarn component is fused, dissolved into a specific solvent or swelled with a specific solvent to adhere to the second yarn component, and thus, the first yarn component of each yarn is bound

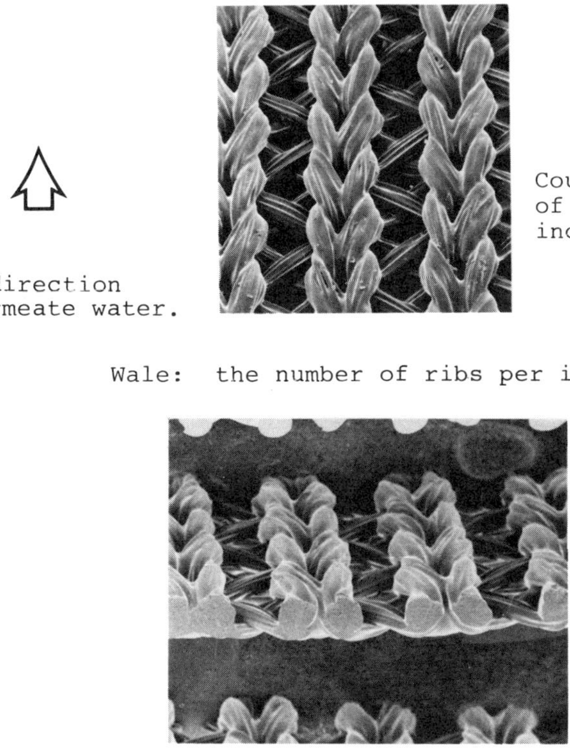

Figure 5. An example of tricot knit fabric.

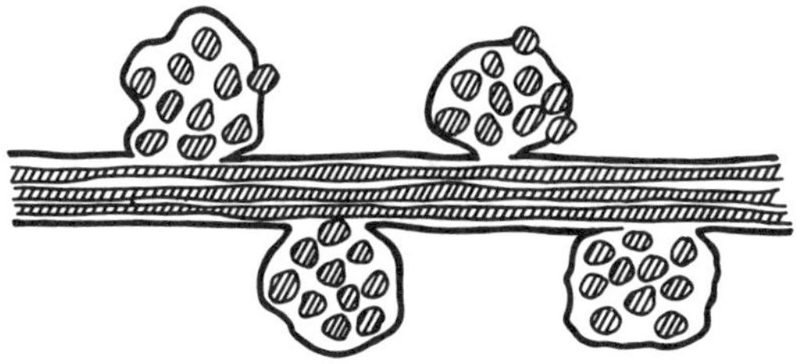

Figure 6. Sectional view of a tricot knit fabric composed of mixed yarns made rigid by heat treatment.

directly to that yarn and/or the other yarns so that the cloth is made rigid. The resultant cloth is rigid enough to resist a high fluid pressure of the feed solution, so that a smooth flow of the permeate water is ensured. And also there is no problem of dissolution in this improved type.

The experimental data for comparing some of the improved type to melamine type is shown in Table II. The flow resistance coefficient H, the so called H-value, means pressure drop per unit flow rate and is given by Equation 1 below.

$$H = \frac{W}{L} \cdot \frac{dp}{Q} \qquad (1)$$

where W, L, dp, Q are the channel width [m], channel length [m], pressure drop [atm] and flow rate [m^3/day] respectively.

And m_H given by Equation 2 means the increase of H-value with increase in operating time, therefore smaller m_H means more durability.

$$m_H = \frac{\log H_t/H_o}{\log t} \qquad (2)$$

where t, H_o, H_t represent operating time, initial H-value, H-value after t has elapsed, respectively. One can see the effect of our improvement from Table II.

On the other hand, we also made a hydrodynamical study on the proper configuration of permeate spacer to make fluid flow resistance less, and reduce spacer thickness at the same time. As the first step, we adopted the plain model of the section configuration shown in Figure 7. If we suppose the membrane is a homogeneous film, its deformation is given by balancing the static load working on the membrane. The schematic cross section of the membrane under operating pressure is shown in Figure 8, where two basic equations are given. From the balance of the force acting vertically.

$$P\ell = T \sin \theta \qquad (3)$$

From the relation between the stress of the membrane and its elongation,

$$\sigma = E \left(\frac{\Delta L - L}{L}\right) \qquad (4)$$

$$= E \left(\frac{\frac{T}{P}\theta - L}{L}\right) \qquad (5)$$

where σ, E, ΔL are the stress [kg/m^2], the Young's Modulus [kg/m^2], elongation [m] of the membrane respectively.
Considering that

$$\sigma = \frac{T}{t} \qquad (6)$$

where t is thickness of the membrane [m], Equation 5 turns into the form of Equation 7 as shown below.

$$\theta = P\ell \left(\frac{1 + m/\ell}{Et} + \frac{1}{T}\right) \qquad (7)$$

Table II. Flow Resistance Characteristics of Some Kinds of Tricot Knit Fabric

Rigid Cloths		Flow Resistance Characteristics H (atm.day/m³)	m_H	Reduction in flow rate after 100 hours (%)
Ex. 1 Polyester conjugate yarns	Heating treatmeent at 240 c	0.78 3.82	0.03 0.15	0 (negligible)
Ex. 2 Polyester mixed yarns	Heating treatmeent at 235 c	1.41	0.04	1.1
Ex. 3 Polyester mixed yarns	Heating treatmeent at 220 c	1.23	0.02	0 (negligible)
Control Polyester yarns with melamine		3.56	0.19	13.0

$$H = \frac{W \cdot dP}{L \cdot Q}$$

H: flow resist. coefficient (atm.day/m3)
W: channel width (m)
L: channel length (m)
dP: pressure drop (atm)
Q: flow rate (m3/day)

$$m_H = \frac{\log Ht/Ho}{\log t}$$ t: time (hour)

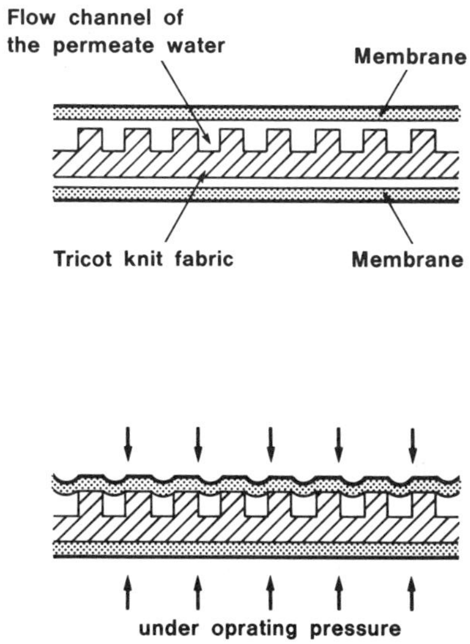

Figure 7. The plain model of the section configuration of the permeate spacer.

The values of θ and T are given by satisfying Equation 3 and Equation 7 simultaneously. In this way, we can calculate the elongation of the membrane and thus, the depth of the bending of the membrane. Hence, we can fix the effective flow channel area of permeate water and the channel resistance H.

The calculated H-value of a certain configuration is shown in Figure 9. In this case, membrane is PEC-1000, operating pressure is 70 kg/cm^2, and groove depth is 0.15 mm. H-value is indicated with the parameter, wale No. This result shows, naturally, that the smaller channel width gives the bigger H-value. The bigger channel width means the smaller H-value but its effect is decreased by the deformation of the membrane. In the case of too big a channel width, H-value could be even increased, when we must consider the limitation in making tricot knit fabric. As for the rib width, the smaller is the better, but about 0.2 mm is the minimum we can reach. The dotted line in Figure 9 means the rib width is 0.2 mm, so the tricot we can make is in the upper area of the dotted line. Another example of calculated H-value is given in Figure 10. The difference from the previous one is channel depth, 0.13 mm in this case. This result shows that only 0.02 mm smaller depth increases H-value greatly.

Such study brought us to the specific design about the most proper configuration which makes the H-value the least. Those are summarized as follows:

1) the bigger channel depth
2) channel width should be in the range from 0.2 mm to 0.3 mm
3) wale number; about 50 is preferable.

We made a lot of samples of tricot whose configurations were similar to the ideal one. We measured the H-value of those samples while comparing them with each other. Figure 11 shows the experimental data. H-value is plotted versus the thickness of tricot. So in this figure, left and lower area means good configuration. It is obvious that our improved tricot has a better configuration than conventional ones. Though the calculated H-value is slightly different from the measured one because of the simplification, experimental data are enough to make us ensure the correctness of our study.

We studied the permeate spacer design to get better performance of the element, mainly by the two ways described above. By applying these results of the development, PEC-1000 membrane elements exhibit excellent high salt rejection at an average of 99.7% and a high permeate flow rate at 56 kg/cm^2 (800 psi) using 3.5% NaCℓ or seawater, as feed at 25°C (77°F) and 13.5% water recovery.

Plant Performance of the PEC-1000 Membrane Element

As stated above, PEC-1000 membrane and element have the characteristics required for single-stage seawater desalination. That means high durability against pressure and temperature, and of course high salt rejection and water flux from the membrane. The reasons for such characteristics are the following:

A. high performance of salt barrier layer,
B. very strong physical structure of the membrane,
C. improved configuration of the element,
D. proper selection of the structural material.

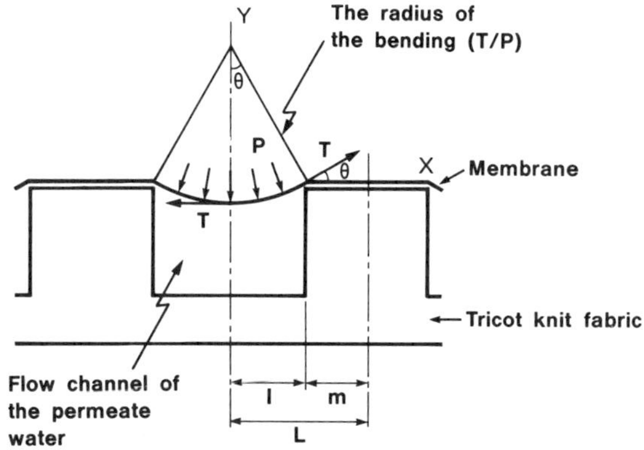

Figure 8. The deformation of the membrane under operating pressure. Key: P, operating pressure (kg/m^2); T, tension of the membrane per unit width (kg/m); θ, angle of deformation (rad.); I, a half of the channel width (m); and m, a half of the rib width (m).

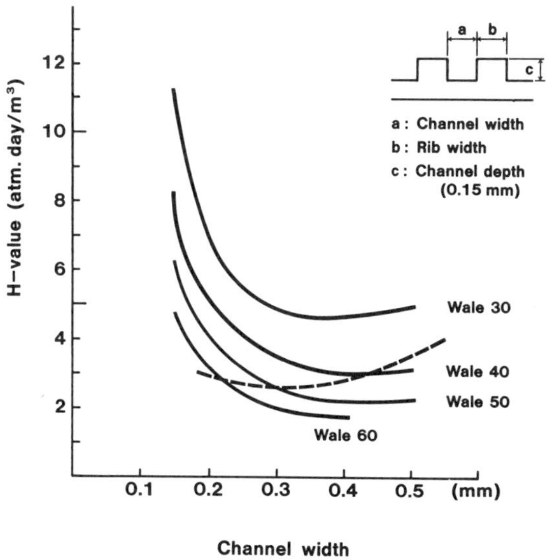

Figure 9. Examples of the calculated H-value (1). Wale 30 means 30 channels per inch of the tricot width. Upper area of the dotted line means that the rib width is bigger than 0.2 mm. Membrane: PEC-1000. Operating pressure: 70 (kg/cm^2).

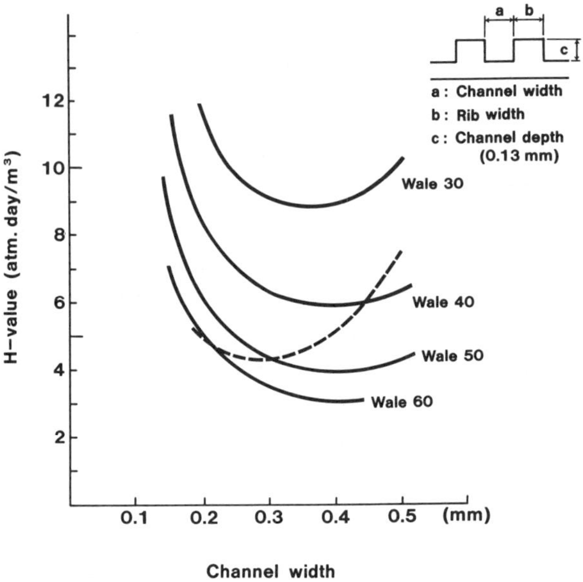

Figure 10. Examples of the calculated H-value (2). Wale 30 means 30 channels per inch of the tricot width. Upper area of the dotted line means that the rib width is bigger than 0.2 mm. Membrane: PEC-1000. Operating pressure: 70 (kg/cm^2).

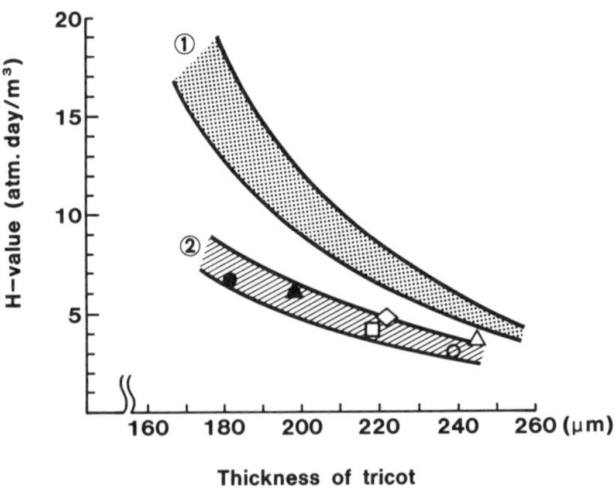

Figure 11. H-value of improved tricot knit fabric. Key: 1, zone of H-value of commonly used tricot: 2, zone of H-value of improved tricot; and △, ▲, ○, ●, □, ◇; actual experimental data of improved tricot.

These characteristics are actually shown in many commercial RO plants (6,7,8,9).

Literature Cited

1. Kurihara, M.; Kanamaru, N.; Harumiya, N.; Yoshimura, K.; Hagiwara, S. Desalination 1980, 32, 13.
2. Kurihara, M.; Watanabe, T.; Inoue, T. U.S. Patent 4366062.
3. Chen, J.Y.; Kurihara, M.; Pusch, W. Desalination 1983, 46, 379.
4. Bairinji, R.; Tanaka, T.; Kurihara, M.; Kanamaru, N.; Tonomura, T. Desalination 1983, 46, 57.
5. Kanamaru, N.; Fujino, H. U.S. Patent 3933646, 1976.
6. Kurihara, M.; Nakagawa, Y.; Takeuchi, H.; Kanamaru, N.; Tonomura, T. Proc. 10th Annual Conferen. Trade Fair of WSIA., 1982.
7. Kurihara, M.; Nakagawa, Y.; Takeuchi, H.; Kanamaru, N.; Tonomura, T. Desalination 1983, 46, 101.
8. Doelle, R.A.; Kallenberg, K.H.; Heyden, W. Poster Presentation No. 29 at First World Congress on Desalination and Water Reuse. Florence Italy, May 26 (1983).
9. Kunisada, Y.; Okada, K.; Sonoda, T.; Setogawa, S.; Ishiwatari, T. Zosui Gijitsu 1983, 9, No. 4, 13.
10. Bairinji, R.; Tanaka, T.; Kawabata, T. G.B. Patent 2000694.

RECEIVED February 22, 1985

9

Analysis of Transport in a Pressure-Driven Membrane Separation Process

STEPHEN W. THIEL, DOUGLAS R. LLOYD, and J. M. DICKSON[1]

Department of Chemical Engineering, University of Texas at Austin, Austin, TX 78712

The analysis of transport through a membrane in a pressure-driven membrane separation process need not include assumptions about the microstructure of the membrane. An analysis of transport is presented which interprets phenomenological relations in terms of the strengths of physicochemical interactions in the solute-solvent-membrane material system. No assumptions about the membrane microstructure are made. The results of this analysis are compared with the results of an analysis of transport through finely porous membranes.

Transport through a membrane is often interpreted in terms of an assumed microstructure for the membrane. Dense, homogeneous membranes are often described using a solution-diffusion analysis (1-3), while membranes with large pores are described by simple Fickian diffusion superimposed on laminar flow (2). Merten (2) suggested that most reverse osmosis and many ultrafiltration membranes are intermediate in structure and behavior between the extremes of dense membranes and large-pored membranes. Merten called these membranes finely porous.

Mason, Wendt, and Bresler noted that a simple pore flow model can be used to correlate membrane performance over a wide range of conditions, including conditions which violate many of the physical assumptions of the model (4). They suggested that the success of the pore flow model can be explained by dimensional analysis. Consequently, the microstructure of the membrane plays a relatively minor role in determining the form of the mathematical relationships that govern membrane permeation. That is, the microstructure of the membrane does not determine the form of the transport equations,

[1] Current address: Department of Chemical Engineering, McMaster University, Hamilton, Ontario, Canada L8S 4L7

0097-6156/85/0281-0113$06.00/0

only the values of the parameters in those equations. Even though structure and transport are related, membrane structure is not always a reliable key to transport modelling and, conversely, transport behavior is not always a reliable indication of membrane structure.

One approach to transport modelling without assumptions about the microstructure of the membrane uses phenomenological relations. These relations correlate fluxes with thermodynamic forces through coefficients which are not necessarily tied to any given physical assumptions. Many investigators have interpreted the phenomenological coefficients in terms of friction (5-8). This friction is a resistance to motion due to attraction between molecules. For example, Jonsson and Boesen (9) derived a frictional model that is formally similar to the phenomenologically-based Spiegler-Kedem model (10).

This paper considers the transport of dilute binary solutions through membranes under the influence of pressure and concentration gradients. Transport is analyzed using phenomenological relations; no assumptions are made about the microstructure of the membrane. The phenomenological coefficients and external forces are interpreted in terms of intermolecular forces. This analysis is compared with the analysis of finely porous membranes in light of physicochemical interactions in the solute-solvent-membrane material system.

Preliminary Assumptions and Definitions

The mathematical analysis that follows applies to a steady-state pressure-driven membrane separation process separating a dilute, binary, isothermal, non-reacting, liquid solution. The membrane is flat. The analysis considers only transport within the membrane; film theory can be used to account for mass transfer effects in the adjacent liquid (11). The system is assumed to be uniform in directions parallel to the surface of the membrane; all transport occurs in the z-direction, normal to the surface of the membrane. The coordinate z increases from the retentate to the permeate.

The membrane phase is chosen to include only the solute and solvent within the polymer matrix. The effects of the membrane material on permeation are accounted for as external forces acting on the solute and solvent. The effective thickness of the membrane phase, τ, includes the effects of tortuosity in the transport path.

The permeating components are partitioned between the membrane phase and the adjacent liquid phase. This equilibrium is described by the equation

$$X_{AM} = K_A X_A \quad (1)$$

The quantity X_{AM} is the solute mole fraction in the membrane phase, X_A is the solute mole fraction in the liquid phase, and K_A is a partition coefficient. The partition coefficient on the retentate side of the membrane, K_{A2}, can be different from the partition coefficient on the permeate side of the membrane, K_{A3}.

General Transport Considerations

Irreversible thermodynamics (12-14) provides a general description of nonequilibrium systems, and so describes processes which advance at a measurably rapid rate. Irreversible thermodynamics suggests that for an n-component system the natural choices for fluxes, J_i, and forces, Y_i, are

$$J_i = CX_i(v_i - V) \tag{2}$$

and

$$Y_i = F_i - \frac{d\mu_i}{dz} \tag{3}$$

where the molar-average velocity, V, is defined as

$$V = \sum_{i=1}^{n} X_i v_i \tag{4}$$

Throughout this paper, C is molar density, X_i is the mole fraction of component i, v_i is the velocity of component i relative to stationary coordinates, F_i is the external force acting on component i, and μ_i is the chemical potential of component i. The flux J_i is the diffusive molar flux of component i relative to the molar-average velocity.

In typical pressure-driven membrane separation processes, the diffusive fluxes cannot be measured; experimental results instead provide values for the fluxes relative to stationary coordinates, N_i. Because the membrane is stationary, stationary coordinates are defined relative to the membrane. The fluxes N_i are defined as

$$N_i = CX_i v_i \tag{5}$$

This definition can be rewritten as

$$N_i = CX_i(v_i - V) + CX_i V = J_i + CX_i V \tag{6}$$

Introducing the total molar flux, N_T, Equation 6 becomes

$$N_i = J_i + X_i N_T \tag{7}$$

Equation 7 expresses the flux of a component relative to stationary coordinates as the sum of two fluxes: the flux due to bulk motion and the flux due to diffusion. The flux due to bulk motion, $X_i N_T$, represents molar-average motion; the inclusion of this term does not imply viscous flow.

Assume that the fluxes and forces are linearly related through the phenomenological relations

$$J_i = \sum_{k=1}^{n} L_{ik} Y_k \tag{8}$$

where the L_{ik} are phenomenological coefficients which relate the flux of component i to the force on component k. DeGroot (12) showed that not all of the L_{ik} are independent, and that in a binary system,

$$L_{AA} = - L_{AB} = - L_{BA} = L_{BB} = L \tag{9}$$

Equation 9 is a generalization of the observation that mass diffusion in a binary system is characterized by one diffusivity. Inserting Equation 9 into Equation 8 provides, for a binary system, the result

$$J_A = L(Y_A - Y_B) \tag{10}$$

The phenomenological coefficient L can be interpreted in terms of intermolecular forces (2):

$$L = \frac{C_M X_{AM}}{\phi_{AB}} \tag{11}$$

where the coefficient ϕ_{AB} represents the resistance to mutual motion of the solute and solvent; ϕ_{AB} is the reciprocal of the mobility of the solute-solvent system. Equation 11 can be inserted into Equation 10 to provide the relationship

$$J_A = \frac{C_M X_{AM}}{\phi_{AB}} (Y_A - Y_B) \tag{12}$$

Equation 12 is central to the analysis of solute transport.

The external force, F_i, represents the effect of chemical and physical interactions between the membrane material and component i on the transport of component i. This external force is assumed to be proportional to the difference between the velocity of component i and the velocity of the membrane material (15). Since the membrane material is stationary, F_i can be expressed as

$$F_i = - \frac{\phi_i}{C_M X_{iM}} N_i = - \frac{\phi_i}{C_M X_{iM}} X_{i3} N_T \tag{13}$$

where C_M is the molar density of the membrane phase. Although the values of the coefficients ϕ_i and ϕ_{AB} are determined by the physicochemical interactions in the solute-solvent-membrane system, the introduction of these coefficients does not require the introduction of a specific microstructure for the membrane (8).

Solute Transport

A relationship between separation and flux can be derived starting with Equation 12. The total force acting on the solute, $Y_A - Y_B$, can be divided into an external force term and a chemical potential term using Equation 3:

$$Y_A - Y_B = (F_A - F_B) - \left(\frac{d\mu_A}{dz} - \frac{d\mu_B}{dz}\right) \tag{14}$$

The external force term can be expanded using Equation 13 to obtain the result

$$F_A - F_B = \left(-\frac{\phi_A}{C_M X_{AM}} X_{A3} N_T\right) - \left(-\frac{\phi_B}{C_M X_{BM}} X_{B3} N_T\right) \tag{15}$$

For dilute solutions, X_{BM} and X_{B3} are approximately unity. Equation 15 then becomes

$$F_A - F_B = \frac{N_T}{C_M X_{AM}} [\phi_B X_{AM} - \phi_A X_{A3}] \tag{16}$$

The chemical potential term can be rewritten using the Gibbs-Duhem equation for the free energy of the system. For an inert, isothermal, binary system, this equation is

$$\bar{V}dp - (X_A d\mu_A + X_B d\mu_B) = 0 \tag{17}$$

where $\bar{V}$ is the molar volume of the system, and p is the pressure of the system. Equation 17 can be rearranged to yield the result

$$d\mu_A - d\mu_B = (d\mu_A - \bar{V}dp)/X_B \tag{18}$$

The differential of the solute chemical potential is given by

$$d\mu_A = \bar{V}_A dp + RTd(\ln X_A) \tag{19}$$

where $\bar{V}_A$ is the partial molar volume of the solute. Since X_B is, by assumption, approximately unity, substituting Equation 19 into Equation 18 provides the result

$$\frac{d\mu_A}{dz} - \frac{d\mu_B}{dz} = (\bar{V}_A - \bar{V}) \frac{dp}{dz} + RT \frac{d(\ln X_A)}{dz} \tag{20}$$

Merten (2) noted that the contribution of pressure to chemical potential, $\bar{V}_A dp$, is often negligible compared to the contribution of composition to chemical potential, $RTd(\ln X_A)$. Similarly, the analysis presented here assumes that $(\bar{V}_A - \bar{V})dp$ is negligible compared to $RTd(\ln X_A)$. Making this approximation, combining

Equations 12, 14, 16, and 20 provides the differential equation

$$J_A = \frac{C_M X_{AM}}{\phi_{AB}} \left[\frac{N_T}{C_M X_{AM}} (\phi_B X_{AM} - \phi_A X_{A3}) - \frac{RT}{X_{AM}} \frac{dX_{AM}}{dz} \right] \quad (21)$$

Combining Equation 21 with Equation 7 yields, after rearrangement,

$$-\frac{N_T \phi_{AB}}{C_M RT} dz = \frac{dX_{AM}}{b_A X_{A3} - b_B X_{AM}} \quad (22)$$

where the b_i are functions defined by the equation

$$b_i = 1 + (\phi_i / \phi_{AB}) \quad (23)$$

The quantities b_A, b_B, C_M, and ϕ_{AB} are assumed to be constant across the membrane. Assuming equilibrium at the membrane-liquid interfaces using Equation 1, Equation 22 can be integrated across the membrane and rearranged to yield the result

$$\frac{X_{A2}}{X_{A3}} = \frac{b_A}{b_B K_{A2}} + \left(\frac{K_{A3}}{K_{A2}} - \frac{b_A}{b_B K_{A2}} \right) \exp \left[-\frac{\phi_{AB} b_B \tau}{C_M RT} N_T \right] \quad (24)$$

Equation 24 indicates that the behavior of the system depends on the relative magnitudes of the solute-membrane, solvent-membrane, and solute-solvent interactions.

Solvent Transport

Since the permeating solution is dilute, the solvent is treated as a pure component. The effects that result from the presence of solute are accounted for as external forces.

Consider first the transport of a pure component. For pure component B, the phenomenological relations expressed in Equation 8 reduce to the single equation

$$J_B = L_{BB} Y_B \quad (25)$$

However, by definition,

$$J_B = C X_B (v_B - V) \quad (26)$$

Since $V = v_B$ for pure component B, the diffusive flux of component B, J_B, is zero. Therefore, the force Y_B must also be zero:

$$Y_B = F_B - \frac{d\mu_B}{dz} = 0 \quad (27)$$

or, more directly,

$$F_B = \frac{d\mu_B}{dz} \tag{28}$$

According to Equation 28, the chemical potential gradient is balanced by external forces at steady-state.

The net force acting on the solvent in a pressure-driven membrane separation process includes not only the force due to the motion of the solvent, but also a force due to the motion of the solute. Consequently, Equation 28 can be expanded to provide the result

$$-\phi_B \frac{N_B}{C_M X_{BM}} X_{BM} - \phi_B' \frac{N_A}{C_M X_{AM}} X_{AM} = \frac{d\mu_B}{dz} \tag{29}$$

where ϕ_B' is a coefficient relating the external force acting on the solvent to the velocity of the solute. Equation 29 can be rearranged to yield the equation

$$N_T = -\frac{C_M/\phi_B}{1 + (\phi_B'/\phi_B - 1)X_{A3}} \frac{d\mu_{BM}}{dz} \tag{30}$$

Assuming constant values for C_M, ϕ_B, and ϕ_B', Equation 30 can be integrated across the membrane to provide the result

$$N_T = -\frac{C_M/\phi_B\tau}{1 + (\phi_B'/\phi_B - 1)X_{A3}} \Delta\mu_{BM} \tag{31}$$

Assuming equilibrium at both membrane-liquid interfaces, $\Delta\mu_{BM} = \Delta\mu_B$. Since the chemical potential difference of the solvent across the membrane can be written as

$$\Delta\mu_B = -\bar{V}_B[(p_2 - p_3) - (\pi_2 - \pi_3)] \tag{32}$$

Equation 31 can be rewritten as

$$N_T = \frac{C_M\bar{V}_B/\phi_B\tau}{1 + (\phi_B'/\phi_B - 1)X_{A3}} [(p_2 - p_3) - (\pi_2 - \pi_3)] \tag{33}$$

Although Equation 33 is formally similar to the equation the for total flux through finely porous membranes (2,9), the derivation of Equation 33 uses no assumptions about the microstructure of the membrane.

Comparison with the Analysis of Transport Through a Finely Porous Membrane

The analysis of transport through finely porous membranes (2) yields the result

$$\frac{X_{A2}}{X_{A3}} = \frac{b_A}{K_{A2}'} + \left(\frac{K_{A3}'}{K_{A2}'} - \frac{b_A}{K_{A2}'}\right) \exp\left[-\frac{\phi_{AB}\tau}{C\varepsilon RT} N_T\right] \tag{34}$$

where ε is the surface porosity of the membrane, and the K_{Ai}' are partition coefficients for equilibrium between free solution and the pore fluid. Equation 34 is similar to Equation 24 derived above. This resemblance is not surprising, since the two analyses share several key assumptions. First, both analyses describe binary diffusion, since both choose a membrane phase which excludes the membrane material. Second, both analyses express the results of intermolecular interactions in terms of partitioning and forces opposing motion. Finally, both analyses include external forces and chemical potential in the thermodynamic force terms.

There is one major difference between this analysis and the analysis of a finely porous membrane: this analysis includes the force acting on the solvent, Y_B, in the phenomenological equation governing solute transport. The chemical potential term in the equation for solute transport is unaffected by the inclusion of the force Y_B because the contribution of pressure to chemical potential is negligible and because the mole fraction of solvent is nearly unity. The inclusion of Y_B does, however, lead to the introduction of the function b_B, thereby influencing the physical interpretation of the parameters in the solute transport equation. The presence of the term b_B indicates a role for solvent-membrane material interactions that the analysis of the finely porous model does not suggest.

This additional role for solvent-membrane material interactions can be clarified by considering the separation achieved as the flux becomes infinitely great. This limiting separation provides a theoretical limit on the performance of the system, a limit which has been observed in laboratory studies (9,16,17). Equation 34 gives the limiting separation as b_A/K_{A2}', while Equation 24 gives the limiting separation as $(b_A/b_B)/K_{A2}$. The quantities b_A/K_{A2}' and $(b_A/b_B)/K_{A2}$ have different relationships to the physicochemical interactions in the solute-solvent-membrane material system.

Consider first the quantity b_A/K_{A2}', obtained from the analysis of a finely porous membrane. The function b_A is a result of the strength of solute-membrane material interactions relative to the strength of solute-solvent interactions, and so increases with increasing solute-membrane material affinity. The partition coefficient K_{A2}' is a result of the strength of solute-membrane material interactions relative to the strength of solvent-membrane material interactions, and so increases with increasing solute-membrane material affinity but decreases with increasing solvent-membrane material affinity. Consequently, b_A/K_{A2}' must increase as solvent-membrane material affinity increases relative to solute-solvent affinity. One cannot determine whether b_A/K_{A2}' increases or decreases with solute-membrane material affinity without more information about b_A and K_{A2}'.

Now consider the quantity $(b_A/b_B)/K_{A2}$. The terms b_A and K_{A2} vary as before. The function b_B is a measure of the strength of solvent-membrane material interactions relative to the strength of solute-solvent interactions, and so increases with increasing solvent-membrane material affinity. Both the ratio b_A/b_B and the coefficient K_{A2} increase as the strength of solute-membrane material interactions increases relative to the strength of solvent-membrane material interactions. Consequently, one cannot determine how the ratio $(b_A/b_B)/K_{A2}$ varies with the physicochemical properties of the solute-solvent-membrane material system without more information about the quantities b_A, b_B, and K_{A2}.

Based on an analysis of the finely porous model, one would expect the limiting separation to increase as solvent-membrane material affinity increases relative to solute-solvent affinity. The analysis presented in this paper shows that the limiting separation does not necessarily increase as the affinity between the solvent and the membrane material increases.

Conclusions

This paper has presented an analysis of transport through membranes which uses no assumptions about the microstructure of the membrane. Since physicochemical interactions among the solute, solvent, and membrane material determine the strength of intermolecular forces, these interactions determine the values of the coefficients which relate separation to flux: ϕ_i, ϕ_{AB}, and K_{Aj}. The values of these coefficients are closely related to both the chemical nature of the system and to the physical properties -- the structure -- of the system. Consequently, this work represents only a partial solution to the problem of transport through membranes. A complete understanding of the permeation process calls not only for an understanding of the transport equations and their fundamental basis, but also for an understanding of the relationship between physicochemical properties and the forces acting in the solute-solvent-membrane material system.

Nomenclature

A	Pure water permeability [$kmol/m^2$-s-kPa]
b_i	A function representing the relative magnitudes of forces due to solute-membrane interactions and solute-solvent interactions [dimensionless]
C	Molar density [$kmol/m^3$]
C_M	Molar density of membrane phase [$kmol/m^3$]
C_{iM}	Molar density of component i in the membrane phase [$kmol/m^3$]
F_i	External force acting on component i [$(kg\text{-}m/s^2)/kmol$]
J_i	Molar diffusive flux of component i relative to the molar-average velocity [$kmol/m^2$-s]
K_{Aj}	Partition coefficient for solute at location j [dimensionless]
K_{Aj}'	Partition coefficient for solute at location j for a porous membrane [dimensionless]

L	Phenomenological coefficient for a binary system $[(kmol/m^2\text{-}s)/((kg\text{-}m/s^2)/kmol)]$
L_{ik}	Phenomenological coefficient relating the force acting on component k to the flux of component i $[(kmol/m^2\text{-}s)/((kg\text{-}m/s^2)/kmol)]$
n	Number of components [dimensionless]
N_i	Molar flux of component i relative to stationary coordinates $[kmol/m^2\text{-}s]$
p	Pressure [kPa]
p_j	Pressure at location j [kPa]
R	Gas law constant [J/kmol-K]
T	Thermodynamic temperature [K]
v_i	Velocity of component i [m/s]
V	Molar-average velocity [m/s]
$\bar{V}$	Molar volume of system $[m^3/kmol]$
$\bar{V}_i$	Partial molar volume of component i $[m^3/kmol]$
X_{ij}	Mole fraction of component i at location j [dimensionless]
Y_i	Thermodynamic force acting on component i $[(kg\text{-}m/s^2)/kmol]$
z	Position within the membrane phase [m]
ε	Surface porosity of the membrane [dimensionless]
μ_i	Chemical potential of component i [J/kmol]
μ_{iM}	Chemical potential of component i in the membrane phase [J/kmol]
π_j	Osmotic pressure at location i [kPa]
ϕ_{AB}	Coefficient related to the diffusivity of the system A--B $[((kg\text{-}m/s^2)/kmol)/(m/s)]$
ϕ_B'	Coefficient relating the solute velocity relative to the membrane to the external force acting on the solvent $[((kg\text{-}m/s^2)/kmol)/(m/s)]$
ϕ_i	Coefficient relating the velocity of component i relative to the membrane to the external force acting on component i $[((kg\text{-}m/s^2)/kmol)/(m/s)]$
τ	Effective thickness of the membrane phase [m]

Subscripts:

A Solute

B Solvent

T Total solution (solute and solvent)

2 Retentate immediately adjacent to the membrane

3 Permeate

Acknowledgments

The authors are grateful for the support provided by the National Science Foundation (Separation Processes Program, grant number CPE-8312671), and also for the support provided by the Separations Research Program, Department of Chemical Engineering, and Graduate School of the University of Texas at Austin.

Literature Cited

1. Lonsdale, H. K.; Merten, U.; Riley, R. L. J. Appl. Polymer Sci. 1965, 9, 1341-1362.
2. Merten, U. In "Desalination by Reverse Osmosis"; U. Merten, Ed.; MIT Press: Cambridge, 1966; p. 15-54.
3. Burghoff, H. G.; Lee, K. L.; Pusch, W. J. Appl. Polym. Sci. 1980, 25, 323-347.
4. Mason, E. A.; Wendt, R. P.; Bresler, E. H. J. Membr. Sci. 1980, 6, 283-298.
5. Laity, R. W. J. Phys. Chem. 1959, 63, 80-83.
6. Smit, J.A.M.; Eijsermans, J. C.; Staverman, A. J. J. Phys. Chem. 1975, 79, 2168-2175.
7. Schmitt, A.; Craig, J. B. J. Phys. Chem. 1977, 81, 1338-1342.
8. Lorimer, J. W. J. Chem. Soc., Faraday Trans. II 1978, 74, 75-83.
9. Jonsson, G.; Boesen, C. E. Desalination 1975, 17, 145-165.
10. Spiegler, K. S.; Kedem, O. Desalination 1966, 1, 311-326.
11. Kimura, S.; Sourirajan, S. AIChE J. 1967, 13, 497-503.
12. DeGroot, S. R. "Thermodynamics of Irreversible Processes"; North-Holland Publishing Company: Amsterdam, 1958.
13. DeGroot, S. R.; Mazur, P. "Non-Equilibrium Thermodynamics"; North-Holland Publishing Company: Amsterdam, 1962.
14. Prigogine, I. "Introduction to Thermodynamics of Irreversible Processes", 3rd ed.; Interscience Publishers: New York, 1967.
15. Spiegler, K. S. Trans. Faraday Soc. 1958, 54, 1408-1428.
16. Jonsson, G. Desalination 1978, 24, 19-37.
17. Baker, R. W.; Eirich, F. R.; Strathmann, H. J. Phys. Chem. 1972, 76, 238-242.

RECEIVED February 22, 1985

10

Physicochemical Interpretation of the Behavior of a Pressure-Driven Membrane Separation Process

STEPHEN W. THIEL, DOUGLAS R. LLOYD, and J. M. DICKSON[1]

Department of Chemical Engineering, University of Texas at Austin, Austin, TX 78712

In the pressure-driven membrane separation of binary aqueous solutions of aromatic hydrocarbons using asymmetric cellulose acetate membranes, the total solution flux is often significantly less than the pure water flux at the same pressure, even in the absence of significant osmotic pressures. This behavior can be explained using a simple model in which adsorbed solute reduces the area available for transport. Flux reduction can be correlated with composition using an adsorption isotherm, and the parameters obtained in that correlation are related to the physicochemical nature of the solute-solvent-membrane system.

In pressure-driven membrane separation processes, the total molar solution flux, N_T, is often significantly less than the molar flux of pure water, N_P, at the same operating pressure, p. One possible reason for this reduction in flux can be found from the results of an analysis of transport through membranes in pressure-driven membrane separation processes. The total flux in a pressure-driven membrane separation process is given by the equation (1)

$$N_T = \frac{C_M \bar{V}_B / \phi_B \tau}{1 + (\phi_B'/\phi_B - 1)X_{A3}} (p - \Delta\pi), \tag{1}$$

where C_M is the molar density of the membrane phase, $\bar{V}_B$ is the molar volume of the solute, ϕ_B is a coefficient representing resistance to solvent flow, τ is the effective thickness of the membrane, $\Delta\pi$ is the osmotic pressure difference across the membrane, ϕ_B' is a

[1] Current address: Department of Chemical Engineering, McMaster University, Hamilton, Ontario, Canada L8S 4L7

0097-6156/85/0281-0125$06.00/0

coefficient representing the resistance to solvent flow due to solute-membrane material interactions, and X_{A3} is the mole fraction of solute in the permeate. Since hydrocarbons are only sparingly soluble in water, aqueous solutions of hydrocarbons are necessarily quite dilute, and so have negligible osmotic pressures; for example, the osmotic pressure difference across a membrane separating an aqueous solution of benzene can be no greater than approximately 56 kPa, much smaller than typical operating pressures. Consequently, osmotic pressure effects cannot explain flux reduction in aqueous hydrocarbon separations. Similarly, the low solubility limits the hydrocarbon mole fraction in the permeate to values much less than unity, so it can be assumed that the denominator of Equation 1 is approximately unity. Thus, the total flux for a system separating an aqueous hydrocarbon solution is given by

$$N_T = (C_M\bar{V}_B/\phi_B\tau)p \tag{2}$$

For dilute solutions, the product $C_M\bar{V}_B$ is approximately constant. Since the thickness of the membrane should be relatively insensitive to the solute concentration, flux reduction must result from a concentration dependence of the coefficient ϕ_B.

Several mechanisms can be hypothesized to explain the concentration dependence of ϕ_B, for example: the adsorption of solute onto pore walls, thereby reducing the area available for transport; the reversible swelling of the polymer matrix by the solute; and the reduction of membrane phase water content due to the dissolution of solute into the membrane phase. Regardless of the view adopted, it is the physicochemical affinity of the solute for the membrane material that results in flux reduction. In this paper, the increase of ϕ_B with solute concentration is attributed to the adsorption of solute onto the membrane. This adsorbed solute reduces the area available for transport by reducing the size of the transport corridors and increasing the resistance to flow. Consequently, it is reasonable to expect flux reduction to be correlated with composition through an adsorption isotherm.

This paper considers flux reduction in the pressure-driven membrane separation of binary aqueous solutions of aromatic hydrocarbons using asymmetric cellulose acetate membranes. Flux reduction is attributed to the adsorption of solute onto the membrane material. These adsorption effects are then correlated with the physicochemical nature of the solute-solvent-membrane system.

Theory

This section develops a theoretically-based adsorption correlation for flux reduction in pressure-driven membrane separation processes. The effects of osmotic pressure and solute-membrane friction are neglected. The development proceeds as follows. First, physical assumptions are presented. Next, an equation relating observable quantities to the amount of solute adsorbed is derived. Finally, adsorption isotherms based on analogies with multilayer adsorption of gases onto solids are presented.

This development assumes that the membrane is a solid film bridged by capillary pores with an effective radius of r. Transport occurs only through these pores. Since pure water fluxes were proportional to the applied pressure, it is assumed that the pore fluid is in laminar flow. Physicochemical interactions between the membrane material and solute molecules in the permeating solution can result in a layer of solute adsorbed onto the pore walls. The effective thickness of this adsorbed layer is h.

The volume of the adsorbed layer in one pore is given by the equation

$$\text{volume adsorbed} = \tau(\pi r^2 - \pi(r-h)^2) \tag{3}$$

The quantity τ is the effective thickness of the membrane including the effects of tortuosity. Rearranging Equation 3 provides the result

$$\text{volume adsorbed} = \tau\pi r^2(2 - (h/r))(h/r) \tag{4}$$

If the thickness of the adsorbed layer is much less than the pore radius (that is, if h is much less than r), the amount of solute adsorbed is proportional to (h/r).

The molar flux of pure water is given by the equation (2)

$$N_P = \frac{C}{8\eta}\,\varepsilon\,\frac{r^2}{\tau}\,p \tag{5}$$

where C is the molar density and η is the viscosity of the permeating liquid. The surface porosity, ε, is the ratio of the total pore area to the apparent surface area of the membrane:

$$\varepsilon = \frac{n\pi r^2}{S} \tag{6}$$

The quantity S is the apparent surface area of the membrane and n is the number of pores. As a result, the pure water flux is proportional to the fourth power of the pore radius:

$$N_P = \frac{\pi n C}{8\eta S\tau}\,r^4\,p \tag{7}$$

The quantity $(\pi n C r^4)/(8\eta S\tau)$ is the pure water permeability, A. When solute from the permeating solution adsorbs to the pore wall, the radius r should be replaced by the reduced pore radius (r-h); for this case, Equation 7 becomes

$$N_T = \frac{\pi n C}{8\eta S\tau}\,(r-h)^4\,p \tag{8}$$

Equations 7 and 8 can be combined and rearranged to obtain the result

$$(h/r) = 1 - (N_T/N_P)^{1/4} \tag{9}$$

giving (h/r), which is proportional to the amount of solute in the adsorbed layer, in terms of the fluxes N_P and N_T.

The amount of solute adsorbed is related to the composition of the system through an adsorption isotherm. Although there are many adsorption isotherms which might correlate flux reduction to composition, this paper will consider only the use of the Langmuir isotherm (3), which assumes less than complete monolayer coverage, and the Brunauer-Emmett-Teller (BET) isotherm (4), which applies to multilayer adsorption. The Langmuir and the BET isotherms were chosen because both are derived from first principles and because both have parameters with physically meaningful interpretations. Although these isotherms were originally derived for the adsorption of gases onto solids, they can, by analogy, be applied to the adsorption of a solute from liquid solution. The analogy represents the amount adsorbed by (h/r), the gas pressure by the solute mole fraction, X_A, and the adsorbent vapor pressure by the solute mole fraction at saturation, X_A*. For adsorption from solution onto pore walls, the Langmuir isotherm is

$$(h/r) = (h/r)_m \frac{K_L X_A}{1 + K_L X_A} \tag{10}$$

while the BET isotherm is

$$\frac{X_A}{(h/r)(X_A* - X_A)} = \frac{1}{(h/r)_m c} + \frac{c-1}{(h/r)_m c} \frac{X_A}{X_A*} \tag{11}$$

The quantity $(h/r)_m$ represents the amount of solute that must be adsorbed for complete monolayer coverage, while K_L and c are constants. The choice between these isotherms ultimately depends on consistency with observation.

Experimental Procedures

This paper uses data presented earlier for the pressure-driven membrane separation of binary aqueous solutions of aromatic hydrocarbons (benzene, toluene, and cumene) (5). Physicochemical properties for these solutes are given in Table I. The experiments were performed using flat, circular membranes in radial flow cells; the apparatus and operating procedures are described elsewhere (5,6). Six asymmetric cellulose acetate membranes were used; the procedures used to cast and heat-treat the membranes are described elsewhere (5,7).

The composition of the solution adjacent to the high-pressure side of the membrane was calculated from the total solution flux. The feed and permeate concentrations were determined using a method described elsewhere (8). The mole fraction of solute adjacent to the membrane, X_{A2}, (that is, the boundary layer concentration) never equalled the solubility limit for the solute, X_A* (values of which are listed in Table I), thereby avoiding phase separation in the boundary layer. Values of the boundary layer concentration are indicated in Figures 1 to 3.

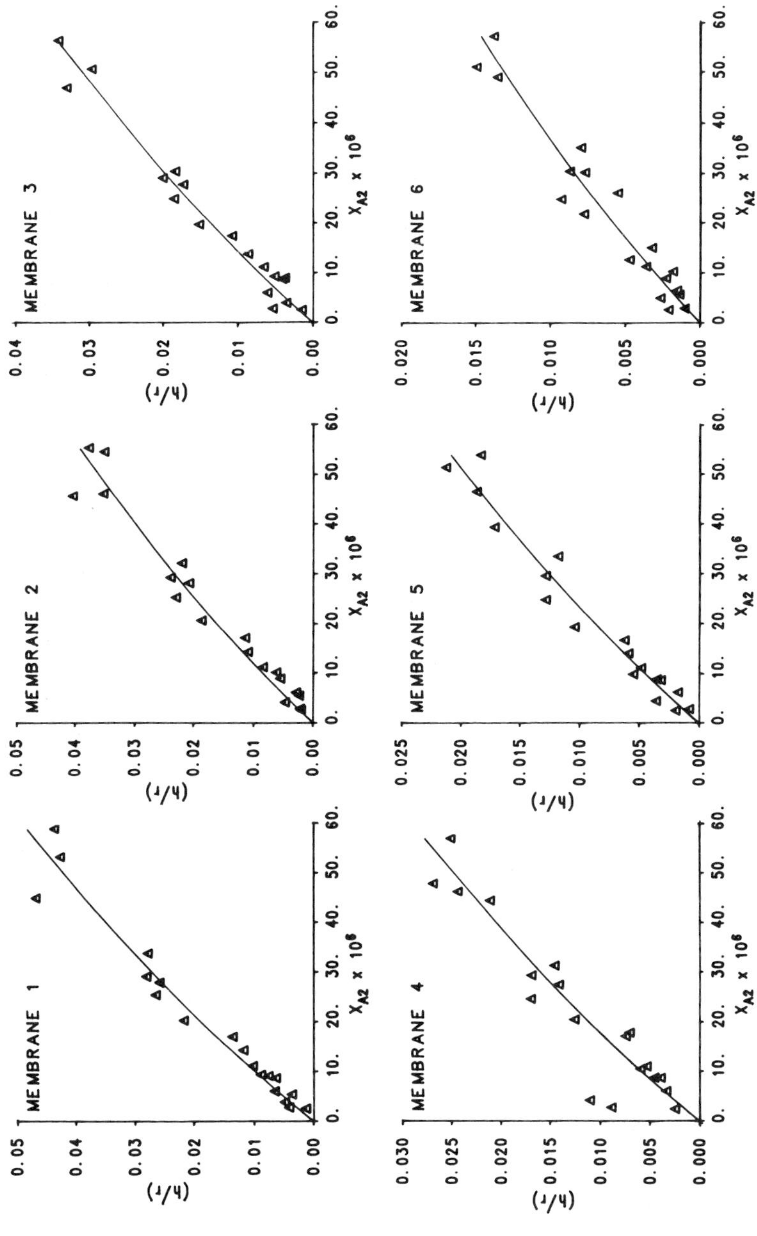

Figure 1: Adsorption correlation for benzene. Symbols represent experimental data; curves are for the best-fit regression.

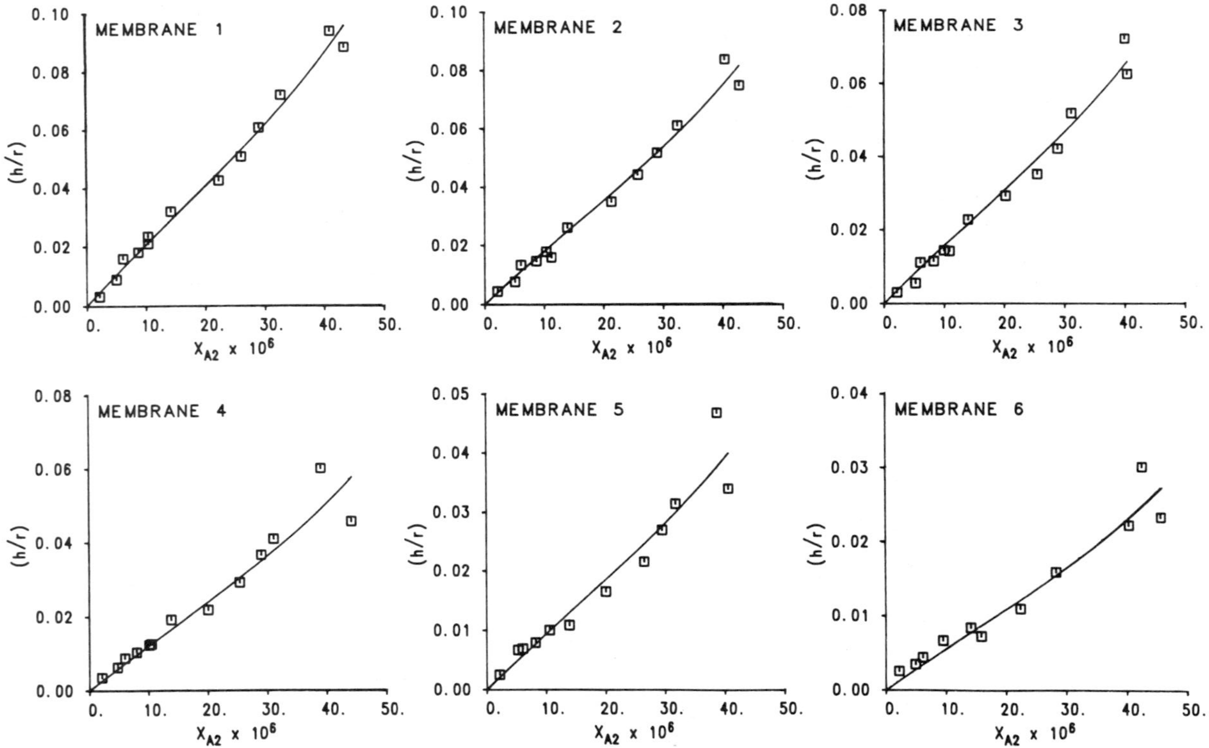

Figure 2: Adsorption correlation for toluene. Symbols represent experimental data; curves are for the best-fit regression.

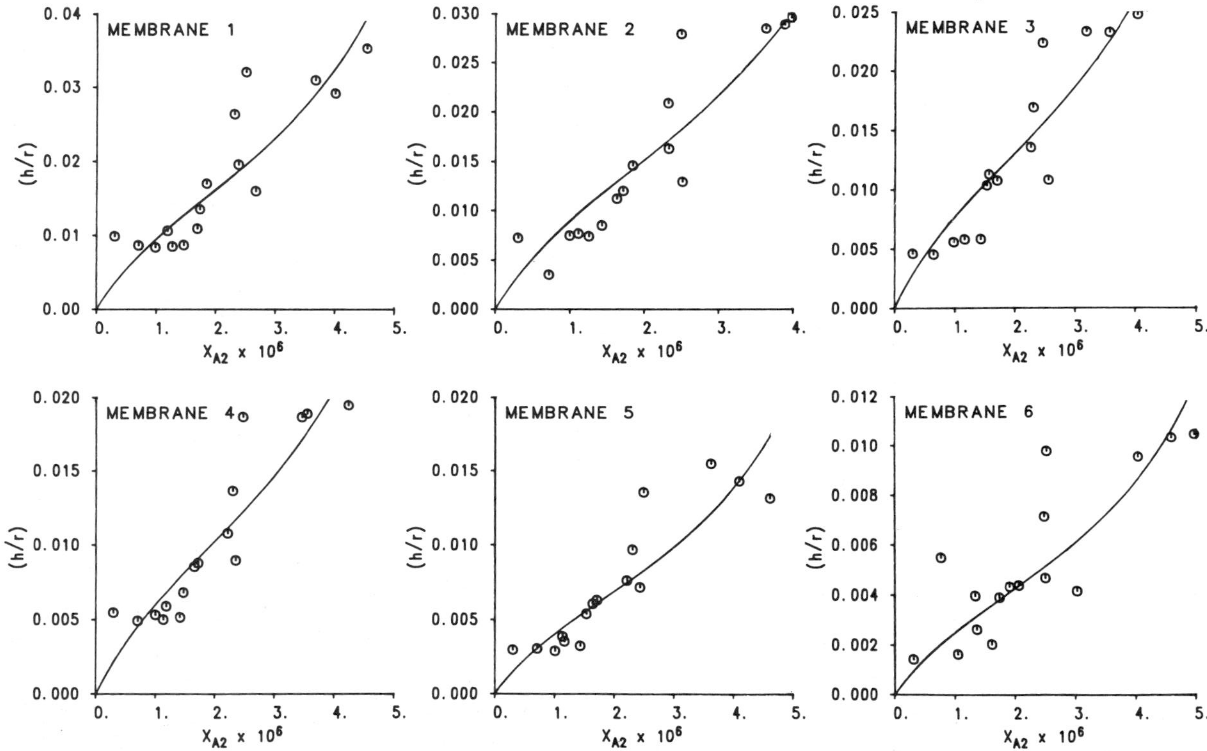

Figure 3: Adsorption correlation for cumene. Symbols represent experimental data; curves are for the best-fit regression.

Table 1. Physicochemical Properties of Solutes

	Benzene	Toluene	Cumene
Molar volume, $m^3/kmol \times 10^3$	88.9	106.8	139.5
$X_A^* \times 10^6$	411.	101.	7.57
Solubility parameters, $MPa^{\frac{1}{2}}$			
δ_d	18.4	18.0	17.0
δ_h	0.0	1.4	0.8
δ_p	2.0	2.0	2.0
Molar cohesive energy, MJ/kmol	30.1	34.6	40.3

Note: Molar volumes from Ref. 9. Saturation mole fractions were calculated from data published by Matsuura and Sourirajan (10). Solubility parameters for benzene and toluene are from Ref. 11. Solubility parameters for cumene were estimated using group contribution methods (12). The molar cohesive energy is the product of the molar volume and the square of the dispersive component of the solubility parameter.

Correlation of Experimental Data

There are several ways to correlate the experimental data. First, the data can be correlated using either the Langmuir isotherm, Equation 10, or the BET isotherm, Equation 11. Second, the concentration X_A can be represented by the retentate concentration, X_{A2}, the permeate concentration, X_{A3}, or the average composition, $(X_{A2}+X_{A3})/2$. Third, the parameters K_L and c can be taken as characteristic either of the solute only or of both the solute and the membrane. Taking all possible combinations of these options gives twelve correlations.

Each of these twelve correlations was performed. Some correlations gave results that were physically unrealistic, yielding values of $(h/r)_m$ which were either negative or greater than unity. These correlations were omitted from further analysis. The correlations which gave physically realistic results were compared to determine which correlation had the minimum residual variance. The residual variances are listed in Table II. The two BET correlations with X_{A2} as the concentration variable fit the data better than any of the other correlations. An analysis of covariance (13) showed that these two correlations were significantly better than the others, and that these two correlations fit the data equally well. Allowing c to vary with both solute and membrane does not significantly improve the descriptive power of the correlation, and so is an additional complication which adds little to the understanding of the adsorption phenomenon. Therefore, this paper considers only the results obtained when c was taken to be a function of solute alone.

Figures 1, 2, and 3 show the adsorption correlations for

benzene, toluene, and cumene; these correlations use the BET isotherm with X_{A2} as the concentration variable and c as a function of solute alone. It can be seen from these figures that flux reduction, as indicated by the ratio (h/r), increases with increasing solute concentration adjacent to the membrane. The values of c obtained from these correlations are listed in Table III, while the values of $(h/r)_m$ obtained for each solute are listed in Table IV.

Discussion

The Variation of $(h/r)_m$ with Solute. For a given membrane, $(h/r)_m$ increases in the order cumene < tolune < benzene. If the simple model of solute physically adsorbed to the pore wall were valid, the thickness of a monolayer, $(h)_m$, would increase with molecular size, and so $(h/r)_m$ would increase with molecular size. Since the molar volume also increases with molecular size, the parameter $(h/r)_m$ should increase as molar volume increases. Thus, the assumption of simple physical adsorption onto pore walls leads to the prediction that $(h/r)_m$ should increase in the order benzene < toluene < cumene. This predicted order is opposite to the observed order, and so the assumption of simple physical adsorption onto pore walls must be rejected.

An alternative explanation is based on physicochemical interactions in the solute-solvent-membrane material system. A solute molecule approaching the pore wall is subject to both attractive and repulsive forces. The solutes are attracted to the membrane material due to their considerable dispersive nature (see Table I). However, the electron-rich conjugated double bonds in the solute are repelled by the electron-rich oxygen, double-bonded oxygen, and double-bonded carbon in the membrane material (cellulose acetate). These forces move the solute molecule to a position where attraction and repulsion are in equilibrium. At equilibrium, the distance from the solute molecule to the membrane is $(h)_m$.

Because benzene, toluene, and cumene each have one aromatic ring, they are subject to nearly equal repulsive forces. Therefore, any differences that exist among these solutes are due to differences in the force of attraction acting between the solute molecules and the membrane material. Increasing the dispersive character of the solute by altering the nature of the pendant group increases the strength of the solute-membrane attraction. The equilibrium distance from the membrane, $(h)_m$, and consequently $(h/r)_m$, should decrease as the solute becomes more dispersive. As a result, $(h/r)_m$ should decrease with increasing molar cohesive energy due to dispersion. Based on the molar cohesive energies of dispersion for these solutes, which are listed in Table I, $(h/r)_m$ should decrease in the order benzene > toluene > cumene. This prediction is consistent with the experimental results: the thickness of the adsorbed layer corresponding to complete monolayer coverage, $(h/r)_m$, decreases as the solute becomes more dispersive.

Table II. Variance of the Residual Error for the Langmuir and BET Isotherms

Langmuir isotherm	Variance x 10^6			Degrees of Freedom
	X_{A2}	X_{A3}	$(X_{A2}+X_{A3})/2$	
$K_L=K_L$(solute)	(a)	46.580	16.556	261
$K_L=K_L$(solute, membrane)	(a)	(a)	(a)	246
BET isotherm	**Variance x 10^6**			**Degrees of Freedom**
	X_{A2}	X_{A3}	$(X_{A2}+X_{A3})/2$	
c = c(solute)	9.087	45.969	17.186	261
c = c(solute, membrane)	8.868	47.792	11.814	246

(a) Correlation results were not physically realistic.

Table III. The C Parameter of the Adsorption Isotherm for Benzene, Toluene, and Cumene

	c
Benzene	4.47
Toluene	2.74
Cumene	5.77

Table IV. Pure Water Permeability and the Amount Adsorbed at Complete Monolayer Coverage, $(h/r)_m$

	A x 10^7	$(h/r)_m$ x 100		
Membrane	kmol/m^2-s-kPa	Benzene	Toluene	Cumene
1	0.72	9.70	8.12	1.76
2	0.88	8.28	7.02	1.65
3	1.25	7.10	6.11	1.42
4	1.87	5.71	4.76	1.12
5	2.30	4.51	3.69	0.72
6	2.52	3.02	2.15	0.47

The Parameter c in the Adsorption Isotherm. In the original derivation of the BET isotherm, c is a function of both the heat of adsorption of the gas to the solid and the heat of liquefaction of the gas being adsorbed. In this work, c should be determined by the free energy change which results from moving the solute molecule from free solution to its equilibrium position. This free energy change should be related to the physicochemical properties of the solute, solvent, and membrane material. Because only one membrane material and only one solvent were used in this study, only the physicochemical nature of the solute need be explicitly considered. As can be seen from the solubility parameters for benzene, toluene, and cumene, which are summarized in Table I, the solutes are not polar and do not hydrogen bond appreciably. The dispersive component of the solubility parameter indicates the strength of dispersion on a volumetric basis. However, the interaction with a single molecule is best expressed on a molar basis, and so the molar cohesive energy calculated from the dispersive component of the solubility parameter is used to indicate the strength of dispersion.

The relationship of the constant c to the strength of solute-membrane and solute-solvent affinity can be shown explicitly by considering the situation of complete monolayer coverage of the membrane. For this situation, $(h/r)=(h/r)_m$ and Equation 11 becomes

$$\frac{(X_{A2})_m}{X_A^* - (X_{A2})_m} = \frac{1}{c} + \frac{c-1}{c}\,\frac{(X_{A2})_m}{X_A^*} \tag{12}$$

where $(X_{A2})_m$ represents the solute mole fraction corresponding to complete monolayer coverage. Solving Equation 12 for c provides the result

$$c = \left[\frac{X_A^*}{(X_{A2})_m} - 1\right]^2 \tag{13}$$

The saturation mole fraction, X_A^*, is determined by the strength of solute-solvent interactions. Since the amount of adsorbed solute required for complete monolayer coverage, reflected by $(h/r)_m$, is determined primarily by the strength of solute-membrane affinity, it is reasonable to assume that the mole fraction corresponding to monolayer coverage is also determined by the degree of solute-membrane affinity. This assumption is borne out by the relationship between $(X_{A2})_m$ and the molar cohesive energy, shown in Figure 4. The less dispersive the solute, the weaker the strength of solute-membrane affinity and the greater the concentration required to obtain complete monolayer coverage. Consequently, the adsorption constant, c, is determined by the strength of solute-solvent interactions relative to the strength of solute-membrane interactions.

The Porous Structure of the Membrane. An earlier paper (8) suggested that the increase in pure water permeability from membrane to membrane was due, at least in part, to an increase in the effective pore radius. The results presented in this paper are consistent with that suggestion. If the equilibrium distance, $(h)_m$, is approximately constant for a given solute, the quantity $(h/r)_m$ should decrease as pore radius, r, increases. The thickness corresponding to complete monolayer adsorption, $(h/r)_m$, is shown as a function of the pure water permeability, A, in Figure 5. This figure shows that $(h/r)_m$ does increase as the pure water permeability increases. Consequently, the increase in pure water permeability from membrane to membrane is due, at least in part, to an increase in pore radius.

Conclusions

Flux reduction in the pressure-driven membrane separation of aromatic hydrocarbons from water using asymmetric cellulose acetate membranes can be correlated using adsorption isotherms. The Brunauer-Emmett-Teller isotherm fits the data better than the Langmuir isotherm. Flux reduction is best correlated as a function of the composition of the retentate immediately adjacent to the membrane rather than as a function of the permeate composition or the average of the permeate and retentate compositions. The behavior of the system is related to the dispersive character of the solute, as indicated by the molar cohesive energy due to dispersion.

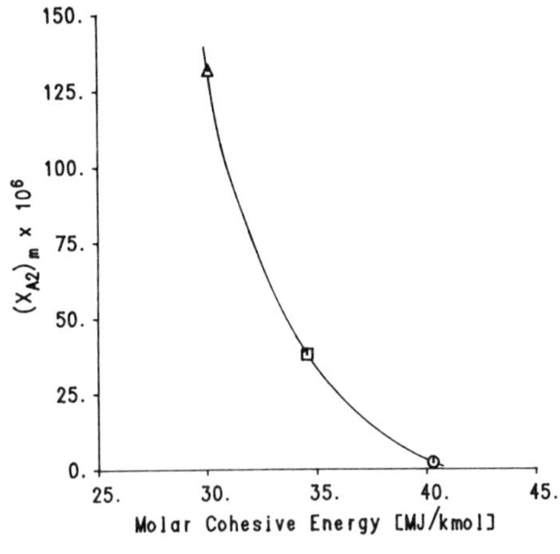

Figure 4: Solute mole fraction for complete monolayer coverage as a function of the molar cohesive energy due to dispersion.

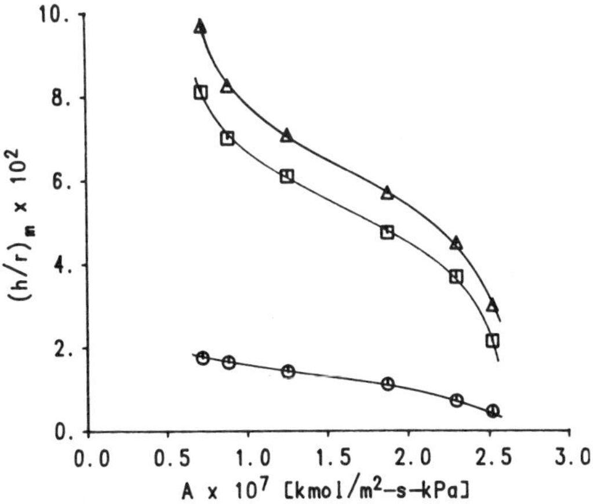

Figure 5: Thickness for complete monolayer coverage, $(h/r)_m$, as a function of pure water permeability, A. △ : benzene; □ : toluene; ○ : cumene

Nomenclature

A	Pure water permeability [kmol/m^2-s-kPa]
c	Parameter in the BET isotherm [dimensionless]
C	Molar density of permeating solution [kmol/m^3]
C_M	Molar density of the membrane phase [kmol/m^3]
h	Thickness of adsorbed layer [m]
K_L	Langmuir adsorption constant [dimensionless]
n	Number of pores [dimensionless]
N_P	Molar pure water flux [kmol/m^2-s]
N_T	Molar solution flux [kmol/m^2-s]
p	Operating pressure [kPa]
r	Radius of pore [m]
S	Apparent surface area of membrane [m^2]
$\bar{V}_B$	Molar volume of water [m^3/kmol]

X_A^* Solute mole fraction at saturation [dimensionless]

X_{Aj} Solute mole fraction at location j [dimensionless]

δ_d Dispersive component of the solubility parameter [$MPa^{\frac{1}{2}}$]

δ_h Hydrogen bonding component of the solubility parameter [$MPa^{\frac{1}{2}}$]

δ_p Polar component of the solubility parameter [$MPa^{\frac{1}{2}}$]

ε Surface porosity [dimensionless]

η Viscosity of permeating solution [kPa-s]

ϕ_B Coefficient relating the solvent velocity relative to the membrane to the external force acting on the solvent [((kg-m/s^2)/kmol)/(m/s)]

ϕ_B' Coefficient relating the solute velocity relative to the membrane to the external force acting on the solvent [((kg-m/s^2)/kmol)/(m/s)]

τ Effective thickness of membrane [m]

$( )_m$ Quantity for complete monolayer coverage

Subscripts:

2 Retentate adjacent to the membrane

3 Permeate

Acknowledgments

The authors are grateful for the support provided by the National Science Foundation (Separation Processes Program, grant number CPE-8312671), the Office of Water Research and Technology (grant number 14-34-0001-0522), and the Separations Research Program, Department of Chemical Engineering, and Graduate School of the University of Texas at Austin.

Literature Cited

1. Thiel, S. W.; Lloyd, D. R., Dickson, J. M., This volume.
2. Merten, U. In "Desalination by Reverse Osmosis"; U. Merten, Ed.; MIT Press: Cambridge, 1966; p. 15-54.
3. Langmuir, I. J. Am. Chem. Soc. 1918, 40, 1361-1403.
4. Brunauer, S.; Emmett, P. H.; Teller, E. J. Am. Chem. Soc. 1938, 60, 309-319.
5. Dickson, J. M.; Babai-Pirouz, M.; Lloyd, D. R. Ind. Eng. Chem. Proc. Des. Dev. 1983, 22, 625-632.
6. Dickson, J. M.; Lloyd, D. R. In "Synthetic Membranes, Vol. II"; A. Turbak, Ed.; ACS: Washington, D.C., 1981; Chapter 18.
7. Pageau, L.; Sourirajan, S. J. Appl. Polym. Sci. 1972, 16, 3185-3206.

8. Babai-Pirouz, M.; Thiel, S. W.; Lloyd, D. R.; Dickson, J. M. J. Membr. Sci. 1984. 21, 21-33.
9. R. C. Weast, Ed. "Handbook of Chemistry and Physics", 55th ed.; CRC Press, Inc.: Cleveland, OH, 1974.
10. Matsuura, T.; Sourirajan, S. J. Appl. Polym. Sci. 1973, 17, 3683-3708.
11. Hansen, C. M.; Beerbower, A. In "Encyclopedia of Chemical Technology: Supplemental Volume"; A. Standen, Ed.; Interscience: New York, 1971.
12. Fedors, R. F. Polymer Sci. Eng. 1974, 14, 147-154.
13. Miller, I.; Freund, J. E. "Probability and Statistics for Engineers", 2nd ed.; Prentice-Hall, Inc.: Englewood Cliffs, NJ, 1977; Chapter 12.

RECEIVED February 22, 1985

11

Effects of Hydrolysis on Cellulose Acetate Reverse-Osmosis Transport Coefficients

S. B. McCRAY and JULIUS GLATER

School of Engineering and Applied Sciences, University of California, Los Angeles, CA 90024

The hydrolysis of cellulose acetate involves the removal of acetyl groups from the cellulose acetate polymer backbone. Since the concentration of acetyl groups in the polymer affects the rate of salt and water transport through the membrane, the performance of the membrane changes upon hydrolysis. Our efforts have shown that the two parameter transport model does not adequately describe the transport of salt and water through hydrolized membranes. Our current work represents the study of the applicability of the tree parameter Kedem-Katchalsky transport model to hydrolyzed cellulose acetate membranes. Data are presented which show that the three parameter model more accurately describes transport through hydrolyzed membranes. This is due to the salt-water coupling term which is neglected in the two parameter model.

It has been known for years that cellulose acetate membranes experience hydrolytic decomposition under adverse feedwater pH conditions. While the optimum pH range for extended membrane life is 4-6 (1,2,3), periodic failure of pH control systems can lead to a decrease in salt rejection due to hydrolysis. A knowledge of the effects of hydrolysis on transport coefficients will allow a prediction of membrane performance with time.

Hydrolysis of cellulose acetate results in the removal of acetyl groups from the polymer. Vos et al. (1) studied the hydrolysis of cellulose acetate membranes between 20-90°C, in a pH range of 2-9. They found that acetyl content decreases with time in a pseudo-first

0097-6156/85/0281-0141$06.00/0

order reaction. The activation energies are 16.4 and 20.1 kcal/mole for the acid (pH 2.0) and base (pH 9.0) catalyzed reactions, respectively.

Vos et al. (4) and Sammon et al. (5) measured changes in membrane performance upon hydrolysis. The salt and water permeability coefficients increased exponentially with time after a short induction period. These were on-line tests under alkaline conditions at constant temperature.

McCray (3) used soak tests to evaluate the effects of hydrolysis on membrane performance. He showed the water flux increased exponentially and the salt rejection decreased linearly with time for both acid and base catalyzed reactions.

Glater and McCray (6) evaluated changes in the Solution-Diffusion transport parameters upon hydrolysis at pH 8.6. The salt and water permeability coefficients increased exponentially with time. Based on their results, it was suggested that the Solution-Diffusion model is not adequate in describing transport through cellulose acetate membranes because it neglects salt-water coupling.

The objective of this study is to evaluate the effects of hydrolysis on the Kedem-Katchalsky transport parameters. Using soak tests, the transport parameters of hydrolyzed membranes were determined from reverse osmosis data. Expressions have been developed which relate changes in transport parameters to hydrolysis time and temperature.

Membrane Transport Equations

The Kedem-Katchalsky transport model is based on nonequilibrium thermodynamics and requires three parameters to characterize salt and water transport through membranes (4,8):

$$J_v = L_p(\Delta P - \sigma \Delta \pi) \tag{1}$$

$$N_s = \omega \Delta \pi + (1-\sigma) J_v C_{s,lm} \tag{2}$$

where J_v is the volumetric flux; ΔP is the hydrostatic pressure difference across the membrane; $\Delta \pi$ is the osmotic pressure difference across the membrane; N_s is the molar salt flux; and $C_{s,lm}$ is the log-mean salt concentration across the membrane.

The water permeability coefficient, L_p, characterizes water flow through the membrane. The salt permeability coefficient, ω, describes the diffusive component of salt flux. The reflection coefficient, σ, is a measure of the salt-water coupling as well as the salt rejection of the membrane. Values of σ range from 0 to 1. For $\sigma = 1$, the membrane is ideal, indicating no salt-water coupling and total salt rejection. For $\sigma = 0$, the membrane shows no semipermeability and thus, no salt rejection.

The Solution-Diffusion model assumes each species must first dissolve into the membrane before diffusing through. All transport is assumed to occur by diffusion. Merten (9,10) gives the equations as follows:

$$J_v = A(\Delta P - \Delta\pi) \qquad (3)$$

$$N_s = B\Delta C_s \qquad (4)$$

where ΔC_s is the salt concentration difference across the membrane; and A and B are the water and salt permeability coefficients, respectively. The Solution-Diffusion model neglects salt-water coupling.

Determination of Transport Parameters

Since the transport parameters are being used to describe transport through reverse osmosis membranes, it was desired to measure them from reverse osmosis data. The salt rejection, r, is defined as:

$$r = 1 - C_{s,p}/C_{s,m} = 1 - N_s/C_{s,m}J_v \qquad (5)$$

where $C_{s,p}$ and $C_{s,m}$ are the salt concentrations of the product stream and at the membrane surface, respectively. Using the transport equations given above, a relationship between salt rejection and volumetric flux can be derived from which the transport parameters can be determined.

Pusch (11) gives the following equations for determining the Kedem-Katchalsky transport parameters:

$$J_v = L_p(\Delta P - r\sigma\pi_m) \qquad (5)$$

$$1/r = 1/r_\infty + \frac{\omega\pi_m}{r_\infty C_{s,m} J_v} = 1/r_\infty + \frac{\omega'}{r_\infty J_v} \qquad (6)$$

where $r_\infty = \sigma$; π_m is the osmotic pressure of the solution at the membrane surface; $\omega' = \omega\pi_m/C_{s,m}$; and r_∞ is the asymptotic salt rejection (the salt rejection at infinite volumetric flow through the membrane). The asymptotic salt rejection is assumed to be equal to the reflection coefficient. The slope of the line formed from 1/r versus $1/J_v$ is ω'/r_∞, while the intercept is $r_\infty = \sigma$. The water permeability coefficient is determined from a plot of J_u versus $(\Delta P - r\sigma\pi_m)$.

The solution-diffusion transport parameters were determined from the following equations:

$$J_v = A(\Delta P - r\pi_m) \tag{7}$$

$$1/r - 1 = B/J_v \tag{8}$$

The salt and water permeabilities are determined from plots of 1/r versus $1/J_v$ and J_v versus $(\Delta P - r\pi_m)$, respectively.

Experimental

The cellulose acetate membranes used in this study were provided by Hydranautics Water Systems of San Diego, California. The membrane was made of refined cellulose acetate. The membrane is thin (3.6 mils = 0.0093 cm wet thickness) and is cast on a paper backing for structural support which was not removed. Untreated, the membrane gave the following transport parameters:

$\sigma = 0.9836$
$L_p = 1.85 \times 10^{-5}$ (cm^3/cm^2-sec-atm)
$\omega = 1.41 \times 10^{-5}$ (cm^3/cm^2-sec)
$A = 1.85 \times 10^{-5}$ (cm^3/cm^2-sec-atm)
$B = 1.92 \times 10^{-5}$ (cm^3/cm^2-sec)

Approximately 150 cm^2 of the membrane were placed into one liter jars and tightly stoppered. The jars contained a pH 8.6 boric acid-sodium hydroxide buffer with a salt concentration of 0.05 molar. The jars were placed into constant temperature baths maintained at the desired temperature to ±0.5°C.

All membrane samples were washed in distilled water and then cut into a 45 cm diameter disc prior to being placed into the radial flow flat plate reverse osmosis test cell described elsewhere (3). The feedwater concentration was constant at 5000 ppm NaCl (0.086 molar) in distilled water at 25 ± 0.5°C. The membranes were pressurized at 5.52×10^3 kPa (800 psig) for 1.5 hours prior to collecting data. The feed and product water concentrations, and volumetric flux were then determined using methods previously described (3). The pressure was then reduced to the desired level and the parameters measured again.

The transport parameters of the membranes were determined using this data and equations 5-8. The salt rejection was corrected for concentration polarization using the formulae given by McCray (12).

Results

Changes in Kedem-Katchalsky Transport Parameters. Figure 1 shows the change in the water permeability coefficient, L_p as a function of time at various hydrolysis temperatures. Since L_p increases exponentially with time, the following equation can be used to describe the changes in L_p upon hydrolysis

$$\ell n\ L_p/L_{po} = k_{Lp}t \qquad (9)$$

where L_{yr} = initial or baseline value of L_p

k_{Lp} = rate constant describing the changes in L_p.

The slopes of the lines found in Figure 1 are values of k_{Lp}, and are given in Table I.

Table I. Rate Constants for Changes in the Kedem-Katchalsky Transport Parameters Upon Hydrolysis at pH 8.6

T °C	k_{LP}x10^6(s^{-1})	k_ωx210^6(s^{-1})	k_σx10^6(s^{-1})
64.7	7.25	33.5	1.69
60.4	3.89	19.1	1.05
54.5	2.03	14.3	0.61
50.0	1.50	8.44	0.36
45.2	0.72	3.61	0.16
E_a (kcal/mole)	23.3	22.7	2.9
E_ax10^3 (J/mole)	97.1	95.0	104.3

Figures 2 and 3 show the value of ω and σ versus time at various hydrolysis temperatures. These transport parameters also show a first order dependence with time. The equations used to describe these changes are

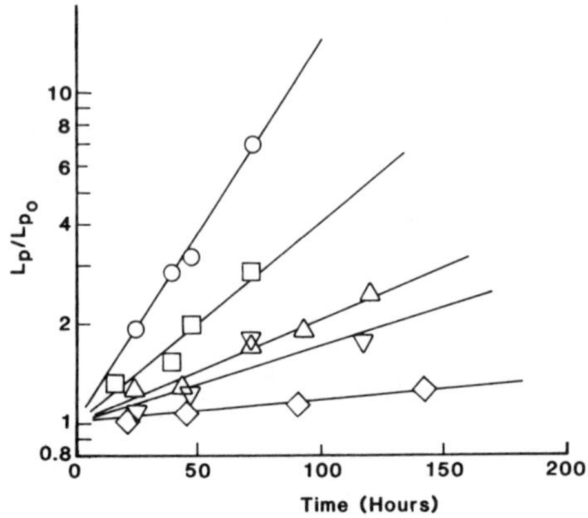

Figure 1. Logarithm Lp/Lp_o versus time for pH 8.6 hydrolysis of β-Hydranautics membranes at various temperatures. Key: ○, 64.7 °C; □, 60.4 °C; △, 54.5 °C; ▽, 50.0 °C; and ◇, 45.2 °C.

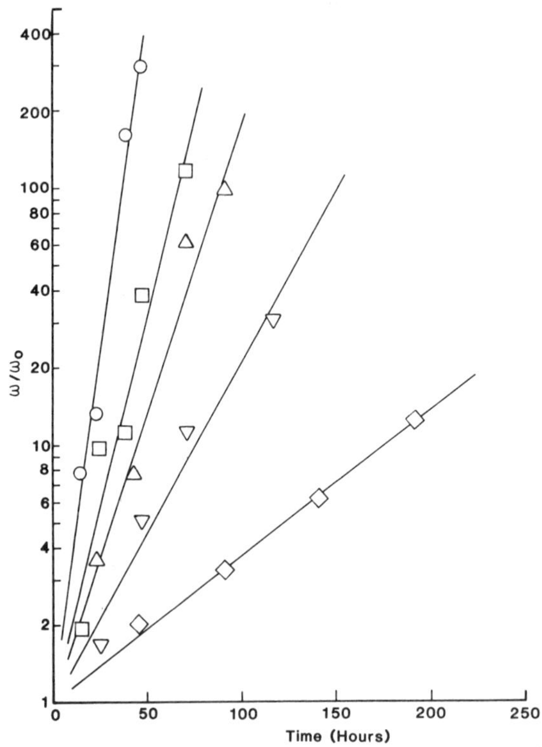

Figure 2. Logarithm ω/ω_o versus time for pH 8.6 hydrolysis of β-Hydranautics membranes at various temperatures. Key: ○, 64.7 °C; □, 60.4 °C; △, 54.5 °C; ▽, 50.0 °C; and ◇, 45.2 °C.

$$\ell n\ \omega/\omega_o = k_\omega t \tag{10}$$

$$\ell n\ \sigma/\sigma_o = -k_\sigma t \tag{11}$$

where subscript o indicates the baseline value, and k_ω and k_σ are the rate constants characterizing changes in ω and σ, respectively. The values of k_ω and k_σ are given in Table I.

Figure 4 shows the rate constants plotted versus inverse temperature. The activation energies calculated from this figure, assuming an Arrhenius expression for the rate constants, are given in Table I. The resultant Arrhenius expressions are

$$k_{Lp}\ (s^{-1}) = 7.41x10^9 e^{-23,300/RT} \tag{12}$$

$$k_\omega\ (s^{-1}) = 1.65x10^{10} e^{-22,700/RT} \tag{13}$$

$$k_\sigma\ (s^{-1}) = 2.60x10^{10} e^{-24,900/RT} \tag{14}$$

Changes in Solution-Diffusion Transport Parameters. The change in the water permeability coefficient, A, with hydrolysis time was essentially the same as the change in the Kedem-Katchalsky water permeability coefficient, L_p. The reasons for this will be discussed below. Values of k_A, determined from the change in A with time are given in Table II. The Arrhenius equation is

$$k_A\ (sec^{-1}) = 7.56x10^9 e^{-23,300/RT} \tag{15}$$

The changes in the salt permeability coefficient, B, were similar to changes in ω. The values of B, however, were somewhat larger than those for ω. Table II gives rate constants and activation energies calculated from changes in B with hydrolysis time. The Arrhenius equation is

$$k_B\ (s^{-1}) = 7.61x10^{10} e^{-23,700/RT} \tag{16}$$

Discussion

The data presented here indicate that hydrolysis can drastically affect membrane performance. Table III gives the time required for a 10% change in the transport parameters for pH 8.6 hydrolysis at 25°C.

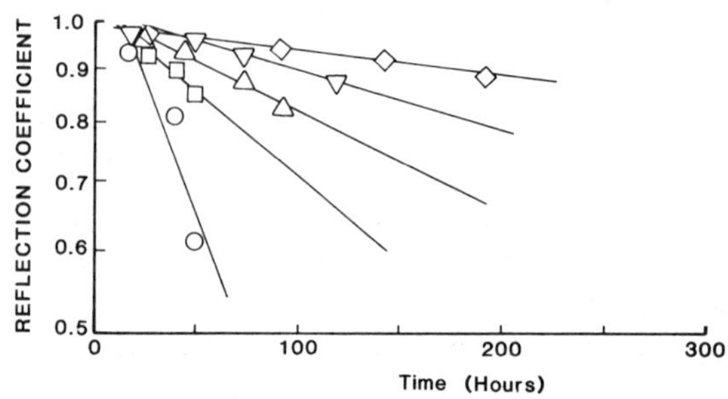

Figure 3. Logarithm reflection coefficient versus time for pH 8.6 hydrolysis of β-Hydranautics membrane at various temperatures. Key: ○, 64.7 °C; □, 60.4 °C; △, 54.5 °C; ▽, 50.0 °C; and ◇, 45.2 °C.

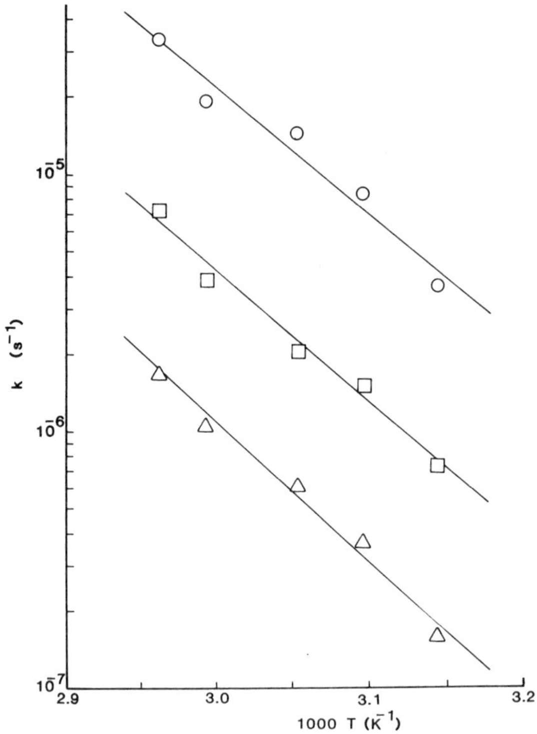

Figure 4. Logarithm k_{Lp}, K_{ω}, and k_{σ} versus inverse temperature for pH 8.6 hydrolysis of β-Hydranautics membranes. Key: ○, k_{ω}; □, k_{Lp}: and △, k_{σ}.

Table II. Rate Constants for Changes in the Solution-Diffusion Transport Parameters Upon Hydrolysis at pH 8.6

T °C	$k_A x10^6 (s^{-1})$	$k_B x10^6 (s^{-1})$
64.7	7.27	36.9
60.4	3.91	21.0
54.5	2.05	14.9
50.0	1.52	8.82
45.2	0.72	3.67
E_a (kcal/mole)	23.3	23.7
$E_a x10^3$ (J/mole)	97.6	99.2

Table III. Time for a 10% Change in Transport Parameters for pH 8.6 Hydrolysis at 25°C

Transport Parameters	$k_i (sec^{-1})$	Time for 10% change (days)
σ	$1.42x10^{-8}$	78
L_p	$6.03x10^{-8}$	18
ω	$3.70x10^{-7}$	3
A	$6.15x10^{-8}$	18
B	$3.15x10^{-7}$	3.5

The salt permeability coefficient is most drastically affected by hydrolysis, requiring only three days for a 10% increase. The reflection coefficient is the least sensitive transport parameter to hydrolysis.

Comparison of equations 12-14 reveals that the activation energies for the rate constants based on changes in the Kedem-Katchalsky transport parameters are within experimental error. Vos et al. (1) found the activation energy at pH 9.0 to be 20.1 kcal/mole, based on changes in acetyl content. Although this value is slightly lower than the ones obtained from changes in transport parameters, the agreement is good considering they were determined by vastly different methods.

Equations 15 and 16 show that the activation energies for changes in the Solution-Diffusion transport parameters are the same as those for the Kedem-Katchalsky parameters. The fact that the activation energies are all within experimental error suggests that changes in transport parameters may be used as a direct measure of the hydrolysis reaction rate.

The changes in transport parameters reported in this study may also be used to examine the differences between the Solution-Diffusion and Kedem-Katchalsky transport models. The changes in the water permeabilities of these models (A and L_p) were essentially the same. Comparison of equations 1 and 3 indicate that they reduce to the same mathematical form for $\sigma = 1$:

$$J_v = L_p(\Delta P-\sigma\Delta\pi) = L_p(\Delta P-\Delta\pi)\big|_{\sigma=1}$$

$$J_v = A(\Delta P-\Delta\pi)$$

At high pressures, the ΔP term of these equations will dominate. Thus, it is reasonable to expect L_p and A to be equivalent.

The values of B were always larger than those of ω. Examination of the salt flux equation shows that they too, become mathematically identical for $\sigma = 1$:

$$N_s = \omega\Delta\pi + (1 - \sigma)C_{s,lm}J_v = \omega'\Delta C_s\big|_{\sigma=1}$$

$$N_s = B\Delta C_s$$

However, as hydrolysis proceeds, the reflection coefficient decreased while the volumetric flux increased. Thus, the coupling term $((1-\sigma)J_vC_{s,lm})$ increased as hydrolysis proceeded. While the Kedem-Katchalsky model provides for both diffusive and convective components of the salt flux, the Solution-Diffusion model neglects coupling and assumes only diffusive flow occurs. One would therefore expect B to be larger than ω. The Solution-Diffusion model thus, overpredicts the diffusive component of the salt flux while the Kedem-Katchalsky model gives a more accurate physical description of the processes occurring.

Conclusions

Transport parameters of cellulose acetate membranes were drastically affected by hydrolysis. The salt and water permeability coefficients of both the Solution-Diffusion and Kedem-Katchalsky models increased exponentially with hydrolysis time. The reflection coefficient decreased exponentially with time.

The Solution-Diffusion transport model does not adequately describe transport of salt through reverse osmosis membranes since it neglects salt-water coupling. Either model will predict water transport, however. Future work will be directed towards further quantifying the differences between the Solution-Diffusion and Kedem-Katchalsky transport models.

Acknowledgments

The authors wish to acknowledge the support for this research provided by the State of California Saline Water Research Funds administered by the Water Resources Center at the University of California, Davis, California. We also express our thanks to Hydranautics Water Systems for providing membranes used in this study.

Literature Cited

1. Vos, K. D.; Burris, F. O., Jr.; Riley, R. L. J. Appl. Poly. Sci., 1966, 10, 825.
2. Spatz, D. D.; and Friedlander, R. H. "Chemical Stability of SEPA Membranes for RP/UF", Osmonics, Inc., Aug. 10, 1977.
3. McCray, Scott B. Masters Thesis, University of California, Los Angeles, 1981.
4. Vos, K. D.; Hatcher, A. P.; Merten, U. Ind. Engr. Chem. Prod. Res. and Dev., 1966, 5(3).
5. Sammon, D. C.; Stringer, B.; Stephen, I. G. 5th International Symposium on Fresh Water from the Sea, 1976, 4, 179-188.
6. Glater, J.; McCray, S. B. Desalination, 1983, 46, 389-397.
7. Kedem, O.; Katchalsky, A. Biochimica et Biophysica Acta, 1958, 27, 229-246.
8. Kedem, O.; Katchalsky, A. Trans. Faraday Soc., 1963, 59, 1941-1953.
9. Merten, U. Ind. Engr. Chem. Fund., 1963, 2, 229-232.
10. Merten, U., in "Desalination by Reverse Osmosis", Merten, U., ed., The MIT Press, Cambridge, MA, 1966.
11. Pusch, W. Berichte: Bunsengesellschaft Fur Physikalishe Chemie, 1977, 81(3), 269-276.
12. McCray, Scott B. Ph.D. Dissertation, University of California, Los Angeles, 1984.

RECEIVED February 22, 1985

12

Application of a Multicomponent Membrane Transport Model to Reverse-Osmosis Separation Processes

MINH-HANG PHAM, P. N. PINTAURO, and K. NOBE

Department of Chemical Engineering, University of California, Los Angeles, CA 90024

A mathematical model, based on the Stefan-Maxwell equations, is used to describe transport in reverse osmosis membranes. The model predicts the salt and solvent fluxes through an asymmetric RO membrane and the concentration and pressure profiles in the dense rejecting layer. For a NaCℓ-water-membrane system, the model contains two equilibrium parameters and three transport parameters. The equilibrium salt and water uptake parameters were determined from experiments using fully dense cellulose acetate membranes. The transport parameters were found by matching experimental RO flux data, using an asymmetric membrane, with the theoretical transport model. Experiments were performed at external NaCℓ concentrations ranging from 0.01 to 0.1 M and the concentration dependence of the resulting parameters was determined.

In an effort to predict the performance of reverse osmosis membranes, a number of investigators have proposed mathematical transport models and developed methods for determining the associated transport parameters. Reid and Breton (1) proposed an RO transport theory based on parallel diffusion and convection processes to characterize the permeability of low flux homogeneous RO membranes. In the years following Loeb and Sourirajan's discovery of a high flux asymmetric cellulose acetate membrane (2), attention was focused on the characteristic solute selectivity of these new membranes and a number of theories were developed to describe RO transport including solution-diffusion models (3,4), pore flow models (5,6), empirical correlations (7,8) and frictional models based on irreversible thermodynamics (9-11). To test the transport models, RO experiments were

0097-6156/85/0281-0153$06.00/0

performed with a variety of asymmetric membranes using aqueous feed solutions containing both organic and inorganic solutes (7,8,12,13). Transport parameters were determined and attempts were made, with some success, to correlate the parameters with the physicochemical properties of the membrane, solute and solvent (4,14).

The transport analysis used in this study is based on the theories of irreversible thermodynamics and is a modified version of the model developed by Pintauro and Bennion (15) for ion exchange membranes. Using computer techniques, the multicomponent Stefan-Maxwell transport equations (16) are integrated across the thin dense rejecting layer of an asymmetric CA membrane and the steady state solute and solvent fluxes are determined. The model includes terms which describe both solute and solvent solubility in the membrane and transport of species through the membrane. The transport parameters, which appear in the model are clearly defined and a method for determining the parameters using experimental data is presented.

Theoretical Transport Model

The fundamental equation describing isothermal multicomponent mass transfer in a membrane can be written as follows:

$$C_i \nabla \mu_i = \sum_j k_{ij}(v_j - v_i) \tag{1}$$

where C_i, μ_i and v_i are the concentration, electrochemical potential and velocity of species i, respectively, and k_{ij} is the friction coefficient describing the interaction between species i and j. The left side of Equation 1 is the driving force per unit volume on species i and the right side is the sum of frictional forces which oppose the driving force. The friction coefficients can be replaced with binary interaction coefficients, D_{ij} by using the relationship (17)

$$k_{ij} = \frac{RTC_iC_j}{C_T D_{ij}} \tag{2}$$

where R and T are the gas constant and absolute temperature and C_T is the total concentration of all species present in the membrane. For a reverse osmosis system with a single binary solute there are three components: j = solute (e), solvent (o) and membrane (m). However, through the application of the Gibbs-Duhem equation (16), there are only two independent transport equations (for this case i = e,o). The analysis assumes the membrane is stationary and $v_m = 0$.

In Equations 1 and 2 for the three component system, there are three independent D_{ij} parameters (D_{eo}, D_{em} and D_{om}) with $D_{ij} = D_{ji}$ (18). To obtain velocities and molar fluxes, N_i, as a function of the electrochemical potential gradients, Equation 1 is inverted to give

$$N_e = -L_{ee}C_e^2\nabla\mu_e - L_{eo}C_eC_o\nabla\mu_o \tag{3}$$

$$N_o = -L_{oe}C_oC_e\nabla\mu_e - L_{oo}C_o^2\nabla\mu_o \qquad (4)$$

There are three independent L_{ij} transport parameters in the above equations with $L_{eo} = L_{oe}$. The functional relationship between the L_{ij} coefficients and concentration, temperature and the three D_{ij} diffusion parameters has been published elsewhere (15). It is desirable to substitute into Equations 3 and 4 expressions which relate μ_e and μ_o to the experimentally measurable quantities of activity (concentration) and pressure

$$\nabla\mu_e = \nu RT\nabla \ell n a_e + \overline{V}_e\nabla P \qquad (5)$$

$$\nabla\mu_o = RT\nabla \ell n a_o + \overline{V}_o\nabla P \qquad (6)$$

where ν is the sum of the number of cations and anions in a neutral salt molecule and $\overline{V}$ is the partial molar volume. The water concentration is eliminated as a variable in Equations 3, 4 and 6 by introducing an equation which defines the dependence of the membrane water concentration on the membrane salt concentration and pressure. Although the dependence may be quite complex, a linear relationship was used in this study (19),

$$C_o = C_o^o - C_e + \alpha(P - P^u) \qquad (7)$$

where P^u is the applied upstream pressure and C_o^o and α are constants which can be determined either experimentally or theoretically. Combining Equations 3-7 and neglecting activity coefficient corrections results in the following membrane flux equations

$$N_e = -L_1\nabla C_e - L_2\nabla P \qquad (8)$$

$$N_o = -L_3\nabla C_e - L_4\nabla P \qquad (9)$$

where

$$L_1 = L_{ee}C_eRT - L_{eo}C_eRT \qquad (10)$$

$$L_2 = L_{ee}C_e^2\overline{V}_e + L_{eo}C_eRT\alpha + L_{eo}C_eC_o\overline{V}_o \qquad (11)$$

$$L_3 = L_{oe}C_o\nu RT - L_{oo}C_oRT \qquad (12)$$

$$L_4 = L_{eo}C_eC_o\overline{V}_e + L_{oo}C_o\alpha RT + L_{oo}C_o^2\overline{V}_o \qquad (13)$$

Equations 8 and 9 are applied to the thin dense rejecting layer (with thickness δ) of an asymmetric cellulose acetate reverse osmosis membrane. This approach assumes that: a) salt rejection and product fluxes are controlled by transport through the dense layer only and

b) there exists a sharp distinction between the dense and porous layers of an asymmetric membrane, thus clearly defining the thickness and homogeneous properties of the rejecting layer. In order to determine the steady state solute and solvent fluxes through an RO membrane and the concentration and pressure profiles in the rejecting layer, Equations 8 and 9 are combined with conservation of species equations,

$$\nabla \cdot N_i = 0 \qquad (i=e,o) \tag{14}$$

Equations 8, 9 and 14 form a set of coupled non-linear partial differential equations with four unknowns, C_e, P, N_e and N_o. The upstream boundary conditions for Equations 8, 9 and 14 are the constant applied pressure and solute concentration,

$$P = P^u \quad \text{and} \quad C_e = C_b K \tag{15}$$

where C_b is the upstream, external solution concentration and K is the partition coefficient, an equilibrium parameter which describes the solute concentration jump at the membrane-solution interface. Downstream, at the dense layer-porous layer interface, the boundary conditions are atmospheric pressure and the downstream salt concentration, which is produced entirely by the ultrafiltration process (20):

$$C_e = \frac{KN_e}{(N_e\overline{V}_e+N_o\overline{V}_o)} \tag{16}$$

Equations 8, 9 and 14 with boundary conditions 15 and 16 were solved in one dimension using the finite-difference numerical techniques outlined by Newman (21). The rejecting layer of thickness δ is divided into a number of mesh points. The differential equations are placed in finite difference form, linearized and solved simultaneously at each mesh point. Since the rejecting layer thickness of an asymmetric membrane is generally not known, the gradient terms ∇C_e and ∇P in the differential equations were made dimensionless with respect to distance from the membrane surface and δ was combined with the transport parameters, resulting in terms of the form D_{ij}/δ.

In order to apply the model to a particular membrane-binary salt-water system, three transport parameters and two equilibrium parameters (the salt partition coefficient and the water concentration, C_o^o, as defined in Equation 7) must be specified. The acetyl content of the rejecting layer of an asymmetric CA membrane (C_m) must also be known. Standard experimental techniques can be used to directly determine K, C_o^o and C_m. The α coefficient in the water concentration equation was estimated to be 6.3×10^{-12} using a simple pressure-stress model (19), along with the bulk modulus of dense cellulose acetate (22). The transport parameters, which appear in the development of the final form flux equations either as D_{ij}/δ, k_{ij}/δ or L_{ij}/δ, are determined indirectly by matching experimental RO flux data with the corresponding predicted quantities determined from the computer transport model.

In this study the three D_{ij}/δ parameters were found by minimizing an objective function, defined as

$$O.F. = \sum_{i=1}^{10} \left(\frac{\Gamma_i^{th} - \Gamma_i^{exp}}{\Gamma_i^{exp}} \right)^2 \qquad (17)$$

where Γ_i^{exp} is a measured downstream salt concentration or product flow rate from an RO experiment and Γ_i^{th} is the corresponding value of Γ_i obtained theoretically by solving the transport model equations with a set of D_{ij}/δ parameters. A computer optimization program, using the direct search techniques of Hooke and Jeeves (23) was used systematically to find the set of D_{ij}/δ yielding the minimum value of the objective function.

According to the theoretical considerations outlined by Pintauro and Bennion (15), the optimum set of experiments for the determination of the D_{ij}/δ parameters would include both dialysis (where the concentration gradient driving force terms in the flux equations dominate) and reverse osmosis (where the ∇P terms dominate). Dialysis experiments, however, were not performed in the present analysis. It has been shown using solution-diffusion models for RO transport (3,4) that the membrane salt flux is primarily dependent on the salt concentration gradient while the membrane water flux is dominated by pressure effects. Thus, by measuring both the downstream salt concentration and product flux, sufficient information regarding both the ∇C_e and ∇P terms in Equations 8 and 9 can be obtained. In addition to this theoretical argument, it was thought that a dialysis experiment would be impractical because the slow diffusion rates would necessitate prohibitively long experiments and the results would be difficult to interpret due to the presence of the porous backing structure.

In principle, only three experimental measurements of downstream salt concentration and product flux at two different upstream pressures are needed to determine a set of three D_{ij}/δ binary interaction parameters. The analysis assumes the transport parameters are independent of pressure, which is reasonable so long as pressure does not affect the physical structure of the membrane during the course of an experiment (i.e., there is no compaction). As will be discussed below, downstream concentrations and product flow rates at five upstream pressures ranging from 400 to 800 psig were measured and inserted into Equation (17). The theory was matched to all five pressure experiments to insure the applicability of the transport model over a wide range of operating pressures. In future work, the use of fewer RO experiments for the D_{ij}/δ determination will be investigated.

Experimental

The transport and equilibrium parameters were determined for a cellulose di,tri-acetate blended membrane. Flux experiments were performed on stock asymmetric membranes (75.3°C cure temperature) supplied by Envirogenics Systems Company, El Monte, California. The equilibrium salt and water uptake experiments were performed on fully dense membranes made from a polymer solution (provided by Envirogenics) whose CA and solvent compositions are identical to those used by Envirogenics to cast their asymmetric membranes. A dense membrane sheet (of thickness 8×10^{-3} cm) was cast on a flat plate at room temperature and allowed to completely air dry. The sheet was then cured in water at 75.3°C for seven minutes, cut into 45 mm circular disks and precompacted at 800 psig for two hours with distilled water. The acetyl content of the fully dense and asymmetric membranes was found experimentally (24) to be 40.33% ($C_m = 1.33 \times 10^{-2}$ moles acetyl/cm^3 wet membrane). The equilibrium and transport parameters were determined as a function of NaCℓ concentration at 25°C by performing experiments in 0.01, 0.03, 0.05, 0.07 and 0.10 M solutions. Reagent grade chemicals and distilled water were used to prepare the salt solutions.

Equilibrium Experiments - In order to determine the salt and water uptake by the dense rejecting layer, three experiments were performed: wet membrane density, membrane salt concentration and membrane water concentration.

Wet Membrane Density - Two dense membrane disks were immersed in a NaCℓ salt solution of known concentration and allowed to equilibrate over a 24 hour period. After 24 hours the membranes were removed from the salt solution and excess electrolyte was wiped off the surface with filter paper. An area outline of the membrane was made on a sheet of paper; then the membrane was weighed and its thickness measured with a micrometer.

Salt Uptake Experiments - Three dense membrane disks were immersed in a NaCℓ solution of known concentration. After 24 hours the membranes were removed from the salt solution, wiped of excess electrolyte, weighed then immersed in 125 mℓ of distilled water for 24 hours. The amount of NaCℓ leached from the disks was determined by atomic absorption spectrometry (Perkin-Elmer Model 303).

Membrane Water Concentration - One membrane disk was immersed in pure distilled water for 24 hours; after which time it was removed, wiped clean of excess surface water and weighed. The membrane was dried in a vacuum oven at 40°C for ca. 48 hours (until its weight became constant) and then reweighed. The difference in weight represents the membrane water content.

Membrane Flux Experiments - These experiments were performed on a radial flow, flat plate reverse osmosis unit designed to accommodate membrane disks of 45 mm diameter. Pressure was supplied to the upstream solution by N_2 gas. The feed container had a total capacity of ca. 5 liters and the brine recycle rate was 600 mℓ/min. The product solution collected downstream was added back to the feed container to restore the NaCℓ concentration to its original value.

Asymmetric membranes were pre-compacted at 800 psig for two hours, after which time the product solution was collected for 10 minutes in a 15 mℓ graduated cylinder and the downstream salt concentration was determined from conductance measurements (Beckman Model 4959 conductivity cell). The applied upstream pressure was then reduced and, after thirty minutes, the product solution flow rate and salinity were measured. Flux experiments were performed sequentially at upstream pressures of 800, 700, 600, 500 and 400 psig with upstream bulk NaCℓ concentrations of 0.01, 0.03, 0.05, 0.07 and 0.1 M. New membranes were used for each upstream concentration to minimize the effects of compaction.

Results and Discussion

The results of two separate wet membrane density (ρ_m) experiments at five external NaCℓ concentrations (C_b) ranging from 0.01 to 0.1 M produced the following linear least-square relationship

$$\rho_m = 1.22 + 15.24C_b \qquad (18)$$

A plot of the partition coefficient, K, for the homogeneous dense membrane vs. the external NaCℓ concentration is shown in Figure 1. The internal molal salt concentration (moles NaCℓ/gram wet membrane) was determined directly from the experimental salt uptake data. These concentrations were then multiplied by the wet membrane density to obtain the membrane concentration in units of moles NaCℓ/cm^3 wet membrane. The dependence of K on C_b can be represented by the following equation

$$K = \frac{C_e}{C_b} = 0.24 - 5.27\text{x}10^3C_b + 33.44\text{x}10^6C_b^2 \qquad (19)$$

The wet homogeneous membrane density was also used to convert the membrane water concentration C_o^o from moles water/gram wet membrane to moles water/cm^3 wet membrane. The resulting value for C_o^o is

$$C_o^o = 7.47\text{x}10^{-3} \text{ moles/cm}^3 \text{ wet membrane} \qquad (20)$$

The steady state product stream volumetric flow rates and salt concentrations from the RO flux experiments are plotted vs. the bulk upstream NaCℓ concentrations in Figures 2 and 3. Each data point was obtained by averaging the results from two experiments.

Transport Parameter Calculations - A set of three D_{ij}/δ binary interaction parameters (D_{eo}/δ, D_{em}/δ and D_{om}/δ) at each external NaCℓ concentration were determined using the theoretical transport model and the data from the equilibrium and transport experiments. The effects of concentration polarization were included in the computer program when analyzing the data (19). The computational technique of determining the D_{ij}/δ parameters assumes that the set of optimized parameters is constant throughout the membrane at that particular upstream NaCℓ concentration. However, due to the fact that RO membranes restrict salt transport, the salt concentration in the rejecting layer changes significantly and the assumption of constant D_{ij}/δ is only valid if the parameters were concentration independent.

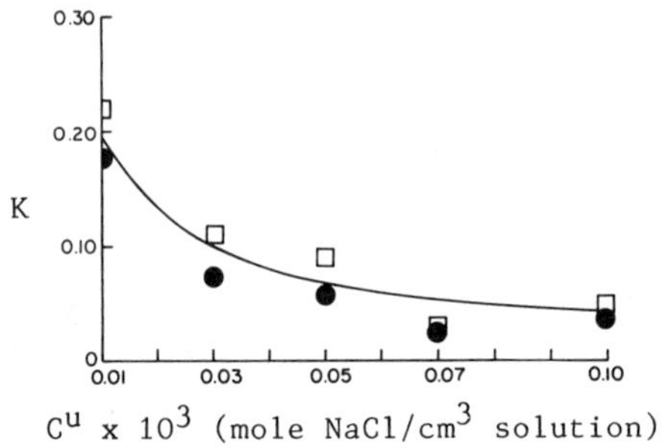

Figure 1. Equilibrium salt partition coefficient versus the upstream NaCl concentration. Key: ●, run 1; and □, run 2. K is mole NaCl inside/cm^3 wet membrane divided by mole NaCl outside/cm^3.

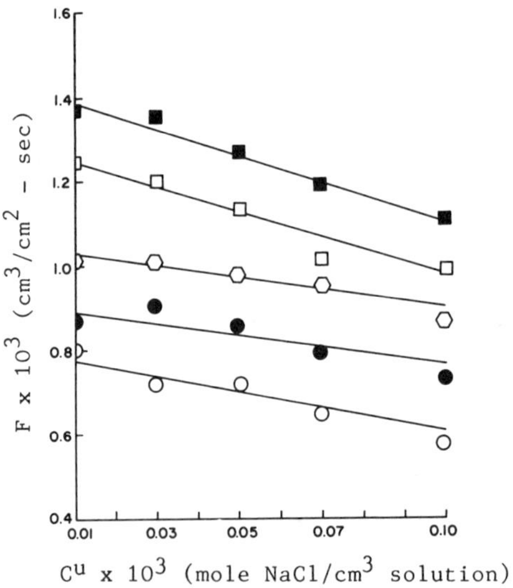

Figure 2. Steady state product stream volumetric flow rate versus the upstream NaCℓ concentration. Key: ■, 800 psig; □, 700 psig; ⬡, 600 psig; ●, 500 psig; and ○, 400 psig.

Table I lists the optimized D_{ij}/δ parameters as a function of the external NaCℓ concentration. These values of D_{ij}/δ, when inserted into the reverse osmosis computer simulation program, were able to predict the experimental fluxes and downstream salt concentrations to within ±12% and ±9%, respectively. Each optimized parameter has an interval of uncertainty of ±1%, meaning that the true minimum in the objective function is somewhere within ±1% of those values listed in Table I. To check for multiple roots of the objective function, the optimization program, with the 0.1 M NaCℓ data, was started using different initial guess values for the D_{ij}/δ parameters. All of the optimization searches resulted in the same set of D_{ij}/δ (within ±1%). The concentration and pressure profiles in the dense rejecting layer for 0.1 M external NaCℓ, computed using the transport model and the optimized D_{ij}/δ, are shown in Figures 4 and 5. Whereas the pressure profiles are linear, the concentration profiles are non-linear and vary slightly with pressure.

Table I. Optimized D_{ij}/δ Parameters for the Envirogenics' RO Membrane-NaCl-Water System

C_b, $mol/cm^3 x10^3$	D_{eo}/δ, $cm/sx10^3$	D_{em}/δ, $cm/sx10^3$	D_{om}/δ, $cm/sx10^2$
0.01	0.854	0.387	1.96
0.03	1.28	1.26	1.90
0.05	1.38	2.09	1.65
0.07	2.04	2.74	1.56
0.10	2.55	3.75	1.41

The D_{ij}/δ results from the 0.1 M optimization program were used as the initial guess parameters when analyzing the 0.07 M data. The results of the 0.07 M search were then used to start the 0.05 M optimization program. This initial guess procedure was used to determine all of the parameters in Table I. Mathematical relationships between the D_{ij}/δ parameters and the rejecting layer salt concentration (C_e) were obtained by combining the results in Table I with Equation 19. The membrane concentration dependence of the parameters is represented by the following least-squares relationships:

$$D_{eo}/\delta = 2.17x10^3 C_e - 1.06x10^3 \tag{21}$$

$$D_{em}/\delta = 4.30x10^3 C_e - 3.26x10^{-3} \tag{22}$$

$$D_{om}/\delta = -7.48x10^3 C_e + 2.62x10^{-2} \tag{23}$$

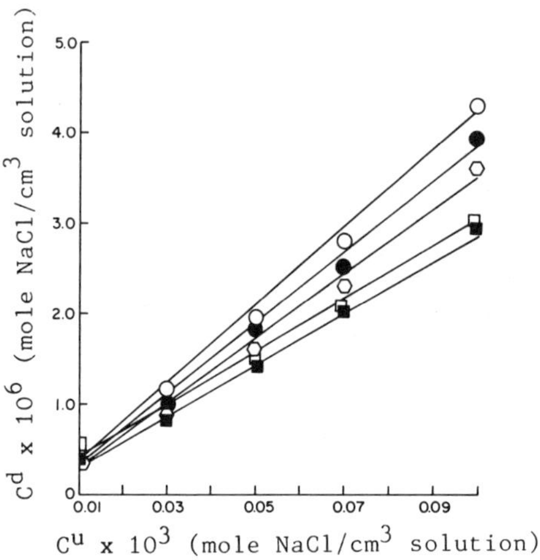

Figure 3. Steady state downstream salt concentration versus the upstream NaCℓ concentration. Key: ■, 800 psig; □, 700 psig; ⬡, 600 psig; ●, 500 psig; and ○, 400 psig.

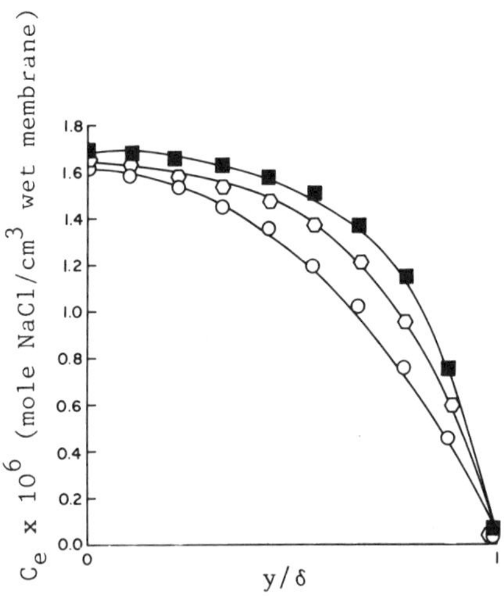

Figure 4. Calculated concentration profiles in the dense rejecting layer for 0.1 M upstream NaCℓ using the D_{ij}/δ parameters in Equations 21-23. Key: ■, 800 psig; ⬡, 600 psig; and ○, 400 psig.

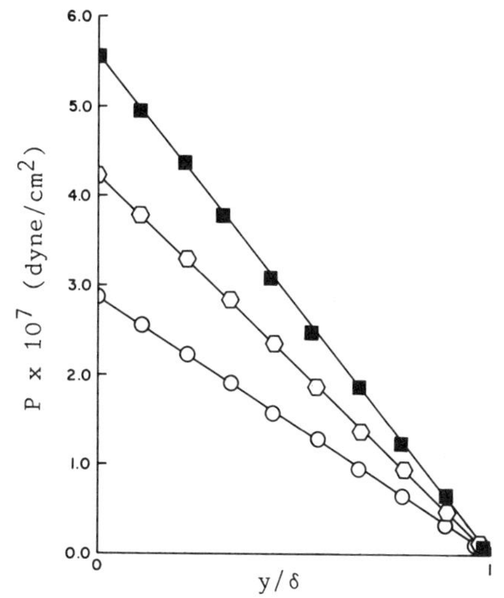

Figure 5. Calculated pressure profiles in the dense rejecting layer for 0.1 M upstream NaCℓ using the D_{ij}/δ parameters in Equations 21-23. Key: ■, 800 psig; ⬡, 600 psig; and ○, 400 psig.

Conclusions

The computer simulation program with Equations 21-23 can be used to predict membrane performance for external NaCℓ concentrations in the range 0.01 M - 0.1 M. Experiments are currently being performed to test the accuracy of the model with the optimized parameters; salt rejection and product flow rates, for upstream NaCℓ concentrations and pressures which differ from those used to determine the D_{ij}/δ parameters will be experimentally determined and compared with the corresponding data predicted by the model. The results of this comparison as well as the application of the model and parameter determination method to other binary and multicomponent salt solutions will be presented in a future publication.

Nomenclature

a_i = Activity of species i inside the membrane.
C_i = Membrane concentration of species i, mol/cm^3 wet membrane.
C_b = Bulk external salt concentration, mole/cm^3 solution.
C_T = Total concentration of all species i inside the membrane, mole/cm^3 wet membrane
D_{ij} = Binary transport parameter describing the interaction of species i and j, cm^2/s.
K = Equilibrium partition coefficient (mol/cm^3 wet membrane)/(mol/cm^3 solution).
k_{ij} = Friction coefficient between species i and j, g/s-cm^3.
L_{ij} = Membrane transport parameters appearing in Equations 3 and 4, g/s cm^3.
N_i = Molar flux of species i, mol/cm^2-s
P = Pressure, dyne/cm^2.
R = Universal gas constant, 8.314×10^7 erg/mol-K.
T = Absolute temperature, K.
v_i = Velocity of species i, cm/s.
$\bar{V}_i$ = Partial molar volume of species i, cm^3/mol.

Greek Symbols

α = Parameter relating membrane pressure and water concentration, mole/dyne-cm.
δ = Membrane rejecting layer thickness, cm.
μ_i = Electrochemical potential of species in inside the membrane, erg/mol.
ν = Number of anions and cations per molecules of electrolyte.
ρ = Wet membrane density, g/cm^3.

Subscripts

e = Membrane salt
o = Solvent (water)
m = Membrane
i,j = Any arbitrary species

Superscripts

d = Downstream
o = Initial or reference state
u = Upstream

Acknowledgments

The authors wish to thank Envirogenics Systems Company for supplying the reverse osmosis membranes used in this study. This work was supported by the University of California Sea Water Desalination Program.

Literature Cited

1. Reid, C. E.; Breton, E. J. J. Appl. Polym. Sci. 1959, 1, 159.
2. Loeb, S.; Sourirajan, S. ADVANCES IN CHEMISTRY SERIES No. 38, American Chemical Society: Washington, D.C., 1963; p. 117.
3. Lonsdale, H. K.; Merten, U.; Riley, R. L. J. Appl. Polym. Sci. 1965, 9, 1341.
4. Hodgson, T. D. Desalination 1970, 8, 99.
5. Gluekauf, E. Proc. 1st Intern. Symp. on Water Desalination 1965, 1, 1943.
6. Sherwood, T. K.; Brian, P. L. T.; Fischer, R. E. Ind. Eng. Chem. Fundam. 1967, 6, 2.
7. Sourirajan, S. "Reverse Osmosis"; Academic: New York, 1970; pp. 1-54.
8. Matsuura, T.; Sourirajan, S. J. Appl. Polym. Sci. 1971, 15, 2905.
9. Kedem, O.; Katchalsky, A. Biochim. Biophys. Acta 1958, 27, 229.
10. Bennion, D. D.; Rhee B. W. Ind. Eng. Chem. Fundam. 1969, 8, 36.
11. Spiegler, K. S. Trans. Faraday Soc. 1958, 54, 1408.
12. Kesting, R. E.; Eberlin, J. J. Appl. Polym. Sci. 1966, 10, 961.
13. Matsuura, Tl; Dickson, D. M.; Sourirajan, S. Ind. Eng. Chem. Process Des. Dev. 1976, 15, 149.
14. Sunu, W. G.; Bennion, D. N. Ind. Eng. Chem. Fundam. 1977, 16, 283.
15. Pintauro, P. N.; Bennion, D. N. Ind. Eng. Chem. Fundam. 1984, 23, 230.
16. Hirschfelder, J. O.; Curtiss, C. F.; Bird, R. D. "Molecular Theory of Gases and Liquids"; Wiley: New York, 1954, p. 714.
17. Newman, J. "Electrochemical Systems"; Prentice-Hall: Englewood Cliffs, NJ; pp. 239-250.
18. Onsager, L. Phys. Rev. 1931, 38, 2265.
19. Pham, Minh-Hang, M.S. Thesis, University of California, Los Angeles, 1984.
20. Lightfoot, E. N. "Transport Phenomena and Living Systems"; Wiley: New York, 1974; pp. 236-265.
21. Newman, J. Eng. Eng. Chem. Fundam. 1968, 1, 514.
22. Reid, C. E.; Kuppers, J. R. J. Appl. Polym. Sci. 1959, 2, 264.
23. Hooke, R.; Jeeves, T. A. J. Assoc. Comput. Mach. 1961, 8 211.
24. Genung, L. B.; Mallatt, R. C. Ind. Eng. Chem. 1941, 13, 369.

RECEIVED March 22, 1985

13

Predictability of Membrane Performance in Reverse-Osmosis Systems Involving Mixed Ionized Solutes in Aqueous Solutions

A General Approach

RAMAMURTI RANGARAJAN, TAKESHI MATSUURA, and S. SOURIRAJAN

Division of Chemistry, National Research Council of Canada, Ottawa, Ontario, Canada K1A 0R9

This paper derives generalized transport equations for the RO system involving solvent water and different completely ionized solutes with different ionic valencies. The analysis imposes no limit to the number of ions or their valencies. The transport equations enable one to predict the separation of individual ions and the permeation rate of product solution at given operating conditions from only a single set of RO primary data for an aqueous sodium chloride reference feed solution.

It was shown in the earlier papers on reverse osmosis (RO) separation of electrolyte mixtures (1-4) that prediction of RO performance data such as the separation of individual ions involved in the mixture, and the membrane permeated product rate, could be successfully accomplished by transport equations based on Kimura-Sourirajan analysis incorporating the necessary modifications arising from ionic equilibria at the solution-membrane interface. It was also shown that the system performance of RO modules by which electrolyte mixtures are separated could also be predicted on the basis of the same fundamental principles (5). The earlier papers cited above have illustrated in no uncertain terms that (1) the basis of Kimura-Sourirajan analysis is sound (2) the concept of free energy parameters which was found useful in predicting RO performance data of single solute systems (6-8) is also relevant in mixed electrolyte solute systems, and (3) the assumptions introduced in the process of the numerical calculations using the transport equations are valid. In this paper, general transport equations are developed, which enable one to predict the performance of RO separations of mixed electrolytes involving any number of ions and ionic valences provided no ionic association takes place in the solution. The system treated in earlier papers are shown to be the special cases of the general expressions derived in this paper.

0097-6156/85/0281-0167$06.00/0
Published 1985, American Chemical Society

Theoretical Analysis and Discussion

Nomenclature. The symbols used (same as those in Part 3 (3) of this series unless otherwise stated) are listed at the end of the paper. To facilitate the reader, a brief note is in order. The subscripts A, B and M refer to salt, water and membrane respectively. All quantities with asterisks (*) refer to ions; with respect to such quantities the first subscript, i, 1, 3, 5... refers to the indicated cations and j, 2, 4, 6... refers to the indicated anions; and, the second subscript M, 1, 2, 3 refers to the indicated phase, namely, membrane, feed solution, concentrated boundary solution on the high pressure side or product solution, respectively. Cations are represented by odd numbers, while anions are represented by even numbers. With respect to the quantities which do not refer to ions specifically, the subscript M, 1, 2, or 3 refers to the indicated phase. Numerical subscripts ij, 1j, i2, 12 refer to single salts with the ions indicated by each number. Thus, for example the quantities X^*_i, X^*_j, X^*_{i1}, X^*_1, X^*_{1M}, X^*_{13}, X_{AM}, X_{AM3}, X_{A3} represent mole fractions of cation i, anion j, cation i in phase 1, cation 1, cation 1 in membrane phase, cation 1 in product solution phase, salt A in membrane phase, salt A in membrane phase in equilibrium with X_{A3} and salt A in product solution phase, respectively; the quantities c_2 and c_{M2} represent molar densities of solutions in phase 2 and in membrane phase in equilibrium with X_{A2}, respectively; and the quantity $(D_{AM}/K\delta)_{12}$ represents solute transport parameter for the single salt 12.

Reverse Osmosis Transport Equations and the General Description of the System. For a RO system involving several ions, under steady state isothermal operating conditions at the gauge pressure P, the flux equations for solvent and ion transport can be written as follows, using the form of the Kimura-Sourirajan analysis (9).

Water transport:

$$N_B = A[P - \pi(\Sigma X^*_{i2} + \Sigma X^*_{j2}) + \pi(\Sigma X^*_{i3} + \Sigma X^*_{j3})] \tag{1}$$

The cation transport:

$$N^*_i = \left(\frac{D^*_{iM}}{K^*_{i2}\delta}\right) c_2 X^*_{i2} - \left(\frac{D^*_{iM}}{K^*_{i3}\delta}\right) c_3 X^*_{i3},\ i = 1,3,5... \tag{2}$$

The anion transport:

$$N^*_j = \left(\frac{D^*_{jM}}{K^*_{j2}\delta}\right) c_2 X^*_{j2} - \left(\frac{D^*_{jM}}{K^*_{j3}\delta}\right) c_3 X^*_{j3},\ j = 2,4,6... \tag{3}$$

$$\text{where } K^*_i = \frac{c\ X^*_i}{c_M\ X^*_{iM}} \tag{4}$$

$$\text{and } K^*_j = \frac{c\ X^*_j}{c_M\ X^*_{jM}} \tag{5}$$

Both Equation 4 and Equation 5 are applicable to concentrated boundary solution on the high pressure side or the dilute product solution on the atmospheric pressure side of the membrane with respect to the second subscript.

In addition to the above equations, the total number of which is equal to (n + 1) where n is the total number of all the ions present, we have the following relations to represent the boundary concentrations of ions:

For cations

$$X^*_{i2} = X^*_{i3} + (X^*_{i1} - X^*_{i3}) \exp \left\{ \frac{N_B + \Sigma N^*_i + \Sigma N^*_j}{k_i\ c} \right\} \qquad (6)$$

and for anions

$$X^*_{j2} = X^*_{j3} + (X^*_{j1} - X^*_{j3}) \exp \left\{ \frac{N_B + \Sigma N^*_i + \Sigma N^*_j}{k_j\ c} \right\} \qquad (7)$$

Furthermore, the product mole fractions of cations and anions are given by,

for cations

$$X^*_{i3} = \frac{N^*_i}{N_B + \Sigma N^*_i + \Sigma N^*_j} \qquad (8)$$

and for anions

$$X^*_{j3} = \frac{N^*_j}{N_B + \Sigma N^*_i + \Sigma N^*_j} \qquad (9)$$

Equations 1 to 9 describe completely the equilibria and transport for all the ionic species. It should, however, be noted that all flux equations have to be considered by suitably subscribing the Equations 2 to 9. The system is also subject to the electroneutrality conditions that prevail in the feed, boundary and the membrane phases:

for the feed solution,

$$\Sigma c_1\ z_i\ X^*_{i1} - \Sigma c_1\ z_j\ X^*_{j1} = 0 \qquad (10)$$

for the concentrated boundary solution,

$$\Sigma c_2\ z_i\ X^*_{i2} - \Sigma c_2\ z_j\ X^*_{j2} = 0 \qquad (11)$$

for the product solution,

$$\Sigma c_3\ z_i\ X^*_{i3} - \Sigma c_3\ z_j\ X^*_{j3} = 0 \qquad (12)$$

for the membrane phase,

$$\Sigma c_M\ z_i\ X^*_{iM} - \Sigma c_M\ z_j\ X^*_{j3} = 0 \qquad (13)$$

and also for the product water flux,

$$\Sigma z_i\ N^*_i - \Sigma z_j\ N^*_j = 0 \qquad (14)$$

Thus, Equations 1 to 14 are a complete general description of a steady state reverse osmosis system with a feed solution consisting of several ions, and our object is to solve the above equations in terms of N_B, N_i^*, N_j^*, X_{i3}^* and X_{j3}^*.

Mathematical Analysis

In order to simplify the solution of the above transport equations, the following assumptions are necessary (1)

i) $c_1 = c_2 = c_3 = c$ (15)

ii) $N_B >> \Sigma N_i^* + \Sigma N_j^*$ (16)

iii) For evaluating N_B using Equation 1, we must have either experimental osmotic pressure data for electrolyte mixtures, or a method to estimate osmotic pressure data for such solutions. In view of the infinite number of solution compositions possible for concentrated boundary solution and product solution phases for a given feed composition, depending upon the operating conditions and the inherent properties of membranes, it is not possible to obtain osmotic pressure data for all possible compositions experimentally. Also there is no theoretical approach to calculate osmotic pressure of solutions containing several ions. For the purpose of our calculations, the osmotic pressure of an electrolyte solution involving several ions is approximated by the equation:

$$\pi\,(\Sigma X_i^* + \Sigma X_j^*) = \Sigma B_i^*(X_i^*)\,X_i^* + \Sigma\,B_j^*(X_j^*)\,X_j^* \tag{17}$$

Equation 17 implies that the osmotic pressure of an electrolyte solution is the result of combination of the contribution from each ion to the total osmotic pressure. Secondly, the osmotic pressure contribution of each ion is proportional to its mole fraction. With these assumptions, sets of $B_i^*(X_i^*)$ and $B_j^*(X_j^*)$ could be evaluated at different mole fractions from the literature data on osmotic pressure of several salts by regression analysis by the method described earlier (3). From $B_i^*(X_i^*)$ and $B_j^*(X_j^*)$ values, for any known composition of a solution, the osmotic pressure of the solution of mixed electrolytes can be calculated.

iv) $k_i^* = k_{NaC\ell} \times (D_i^*/D_{NaC\ell})^{2/3}$ (18)

where k_i^* and $k_{NaC\ell}$ are mass transfer coefficients for ion i (calcd) and for NaCℓ respectively. The quantity $k_{NaC\ell}$ is obtained from Kimura-Sourirajan analysis of the experimental data with NaCℓ feed. The quantities D_i^* and $D_{NaC\ell}$ are the self-diffusion coefficients of ion i and NaCℓ in water, both at infinite dilution.

v) The ratio of diffusivity through the membrane to that in water for all species is constant, so that

$$\frac{D_{AM}}{D_{AB}} = \frac{D_{iM}^*}{D_i^*} = \frac{D_{jM}^*}{D_j^*} = \text{const} \tag{19}$$

With the above assumptions the transport equations 1 to 14 can be simplified as follows:

Corresponding to Equation 1,

$$N_B = AP - A\left[\left\{\Sigma B^*_{i2}(X^*_{i2})\,X^*_{i2} + \Sigma B^*_{j2}(X^*_{j2})\,X^*_{j2}\right\} - \left\{\Sigma\, B^*_{i3}(X^*_{i3})\,X^*_{i3} + \Sigma\, B^*_{j3}(X^*_{j3})\,X^*_{j3}\right\}\right] \qquad (20)$$

Corresponding to Equation 2 and Equation 3,

$$N^*_i = \left(\frac{D^*_{iM}}{K^*_{i2}\delta}\right) c\, X^*_{i2} - \left(\frac{D^*_{iM}}{K^*_{i3}\delta}\right) c\, X^*_{i3} \quad i = 1,3,5... \qquad (21)$$

$$N^*_j = \left(\frac{D^*_{jM}}{K^*_{j2}\delta}\right) c\, X^*_{j2} - \left(\frac{D^*_{jM}}{K^*_{j3}\delta}\right) c\, X^*_{j3} \quad j = 2,4,6... \qquad (22)$$

Corresponding to Equation 6

$$X^*_{i2} = X^*_{i3} + (X^*_{i1} - X^*_{i3})\,\alpha_i \qquad (23)$$

where

$$\alpha_i = \exp\,(N_B/k^*_i\, c) \qquad (24)$$

Corresponding to Equation 7

$$X^*_{j2} = X^*_{j3} + (X^*_{j1} - X^*_{j3})\,\alpha_j \qquad (25)$$

where $\alpha_j = \exp\,(N_B/k^*_j\, c)$ (26)

Corresponding to Equation 8 and Equation 9

$$X^*_{i3} = N^*_i/N_B \qquad (27)$$

and

$$X^*_{j3} = N^*_j/N_B \qquad (28)$$

<u>Expressions for Interfacial Equilibrium Constants and Corresponding Ionic Solute Transport Parameters.</u> It has been shown in the earlier paper (4) that the following equilibrium constants for the salt and ions comprising it are applicable and they are related to each other as follows:

$$\underset{\sim}{K_{ij}} = \frac{c\, \underset{\sim}{X_{ij}}}{c_M\, (\underset{\sim}{X^*_{ij}})_M} \qquad (29)$$

$$\underset{\sim}{K^{\pm}_{ij}} = \frac{(c\, X^*_i)^{z_j}\,(c\, X^*_j)^{z_i}}{(c_M\, X^*_{iM})^{z_j}\,(c_M\, X^*_{jM})^{z_i}} \qquad (30)$$

$$\text{and } K_{ij}^{\pm} = K_{ij}^{(z_i + z_j)} = (K_i^*)^{z_j} (K_j^*)^{z_i} \tag{31}$$

Expressions for the Ionic Equilibrium Constants, K_i^* and $D_{iM}^*/K_i^*\delta$. It can be seen from Equation 2 that $(D_{iM}^*/K_i^*\delta)$ has to be known for evaluating N_i^*. In this section an attempt is made to evaluate K_i^* first, and then to evaluate $(D_{iM}^*/K_i^*\delta)$ by using assumption (v). In developing expressions for K_i^*, we have to keep in mind that:
(1) when an ion is in equilibrium between the aqueous phase and the membrane, we have the composition dependent, K_i^*, given by Equation 4; once $(c_M X_{iM}^*)$ is obtained in terms of the composition $(c X_i^*)$ with which membranes are in contact K_i^* can be evaluated; and
(2) electroneutrality should be maintained in the membrane phase (Equation 13). After expanding and rearranging, Equation 13 becomes

$$c_M z_1 X_{1M}^* \{1 + \frac{c_M z_3 X_{3M}^*}{c_M z_1 X_{1M}^*} + \frac{c_M z_5 X_{5M}^*}{c_M z_1 X_{1M}^*} + \ldots\}$$

$$= c_M z_2 X_{2M}^* \{1 + \frac{c_M z_4 X_{4M}^*}{c_M z_2 X_{2M}^*} + \frac{c_M z_6 X_{6M}^*}{c_M z_2 X_{2M}^*} + \ldots\}$$

or

$$c_M z_1 X_{1M}^* \{1 + \sum_{i=3} \frac{z_i}{z_1} \frac{c_M X_{iM}^*}{c_M X_{1M}^*}\} = c_M z_2 X_{2M}^* \{1 + \sum_{j=4} \frac{z_j}{z_2} \frac{c_M X_{jM}^*}{c_M X_{2M}^*}\} \tag{32}$$

Further, using the general equation given by Equation 30, general expressions for the concentration ratios of cations and anions can be obtained as:

$$\frac{c_M X_{iM}^*}{c_M X_{1M}^*} = \left(\frac{K_{12}^{\pm}}{K_{i2}^{\pm}}\right)^{1/z_2} \left(\frac{c X_i^*}{c X_1^*}\right) \left(\frac{c X_2^*}{c_M X_{2M}^*}\right)^{(z_i - z_1)/z_2} \tag{33}$$

for cations, and

$$\frac{c_M X_{jM}^*}{c_M X_{2M}^*} = \left(\frac{K_{12}^{\pm}}{K_{1j}^{\pm}}\right)^{1/z_1} \left(\frac{c X_j^*}{c X_2^*}\right) \left(\frac{c_M X_{1M}^*}{c X_1^*}\right)^{(z_2 - z_j)/z_1} \tag{34}$$

for anions. Inserting Equation 33 and Equation 34 in Equation 32 we obtain

$$z_1 c_M X_{1M}^* \{1 + \sum_{i=3} \left(\frac{K_{12}^{\pm}}{K_{i2}^{\pm}}\right)^{1/z_2} \frac{z_i}{z_1} \left(\frac{c X_i^*}{c X_1^*}\right) \left(\frac{c X_2^*}{c_M X_{2M}^*}\right)^{(z_i - z_1)/z_2} \}$$

$$= z_2 c_M X_{2M}^* \{1 + \sum_{j=4} \frac{z_j}{z_2} \left(\frac{K_{12}^{\pm}}{K_{1j}^{\pm}}\right)^{1/z_1} \left(\frac{c X_j^*}{c X_2^*}\right) \left(\frac{c_M X_{1M}^*}{c X_1^*}\right)^{(z_2 - z_j)/z_1} \} \tag{35}$$

Also, using Equation 30 in terms of i = 1 and j = 2, Equation 35 becomes,

$$z_1(c_M X^*_{IM})^{(1 + z_2/z_1)}\left[1 + \sum_{i=3} \frac{z_i}{z_1} (K^{\pm}_{12})^{z_i/z_1 z_2} \left(\frac{1}{K^{\pm}_{i2}}\right)^{1/z_2} \left(\frac{c\ X^*_i}{c\ X^*_1}\right)\right.$$

$$\left.\times \left(\frac{c_M\ X^*_{IM}}{c\ X^*_1}\right)^{(z_i - z_1)/z_2}\right] =$$

$$z_2 \frac{(c\ X^*_1)^{z_2/z_1}}{(K^{\pm}_{12})^{1/z_1}} (c\ X^*_2) \left[1 + \sum_{j=4} \frac{z_j}{z_2} \left(\frac{K^{\pm}_{12}}{K^{\pm}_{1j}}\right)^{1/z_1} \left(\frac{c\ X^*_j}{c\ X^*_2}\right)\right.$$

$$\left.\times\left(\frac{c_M\ X^*_{IM}}{c\ X^*_1}\right)^{(z_2 - z_j)/z_1}\right] \quad (36)$$

Equation 36 is a general expression, which is a power series in terms of $(c_M X^*_{IM})$. It has to be noted that special cases of Equation 36 appear as Equation 50 in (1), Equations 52-54 in (2), Equation 44 in (3) and Equation A-21 in (4) with numerical values of z_1, z_2, z_i and z_j appropriate for the respective cases.

Such approach to the evaluation of the roots (in terms of $c_M X^*_{IM}$) of Equation 36 is straight forward, provided values of K^*_{12}, $K^{\pm}_{i2}$, and $K^{\pm}_{1j}$ are available. If they are not available, they are obtainable using the assumption number (v) in the following way:

From Equation 19,

$$\frac{1}{D^*_i} \frac{D^*_{IM}}{\delta} = \frac{1}{D^*_j} \frac{D^*_{jM}}{\delta} = \frac{1}{D_{AB}} \frac{D_{AM}}{\delta} \left(\equiv \frac{1}{D_{ij}} \frac{(D_{ij})_M}{\delta}\right) = \ldots = \text{const} = \phi \quad (37)$$

Then,

$$\frac{D^*_{IM}}{K^*_i \delta} = \frac{D^*_i \phi}{K^*_i}, \qquad \frac{D^*_{jM}}{K^*_j \delta} = \frac{D^*_j \phi}{K^*_j}, \qquad \frac{(D_{ij})_M}{K_{ij}\delta} = \frac{D_{ij}\phi}{K_{ij}} \quad (38)$$

therefore,

$$\frac{1}{K^*_i} = \frac{(D^*_{IM}/K^*_i\delta)}{D^*_i\phi}, \qquad \frac{1}{K^*_j} = \frac{(D^*_{jM}/K^*_j\delta)}{D^*_j\phi}, \qquad \frac{1}{K_{ij}} = \frac{(D_{AM}/K\delta)_{ij}}{D_{ij}\phi} \quad (39)$$

Replacing $K^{\pm}_{ij}$ in Equation 36 by K_{ij} using Equation 31 and further using the relation given by Equation 39, and rearranging we obtain

$$z_1\ (\phi\ c_M\ X^*_{IM})^{(1 + z_2/z_1)} \left[1 + \sum_{i=3} \frac{z_i}{z_1} \left\{\frac{(D_{AM}/K\delta)_{i2}}{D_{i2}}\right\}^{(1 + z_i/z_2)}\right.$$

$$\times \left\{\frac{D_{12}}{(D_{AM}/K\delta)_{12}}\right\}^{z_i(1/z_1 + 1/z_2)} \left\{\frac{c\ X_i^*}{(c\ X_1^*)^{z_i/z_1}}\right\}$$

$$\times (\phi\ c_M\ X_{1M}^*)^{(z_i - z_1)/z_1}] = z_2\ (c\ X_1^*)^{z_2/z_1}\ (c\ X_2^*)$$

$$\times \left\{\frac{(D_{AM}/K\delta)_{12}}{D_{12}}\right\}^{(1 + z_2/z_1)} \left[1 + \sum_{j=4} \frac{z_j}{z_2} \left\{\frac{(D_{AM}/K\delta)_{1j}}{D_{1j}}\right\}^{(1 + z_j/z_1)}\right.$$

$$\times \left\{\frac{D_{12}}{(D_{AM}/K\delta)_{12}}\right\}^{(1 + z_2/z_1)} \left(\frac{c\ X_j^*}{c\ X_2^*}\right) \left(\frac{1}{c\ X_1^*}\right)^{(z_2 - z_j)/z_1}$$

$$\left. \times (\phi\ c_M\ X_{1M}^*)^{(z_2 - z_j)/z_1}\right] \quad (40)$$

Equation 40 is a polynomial function in terms of $(\phi\ c_M\ X_{1M}^*)$ which is readily soluble when all required numerical values of $(D_{AM}/K\delta)_{ij}$, D_{ij}, and the solution compositions are given. The method of obtaining numerical values for $(D_{AM}/K\delta)_{ij}$ and D_{ij} will be shown later. Suppose such a solution is ψ_1, then, since $\phi = D_{1M}^*/D^*\delta$ (Equation 37) and $c_M\ X_{1M}^* = (c\ X_1^*)/K_1^*$ (Equation 4) ψ_1 can be written as

$$\psi_1 = \left(\frac{D_{1M}^*}{D^*\delta}\right)_1 \left(\frac{c\ X_1^*}{K^*}\right)_1 \quad (41)$$

by rearranging

$$\left(\frac{D_{1M}^*}{K^*\delta}\right)_1 = \frac{D_1^*\psi_1}{c\ X_1^*} \quad (42)$$

It can be seen that Equation 40 and Equation 41 do not specify the phase. For the solutions in the concentrated boundary layer phase (phase 2) and the product solution phase (phase 3) appropriate roots, ψ_1, have to be found corresponding to the ionic mole fractions X_i^* and X_j^* in the respective phases. The roots which correspond to phase 2 and phase 3 are designated as ψ_{12} and ψ_{13}, respectively. Furthermore, using Equations 42, 30 and 37, we can deduce,

$$\left(\frac{D_{jM}^*}{K_j^*\delta}\right) = D_j^* \left\{\frac{(D_{AM}/K\delta)_{1j}}{D_{1j}}\right\}^{(1 + z_j/z_1)} \left(\frac{c\ X_1^*}{\psi_1}\right)^{z_j/z_1} \quad (43)$$

Similarly, we can obtain

$$\left(\frac{D_{iM}^*}{K_i^*\delta}\right) = D_i^* \left\{\frac{(D_{AM}/K\delta)_{i2}}{D_{i2}}\right\}^{(1 + z_i/z_2)} \left\{\frac{D_{12}}{(D_{AM}/K\delta)_{12}}\right\}^{(z_i/z_1 + z_i/z_2)}$$

$$\times \left(\frac{\psi_1}{c\ X^*_1}\right)^{z_i/z_1} \tag{44}$$

Both Equations 43 and 44 are applicable for the respective phases, namely

$$\left(\frac{D^*_{iM}}{K^*_{i2}\delta}\right) = D^*_i \left\{\frac{(D_{AM}/K\delta)_{i\ell}}{D_{i\ell}}\right\}^{(1 + z_i/z_2)} \left\{\frac{D_{\ell\ell}}{(D_{AM}/K\delta)_{\ell\ell}}\right\}^{(z_i/z_1 + z_i/z_2)}$$

$$\times \left(\frac{\psi_{12}}{c\ X^*_{12}}\right)^{z_i/z_1} \tag{45}$$

$$\left(\frac{D^*_{iM}}{K^*_{i3}\delta}\right) = D^*_i \left\{\frac{(D_{AM}/K\delta)_{i\ell}}{D_{i\ell}}\right\}^{(1 + z_i/z_2)} \left\{\frac{D_{\ell\ell}}{(D_{AM}/K\delta)_{\ell\ell}}\right\}^{(z_i/z_1 + z_i/z_2)}$$

$$\times \left(\frac{\psi_{13}}{c\ X^*_{13}}\right)^{z_i/z_1} \tag{46}$$

$$\left(\frac{D^*_{jM}}{K^*_{j2}\delta}\right) = D^*_j \left\{\frac{(D_{AM}/K\delta)_{\ell j}}{D_{\ell j}}\right\}^{(1 + z_j/z_1)} \left(\frac{c\ X^*_{12}}{\psi_{12}}\right)^{z_j/z_1} \tag{47}$$

$$\left(\frac{D^*_{jM}}{K^*_{j3}\delta}\right) = D^*_j \left\{\frac{(D_{AM}/K\delta)_{\ell j}}{D_{\ell j}}\right\}^{(1 + z_j/z_1)} \left(\frac{c\ X^*_{13}}{\psi_{13}}\right)^{z_j/z_1} \tag{48}$$

By using Equation 45 to Equation 48, the flux equations given by Equation 2 and Equation 3 can be written as

$$N^*_i = D^*_i \left[\left\{\frac{(D_{AM}/K\delta)_{i\ell}}{D_{i\ell}}\right\}^{(1 + z_i/z_2)} \left\{\frac{D_{\ell\ell}}{(D_{AM}/K\delta)_{\ell\ell}}\right\}^{(z_i/z_1 + z_i/z_2)}\right]$$

$$\times \left\{\left(\frac{\psi_{12}}{c\ X^*_{12}}\right)^{z_i/z_1} - \left(\frac{\psi_{13}}{c\ X^*_{13}}\right)^{z_i/z_1}\right\} \tag{49}$$

for cations (i = 1, 3, 5...), and

$$N^*_j = D^*_j \left\{\frac{(D_{AM}/K\delta)_{\ell j}}{D_{\ell j}}\right\}^{(1 + z_j/z_1)} \left\{\left(\frac{c\ X^*_{12}}{\psi_{12}}\right)^{z_j/z_1} - \left(\frac{c\ X^*_{13}}{\psi_{13}}\right)^{z_j/z_1}\right\} \tag{50}$$

for anions (j = 2,4,6...).

Method of Prediction of Separation of Ions and Product Rate for a Given Membrane for Which RO Data with NaCℓ Feed Solution are Available. From the RO data with NaCℓ feed solution, by Kimura-Sourirajan analysis (9), A, $(D_{AM}/K\delta)_{NaC\ell}$ and $k_{NaC\ell}$ are determined.

From $(D_{AM}/K\delta)_{NaCl}$, C^*_{NaCl} as well as all pertinent $(D_{AM}/K\delta)_{ij}$ necessary for solution of Equation 49 and Equation 50, are evaluated using $(-\Delta\Delta G/RT)^*_i$ values available in the literature (6, 7, and 8) and the following equations,

$$\ln C^*_{NaCl} = \ln (D_{AM}/K\delta)_{NaCl} - (-\Delta\Delta G/RT)^*_{Na^+} - (-\Delta\Delta G/RT)^*_{Cl^-} \quad (51)$$

$$\ln (D_{AM}/K\delta)_{ij} = \ln C^*_{NaCl} + \left\{ n_c \left(\frac{-\Delta\Delta G}{RT}\right)_{cation} + n_a \left(\frac{-\Delta\Delta G}{RT}\right)_{anion} \right\} \quad (52)$$

The mass transfer coefficients of all ions are evaluated using the following relationship (10),

$$k^*_i = k_{NaCl} \, (D^*_i/D_{NaCl})^{2/3} \quad (53)$$

The diffusivities of different ions can be obtained from the literature (11).

In addition to $(D_{AM}/K\delta)_{ij}$, k^*_i, etc., the other necessary information to be used in the transport analysis are values of B^*_i versus X^*_i for different ions. This is obtainable from the literature values of osmotic pressure for different salts and listed in the literature (3,4) for all ions involved in this study. The procedure to compute the pertinent B^*_i values is illustrated in our previous work (3).

Thus, from the values of A, $\ln C^*_{NaCl}$, $(-\Delta\Delta G/RT)^*_i$, k^*_i and B^*_i versus X^*_i, the separation of all the ions and the product rate, [PR], can be computed by the standard procedure of solving the simultaneous equations including Equation 20 and Equations 49 and 50 (3).

Experimental Verification of the Prediction Technique. The prediction technique described above was tested in our previous work for systems including 1:1 and 1:1 electrolyte mixtures, 1:1 and 2:1 electrolyte mixtures, 1:1:1 and 1:1:1 electrolyte mixtures, and electrolyte mixtures simulating the sea water. The test also included the cases where common anions or cations were involved in the mixed electrolytes. Some examples of the feed solutions tested are given in Table I. All numerical values for $-\ln C^*_{NaCl}$, A, k, $B^*_i(X^*_i)$, $(-\Delta\Delta G/RT)_i$, D^*_i and D_{ij}, which are necessary for the calculation, are given in the literature (1-4). A comparison of the calculated and experimental results on ion separations and product rates are shown in Table II. The agreement is excellent, indicating the validity and practical utility of the prediction technique developed above.

Conclusion

The usefulness of the prediction technique based on membrane equilibria, free energy parameters, and Kimura-Sourirajan analysis, has been experimentally substantiated in Part 1-4 of this series and the method has been extended to the most general case where no limit to the number of ions and valences is imposed. Thus, this approach presents a simple prediction technique of practical value and also lends support to the preferential sorption-capillary flow mechanism for RO, which forms the basis for this mathematical treatment.

Table I. Description of Experimental Conditions Used in the RO Runs with Mixed Electrolytes

Run No.	Operating Pressure, kPag	Composition of Feed Solution, Molality			
		NaCl	KNO_3		
1	10342	1.38	1.31		
2	10342	1.38	1.31		
3	10342	0.49	2.73		
4	6895	0.31	0.21		
		$NaNO_3$	$NaC\ell$		
5	10342	0.82	0.26		
6	10342	1.014	1.014		
		$NaC\ell$	$Ca(NO_3)_2$		
7	6895	0.076	0.035		
8	6895	0.250	0.310		
9	6895	0.763	0.326		
10	6895	0.500	0.615		
		NaBr	$MgC\ell_2$		
11	6895	0.254	0.094		
12	6895	0.113	0.202		
13	6895	0.113	0.202		
14	6895	0.073	0.400		
		$NaC\ell$	$MgC\ell_2$		
15	6895	0.42	0.11		
16	6895	0.20	0.32		
		$NaC\ell$	KI	LiBr	
17	6895	0.07	0.07	0.34	
18	6895	0.15	0.14	0.69	
		$NaC\ell$	KBr	$LiNO_3$	
19	6895	0.32	0.67	0.70	
20	10342	1.30	0.32	0.20	
		mole fraction x 10^3			
		Na^+	K^+	Mg^{++}	Ca^{++}
21	10342	10.045	0.222	1.259	0.223
22	10342	10.045	0.222	1.259	0.223

Continued on next page.

Table I. Continued

Run No.	Operating Pressure, kPag	Mole Fraction x 10^3				
		Sr^{++}	$C\ell^-$	Br^-	HCO_3^-	$SO_4^=$
21	10342	0.005	11.795	0.015	0.043	0.695
22	10342	0.005	11.795	0.015	0.043	0.695

Table II. Comparison of Experimental and Predicted RO Performance of Cellulose Acetate Membranes with Different Mixed Electrolyte Feeds

Run No.	(PR)[e] x 10^3 kg/h	Solute Separation, %			
		1-1, 1-1 electrolyte mixtures[a]			
		Na^+	K^+	$C\ell^-$	NO_3^-
1	10.3 (10.3)	68.3 (65.5)	60.6 (61.4)	75.0 (78.0)	52.6 (48.3)
2	60.0 (61.3)	35.5 (30.2)	24.6 (26.8)	44.3 (41.3)	15.3 (15.2)
3	96.1 (97.8)	17.2 (16.7)	12.7 (14.4)	36.2 (36.6)	9.2 (10.8)
4	34.9 (34.3)	80.2 (78.4)	74.3 (75.0)	83.5 (86.0)	69.5 (64.0)
		$NaNO_3$	$NaC\ell$		
5	28.9 (26.4)	94.3 (95.3)	99.6 (98.5)		
6	97.7 (105.8)	21.4 (16.2)	32.3 (35.2)		
		1-1, 2-1 electrolyte mixtures[b]			
		Na^+	Ca^{++}	$C\ell^-$	NO_3^-
7	48.6 (48.7)	81.7 (85.8)	97.2 (98.6)	93.9 (95.2)	82.3 (87.1)
8	138.1 (152.3)	0.5 (-2.0)	11.9 (18.5)	15.3 (16.1)	5.9 (11.2)

Table II. Continued

Run No.	(PR)[e] x 10^3 kg/h	Solute Separation, %			
		1-1, 2-1 electrolyte mixtures[b]			
		Na^+	Ca^{++}	Cl^-	NO_3^-
9	18.0 (16.5)	51.3 (52.7)	91.5 (91.6)	77.3 (84.9)	61.1 (54.0)
10	14.3 (13.8)	29.3 (29.1)	88.1 (88.0)	80.9 (85.2)	67.0 (65.0)
		Na^+	Mg^{++}	$C\ell^-$	Br^-
11	107.1 (102.6)	66.9 (65.7)	91.5 (85.2)	78.4 (80.2)	68.9 (69.3)
12	49.8 (48.7)	82.3 (83.0)	97.9 (97.1)	95.4 (94.7)	90.3 (91.6)
13	91.8 (93.0)	55.3 (56.0)	90.8 (93.9)	86.3 (88.0)	73.9 (77.2)
14	66.7 (70.5)	34.7 (42.0)	88.9 (89.8)	83.9 (86.6)	80.0 (77.4)
		$NaC\ell$	$MgC\ell_2$		
15	32.4 (29.8)	82.4 (83.6)	96.9 (99.4)		
16	45.0 (42.8)	50.7 (54.4)	86.7 (91.2)		
		1-1-1, 1-1-1 electrolyte mixtures[c]			
		Na^+	K^+	Li^+	$C\ell^-$
17	22.6 (23.3)	93.5 (93.9)	92.3 (94.7)	94.7 (92.8)	96.2 (95.1)
18	47.9 (51.5)	30.0 (31.8)	28.2 (30.9)	31.9 (28.7)	42.1 (33.0)

Continued on next page.

Table II. Continued

Run No.	(PR)[e] x 10^3 kg/h	Solute Separation, %			
		1-1-1, 1-1-1 electrolyte mixtures[c]			
		I^-	Br^-		
17	22.6 (23.3)	89.6 (91.6)	95.0 (93.3)		
18	47.9 (51.5)	14.3 (13.9)	29.9 (31.8)		
		Na^+	K^+	Li^+	$C\ell^-$
19	27.7 (28.8)	32.3 (33.3)	34.9 (36.4)	36.8 (37.0)	35.0 (33.6)
20	50.6 (52.0)	52.5 (51.9)	56.3 (56.1)	50.0 (56.9)	55.8 (52.7)
		Br^-	NO_3^-		
19	27.7 (28.8)	34.3 (37.3)	33.3 (36.0)		
20	50.6 (52.0)	50.0 (56.2)	52.6 (51.3)		
		electrolyte mixture of sea water desalination[d]			
		Na^+	K^+	Mg^{++}	Ca^{++}
21	60.2 (61.7)	96.4 (96.9)	95.2 (96.6)	99.8 (99.9)	99.2 (99.9)
22	37.0 (38.7)	97.4 (98.1)	97.0 (97.9)	99.4 (~100)	99.0 (~100)

Table II. Continued

Run No.	(PR)[e] x 10^3 kg/h	Solute Separation, %				
		electrolyte mixture of sea water desalination[d]				
		Sr^{++}	$C\ell^-$	Br^-	HCO_3^-	$SO_4^=$
21	60.2 (61.7)	99.8 (~100)	96.5 (95.9)	- (94.3)	99.7 (99.2)	99.8 (~100)
22	37.0 (38.7)	99.4 (~100)	97.7 (97.6)	- (96.6)	99.7 (99.5)	99.4 (~100)

Note: Numbers without parentheses are experimental values. Numbers inside the parentheses are calculated values.

[a]From literature (1).

[b]From literature (2).

[c]From literature (3).

[d]From literature (4).

[e]Effective film area = 13.2×10^{-4} m^2.

Nomenclature

A = pure water permeability constant, $kmol/m^2 \cdot s \cdot kPa$

B_i^* = $B_i^*(X_i^*)$, osmotic pressure coefficient, kPa (per unit mole fraction of cation i)

B_j^* = $B_j^*(X_j^*)$, osmotic pressure coefficient, kPa (per unit mole fraction of anion j)

B_{i2}^*, B_{i3}^* = osmotic pressure coefficient of cation i in the boundary of high pressure side of the solution and the product solution, respectively

B_{j2}^*, B_{j3}^* = osmotic pressure coefficient of anion j in the boundary of high pressure side of the solution and the product solution, respectively

C_{NaCl}^* = constant characterizing the porous structure of the membrane defined by Equation 51

c = molar density of the solutions, mol/m^3

c_1, c_2, c_3, c_M = molar density, c, in solution phase 1, phase 2, phase 3, and the membrane phase, respectively

c_{M2}, c_{M3} = molar density, c, in membrane phase in equilibrium with X_{A2}, and X_{A3}, respectively

D_{AB}, D_{AM} = diffusion coefficient of solute in water and membrane phase, respectively, m^2/s

D_i^*, D_{iM}^* = diffusion coefficient of cation i, in water and in membrane phase respectively, m^2/s

D_j^*, D_{jM}^* = diffusion coefficient of anion j in water and in membrane phase respectively, m^2/s

$D_{\underset{\sim}{ij}}$ = diffusion coefficient of salts in water, m^2/s

$(D_{AM}/K\delta)$ = solute transport parameter, m/s

$(D_{AM}/K\delta)_{\underset{\sim}{ij}}$ = solute transport parameter for salt, m/s

f = fraction solute separation

$(-\Delta\Delta G/\underset{\sim}{R}T)_i^*$ = free energy parameter for cation i

$(-\Delta\Delta G/\underset{\sim}{R}T)_j^*$ = free energy parameter for anion j

$K_{\underset{\sim}{ij}}$ = interfacial equilibrium constants for single solute, ij, defined by Equation 29

$K^{\pm}_{ij}$ = interfacial equilibrium constants for single solute, ij, defined by Equation 30

$K^*_i, K^*_{i2}, K^*_{i3}$ = equilibrium distribution coefficients for cation i, cation i in solution phase 2, and cation i in solution phase 3, respectively

$K^*_j, K^*_{j2}, K^*_{j3}$ = equilibrium distribution coefficients for anion j, anion j in solution phase 2, and anion j in solution phase 3, respectively

k, k^*_i, k^*_j = mass transfer coefficients on the high pressure side of the membrane; k for cation i and anion j, respectively, m/s

k_{NaCl} = mass transfer coefficient of NaCl as determined by experiment, m/s

N_A = solute flux through the membrane, $kmol/m^2 \cdot s$

N_B = solvent flux through the membrane, $kmol/m^2 \cdot s$
N^*_i = ionic flux of cation i through the membrane, $kmol/m^2 \cdot s$
N^*_j = ionic flux of anion j through the membrane, $kmol/m^2 \cdot s$

n_a, n_c = number of moles of anions and cations, respectively, arising from dissociation of one mole of solute

P = operating pressure, kPa

(PR) = product rate through a given area of the membrane surface, kg/h

(PWP) = pure water permeation rate through a given area of the membrane surface, kg/h

R = gas constant

T = absolute temperature

X = mole fraction

$X_{A1}, X_{A2}, X_{A3}, X_{AM}$ = mole fraction of solute A, in solution phase 1, solution phase 2, solution phase 3, and membrane phase, respectively

X_{AM2}, X_{AM3} = mole fraction of solute in membrane phase in equilibrium with X_{A2}, and X_{A3}, respectively

$X^*_i, X^*_{i1}, X^*_{i2}, X^*_{i3}, X^*_{iM}$ = mole fraction of cation i, cation i in solution phase 1, cation i in solution phase 2, cation i in solution phase 3 and cation i in membrane phase, respectively

$X_j^*, X_{j1}^*, X_{j2}^*, X_{j3}^*, X_{jM}^*$ = mole fraction of anion j, anion j in solution phase 1, anion j in solution phase 2, anion j in solution phase 3 and anion j in membrane phase, respectively

X_{ij}, $(X_{ij})_M$ = mole fraction of salt ij, X_{ij} in membrane phase respectively

z_i, z_j = valence of cation i and anion j, respectively

Greek Letters

α_i, α_j = quantities defined by Equation 24 and Equation 26, respectively

δ = effective thickness of the membrane, m

$\pi(\Sigma X_i^* + \Sigma X_j^*)$, $\pi(\Sigma X_{i2}^* + \Sigma X_{j2}^*)$, $\pi(\Sigma X_{i3}^* + \Sigma X_{j3}^*)$ = osmotic pressure of the mixed electrolyte system, that in the boundary phase, and that in the product phase, respectively, kPa

ϕ = quantity defined by Equation 37

ψ_1 = quantity defined by Equation 41

Acknowledgment

This paper was issued as NRC No. 24034.

Literature Cited

1. Rangarajan, R.; Matsuura, T.; Goodhue, E.C.; Sourirajan, S. Ind. Eng. Chem. Process Des. Dev. 1978, 17, 46.

2. Rangarajan, R.; Matsuura, T.; Goodhue, E.C.; Sourirajan, S. Ind. Eng. Chem. Process Des. Dev. 1979, 18, 278.

3. Rangarajan, R.; Baxter, A.G.; Matsuura, T.; Sourirajan, S. Ind. Eng. Chem. Process Des. Dev. 1984, 23, 367.

4. Rangarajan, R.; Mazid, M.A.; Matsuura, T.; Sourirajan, S. Ind. Eng. Chem. Process Des. Dev. 1985, in press.

5. Matsuura, T.; Sourirajan, S. Ind. Eng. Chem. Process Des. Dev. 1985, in press.

6. Matsuura, T.; Pageau, L.; Sourirajan, S. J. Appl. Polym. Sci. 1975, 19, 179.

7. Rangarajan, R.; Matsuura, T.; Goodhue, E.C.; Sourirajan, S. Ind. Eng. Chem. Process Des. Dev. 1976, 15, 529.

8. Rangarajan, R.; Matsuura, T.; Goodhue, E.C.; Sourirajan, S. Ind. Eng. Chem. Process Des. Dev. 1978, 17, 71.

9. Sourirajan, S. "Reverse Osmosis", Academic: New York, 1970; Chap. 3.

10. Reid, R.C.; Sherwood, T.K. "The Properties of Gases and Liquids", McGraw-Hill: New York, N.Y., 1958, p. 295.

11. Parsons, R. "Handbook of Electrochemical Constants", Butterworths: London, 1959; Table 73.

RECEIVED February 22, 1985

Solute Separation and Transport Characteristics Through Polyether Composite (PEC)-1000 Reverse-Osmosis Membranes

Y. NAKAGAWA[1], K. EDOGAWA[1], M. KURIHARA[2], and T. TONOMURA[1]

[1]Technical Development Department, Toray Industries, Inc., 3-Chome, Sonoyama, Otsu, Shiga 520, Japan
[2]Pioneering Research and Development Laboratories, Toray Industries, Inc., 3-Chome, Sonoyama, Otsu, Shiga 520, Japan

Solute separation and transport characteristics through PEC-1000 composite membrane were compared with those of asymmetric cellulose acetate membrane and polyamide composite membrane using monovalent and divalent ions. Cation rejection becomes worse and anion rejection, on the other hand, becomes higher as feedwater salinity becomes lower in the case of PEC-1000 membrane. This membrane was confirmed to be negatively charged by membrane potentials and ion-exchange capacities. Permeate flow rate through this membrane was quantitatively related with the molar ratio of divalent metal ions and monovalent metal ions in the feedwater. "Donnan-membrane effects" illustrates that this membrane may be close to the most perfect semipermeable membrane among the three membranes evaluated.

In recent years, reverse osmosis (RO) processes have risen to the stage of practical use even in the field of seawater desalination as well as brackish water desalination because of the advantages in saving energy, low capital and operating cost, quick delivery, easy operation and easy maintenance (1, 2).

PEC-1000 membrane manufactured by Toray Industries Inc., among the various kinds of RO membranes, is reported to exhibit excellently high RO performance (especially in solute rejection) not only in seawater desalination but also in the field of concentration and recovery of valuable materials (3 - 7).

In the former case, for example, complete single stage seawater RO (SWRO) in the Middle East is suggested and examined at a demonstration and practical plants in Kingdom of Saudi Arabia. Permeate salinity low enough for world health organization (WHO) standard (500 ppm TDS) is obtained even at high water recovery rate such as 35 - 40% (8, 9). One of the examples of practical SWRO plants at the Arabian Gulf using PEC-1000 membrane elements is also reported (9).

0097-6156/85/0281-0187$06.00/0

This membrane designated PEC-1000 is a composite one, where the active layer is prepared from cross-linked polyether on the microporous polysulfone substrate (3, 10). This membrane is expected to belong to negatively charged membranes from the view points of its preparation method (11). Solute separation and transport characteristics through this membrane has been investigated and studied. At the same time, other representative RO membranes, that is, cellulose acetate (CA) asymmetric membrane and polyamide (PA) composite membrane have been examined, the former of which may be expected to be almost neutrally charged membrane and the latter of which may be expected to be positively charged one.

In this paper, some of the experimental results on the solute separation and transport characteristics through these membranes described above, especially through PEC-1000 membrane are mainly reported.

Experimental

Membranes selected. Three kinds of RO membranes were selected. Commercially available PEC-1000 membrane, CA (SC-3000) membrane manufactured by Toray Ind., Inc. and a PA composite membrane prepared as described in the reference (12) and presumed to be similar enough to the commercial FT-30 membrane were employed.

Analytical Methods. Salt concentration in the feedwater and the permeate was determined by measuring the electro-conductivity, ion concentration of which was measured by means of ion chromatography of a Yokogawa Hokushin Electric Ion chromatographic analyzer, Model IC 100.

Electron charge density of the membrane was determined by ion exchange method or by measuring membrane potentials. In the case of ion-exchange method for PEC-1000 membrane, cross-linked polyether resin forming the ultrathin active layer was extracted out from membrane itself as completely as possible with dimethylformamide (DMF).

Test Results

Effects of feedwater salt concentration. Effects of feedwater salt concentration were examined for three kinds of membranes. Sodium chloride (NaCl) concentration in the feedwater was varied from 100 to 35000 ppm. Figure 1 shows the effects of feedwater salt concentration on the permeate through these three membranes. Here, salt concentration in permeate is computed from the relationship between electro-conductivities and NaCl concentrations, assuming that salt in the permeate is composed of sodium ion (Na^+) and chloride ion (Cl^-) in the same molar ratio.

There is a big difference between the behaviors of CA membrane and other two membranes, especially PEC-1000 membrane. In the former case, salinity in the permeate decreases in proportion to that of feedwater salinity, which means that salt rejection remains constant in the range described above. In the case of the PEC-1000 membrane, on the other hand, the salinity in the permeate seems to come to a certain asymptotic value in the lower salinity range of feedwater containing below several thousand ppm of NaCl, which suggests that salt rejection becomes worse as feedwater salinity becomes lower. The behavior of the PA composite membrane seems to be in the middle of these two membranes.

Ion transport characteristics. Variations of ion concentrations in the permeate were examined with these three kind of membranes in relation to feedwater salt concentrations using NaCl solution of 100 to 10000 ppm under such evaluation conditions as 30 Kg/cm^2, pH 6.5 and 25°C as those in Figure 1. Figure 2 indicates the ion transport characteristics of Na^+ and Cl^- through these membranes. In the case of PEC-1000 membrane, Cl^- concentration becomes lower in proportion as feedwater salinity becomes lower. On the other hand, Na^+ concentration seems to reach a certain asymptotic value, that is 7×10^{-6} mol/liter, even when feedwater salinity becomes lower in this range of salt concentration. This tendency occurs at the feedwater salt concentrations below several thousand ppm or 10^{-1} mol/liter. PA composite membrane exhibited the opposite tendency to that of PEC-1000 membrane, that is, Cl^- concentration is apt to become difficult to be rejected and Na^+ is more preferably rejected when feedwater salt concentration is lower than or 10^{-2} mol/liter. CA membrane was not affected by feedwater salinity in the range of 100 to 10000 ppm or 10^{-3} to 10^{-1} mol/liter of NaCl solution and there is little difference between the variation behaviors of Na^+ and Cl^- concentrations in the permeate.

The reason of the behavior differences between these three kinds of membranes in Figure 1 is explained from the concentration behaviors of Na^+ and Cl^- ions in the permeate in this figure. Salinity in the permeate through PEC-1000 membrane in the feedwater salinity below 10^{-2} mol/liter of NaCl solution is mainly composed of a certain asymptotic concentration of Na^+ and little amount of Cl^-, which produces weak alkaline solution. On the other hand, permeate through PA composite membrane is mainly composed of a certain asymptotic concentration of Cl^- instead of Na^+ in the feedwater salinity below 10^{-3} mol/liter NaCl solution, which makes the permeate rather acidic. In the case of CA membrane, there is no such asymptotic ion concentration above 10^{-3} mol/liter NaCl solution of feedwater, which shows almost neutral pH value.

Electron charge density of PEC-1000 membrane. The PEC-1000 membrane, or more correctly the active layer of this membrane is presumed to involve anionic groups, such as sulfonic acid, as shown in the literature (11). Elementary analysis of the model resin, which were produced in the similar reaction condition as in the membrane fabrication and washed as completely as possible with deionized water, suggests that the active layer of the membrane possesses 0.7 meq/g of negatively charged electron density (See Table I). The membrane active layer, which was produced by extracting out the polysulfone substrate with DMF and washed as completely as possible also with deionized water, was soaked in two kinds of lithium chloride (LiCl) solutions (1 N and 1/100 N) and protons in the anionic groups were exchanged by Li^+. It was confirmed that active layer of the membrane possess 0.6 meq/g of negatively charged electron density from this experiment (See Table I).

In order to determine the electron charge densities of these membranes from another kind of analytical approach, membrane

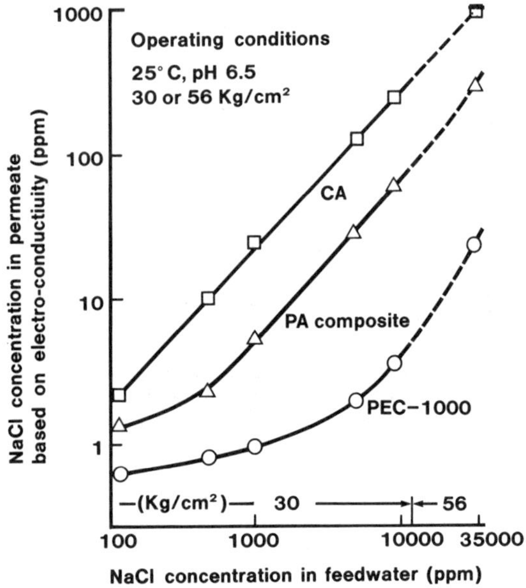

Figure 1. Effects of feedwater salt concentration on the permeate through membranes.

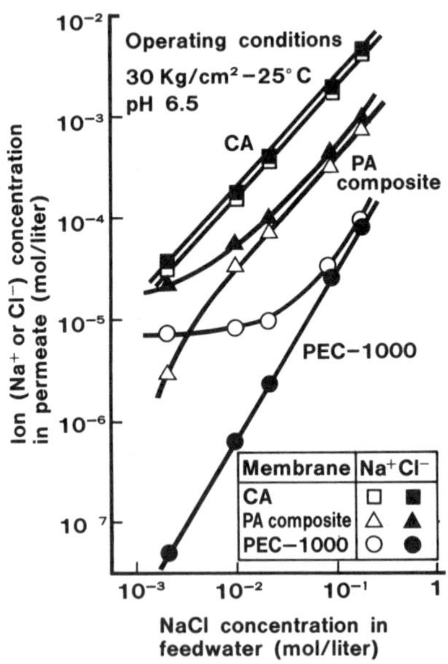

Figure 2. Ion (Na^+ & Cl^-) transport characteristics through membranes in relation to the feedwater salt concentration.

Table I. Electron-Charge Density of PEC-1000 Membrane

	From elementary analysis	From ion-exchange method*)
Electron-charge density (meq/g)	0.7	0.6

*) LiCl (1N, 1/100N) solution was used.

potentials through these three kind of membranes were measured by using two different concentration of potassium chloride (KCl) solutions according to the literature (13) (See Figure 3). The ratio of KCl concentrations in both solutions through the membranes is kept 4 and the concentration in the higher one is varied 10^{-4} to 0.5 mol/liter, variation behaviors in the membrane potentials for these membranes are shown in Figure 4. It is also suggested that PEC-1000 membrane is negatively charged and PA composite membrane possesses a slight positive charge from this figure.

Effects of divalent cations on RO performance of PEC-1000 membrane. Table II shows typical rejections of representative ions involved in the seawater as feedwater when RO performance of PEC-1000 membrane element was evaluated under such operating conditions at 4.9% seawater, 63 Kg/cm^2, 35°C, pH 6.5 and 25% water recovery. Rejections of divalent cations, such as calcium ion (Ca^{2+}) or magnesium ion (Mg^{2+}), are nearly ten times higher than those of monovalent cations, such as Na^+ or K^+. Similar results are obtained between divalent anions, such as sulfate ion ($SO_4{}^{2-}$), and monovalent anions, such as Cl^-.

Table II. Typical Water Quality and Performance of PEC-1000 Membrane Element in SWRO

Constituent	Feedwater	Permeate	Rejection (%)**
TDS (ppm)	49,000	206	99.64
Ca^{2+} (mg/l)	520	0.3	99.95
Mg^{2+} (mg/l)	1,740	1.6	99.92
Na^+ (mg/l)	14,800	75	99.57
K^+ (mg/l)	568	3.7	99.44
$SO_4{}^{2-}$ (mg/l)	3,670	3.4	99.92
Cl^- (mg/l)	26,300	109	99.64

* Operating conditions: 63 Kg/cm^2, 35°C, 25% water recovery
** Calculated at the base of average seawater ionic concentration

RO performance may be analytically expressed by A and B values based on the solution-diffusion theory (14).

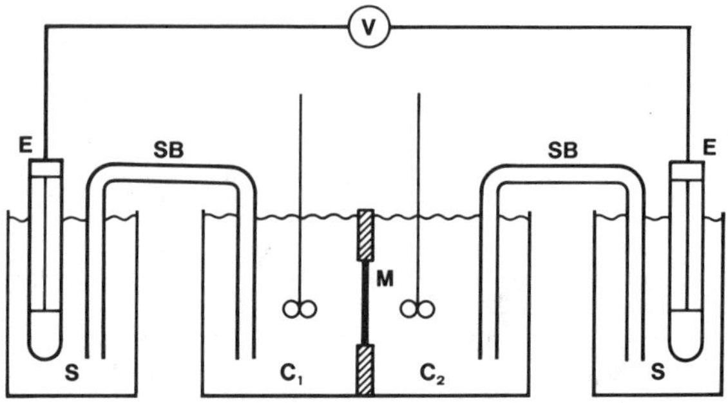

Figure 3. Apparatus for the measurements of membrane potential: C_1,C_2, KCl solution; S, aqueous solution saturated with KCl; SB, agar-bridge saturated with KCl; V, potentiometer; M, membrane; and E, electrodemeter.

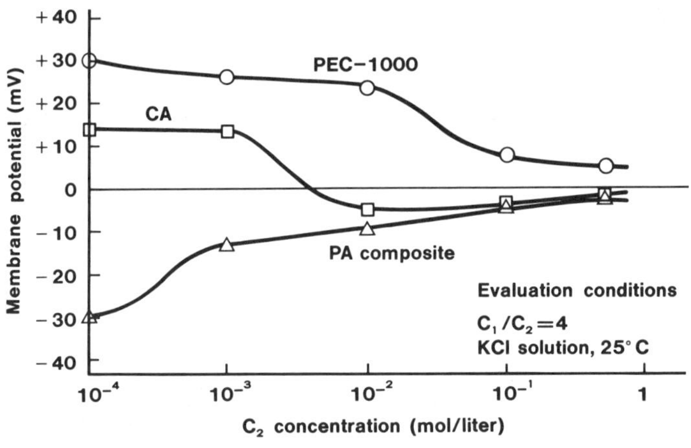

Figure 4. Variation curves in membrane potentials in relation to concentration of lower KCl solution (C_2).

$$J_V = A\ (\Delta P - \Delta\pi) \tag{1}$$

$$J_S = B \cdot \Delta C \tag{2}$$

where,

J_V: volume flux
J_S: solute flux

Figure 5 shows the relationships between molar ratio of divalent cations such as Ca^{2+} or Mg^{2+} to Na^+ as monovalent ion and relative A value for the membranes in the combination system composed of 1.0 % by weight of NaCl solution and some calcium chloride ($CaCl_2$) or magnesium chloride ($MgCl_2$). Here, relative A value means the ratio of A value to the initial A value (A_0) at the evaluation conditions in the absence of divalent cations. In regard to both of CA membrane and PA composite membrane, little variation in relative A value was observed. On the other hand, there is a decrease in relative A value in the case of PEC-1000 membrane when divalent cations, such as Ca^{2+} or Mg^{2+}, were added to the NaCl solution, degrees of which were determined quantitatively by the molar ratio of divalent cations to monovalent ion. This tendency was not observed even in the case of the PEC-1000 membrane when SO_4^{2-} as divalent anion was dosed as Na_2SO_4 salt. These results suggest that there is some interaction between the functional sulfonic acid groups of PEC-1000 membrane and divalent metal ions, such as Ca^{2+} or Mg^{2+}.

Relationships between the molar ratio of Mg^{2+} to Na^+ and Na^+ rejections for such three membranes is exhibited in Figure 6, where $MgCl_2$ was dosed to NaCl solution in the range of molar ratio of 0 to 3.0 under the initial evaluation conditions of 0.13 % NaCl, 30 Kg/cm^2, 25°C and pH 6.5. PEC-1000 membrane exhibits highest ion rejections among these three kinds of membranes and CA membrane exists in the lowest position in rejection. The variation behaviors of ion rejections, however, seems to be similar in all membranes and Na^+ rejection becomes worse as molar ratio of Mg^{2+} to Na^+ becomes higher on every kind of membrane, degrees of which are different each other, though. The variation behaviors of each ion rejection, for example, is illustrated in Figure 7 in the case of PEC-1000 membrane at a similar operating conditions as those in Figure 6 except that the total feedwater salinity is 0.03 mol/liter.

Discussion

Characteristics of charged membrane. When solid material such as membrane or resin which are negatively charged, for example, are contacted with saline water, there exists theoretical relationships exhibited in the following equations as the result of "Donnan equilibrium" under atmospheric condition (14).

$$\frac{Ai}{X^-} = \frac{1}{2}\left[\sqrt{1 + \left(\frac{2C_0}{X^-}\right)^2 \frac{\gamma_{+o}\ \gamma_{-o}}{\gamma_{+i}\ \gamma_{-i}}} - 1\right] \tag{3}$$

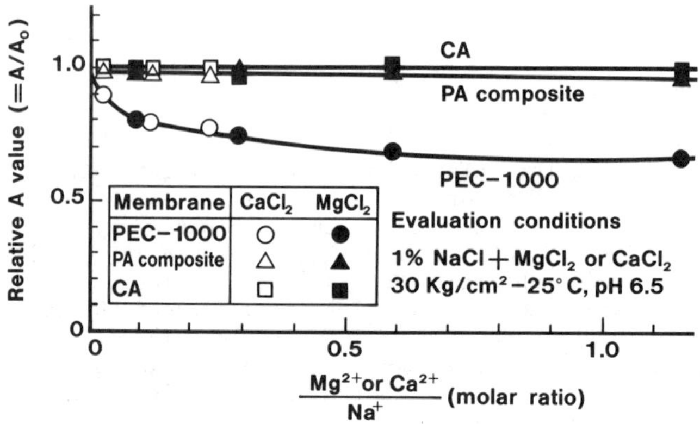

Figure 5. Relationships between molar ratio of divalent cation (Mg^{2+} or Ca^{2+}) to monovalent cation (Na^+) and relative A values.

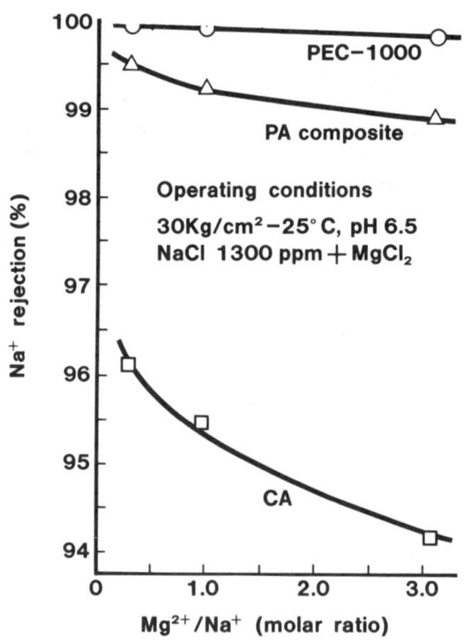

Figure 6. Relationships between molar ratio of divalent cation (Mg^{2+}) to monovalent cation (Na^+) and Na^+ rejection.

$$\frac{B_i}{X^-} = \frac{1}{2}\left[\sqrt{1 + \left(\frac{2C_0}{X^-}\right)^2 \frac{\gamma_{+o}}{\gamma_{+i}}\frac{\gamma_{-o}}{\gamma_{-i}}} + 1\right] \quad (4)$$

where,

X^-: fixed charge density
A_i: anion concentration in the solid phase
B_i: cation concentration in the solid phase
C_0: salt concentration in the liquid phase
γ: activity coefficient

When membranes are negatively charged, distinctive characteristics of charged membranes come into appearance in the salinity range of $X^- \gg C_0$ because it approaches to asymptotic values of $A_i \to 0$ and $B_i \to X^-$ according to equations (3) and (4) (See Figure 8). There exists quite opposite relationship for positively charged membranes. These relationships, however, are strictly achieved only when equilibrium exists in the system without hydrodynamic forces, such as an applied pressure in RO evaluation system. In spite of the existence of such restriction mentioned above, similar relationships are obtained in our experiment as already shown in Figure 2 even in the system where hydrodynamic force exists; in RO evaluation system.

Therefore, from these ion separation and transport characteristics through these three membranes described in Figure 2 and Figure 8, schematic transport characteristics through these three kinds of representative RO membranes may be illustrated in Figure 9.

Donnan-membrane effects. In the hyperfiltration or RO which hydrodynamic forces exist through membrane, it is reported that the presence of a membrane-impermeable ion in the pressurized solution can markedly affect the transport characteristics of co-ions through the membranes and this effect, at the same time, is general for any kind of membrane (16). Lonsdale et al. show a following equation from theoretical point of view, when the impermeable ion, for example, is anionic ion such as citrate and permeable ion is Cl^- (16).

$$1 - S_- = [(1 - S_0)/\{1 - (1 - B/B')S_0\}]\sqrt{(1 + C_1'/C_-')} \quad (5)$$

where,

S_-: rejection for membrane-permeable anion
S_0: salt rejection in the absence of membrane-impermeable anion
B: solute permeability coefficient in the absence of membrane-impermeable anion
B': solute permeability coefficient in the presence of membrane-impermeable anion
C_1': concentration of membrane-impermeable anion in the feedwater
C_-': concentration of membrane-permeable anion in the feedwater

Equation 5 leads to following relationships (See Figure 10).

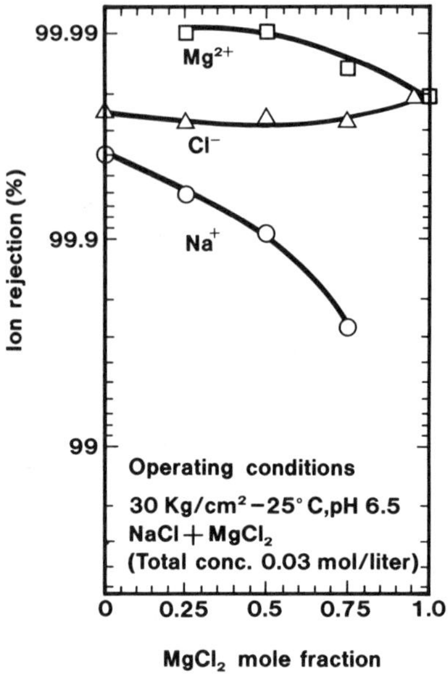

Figure 7. Relationships between mole fraction of $MgCl_2$ and each ion rejection in the case of PEC-1000 membrane.

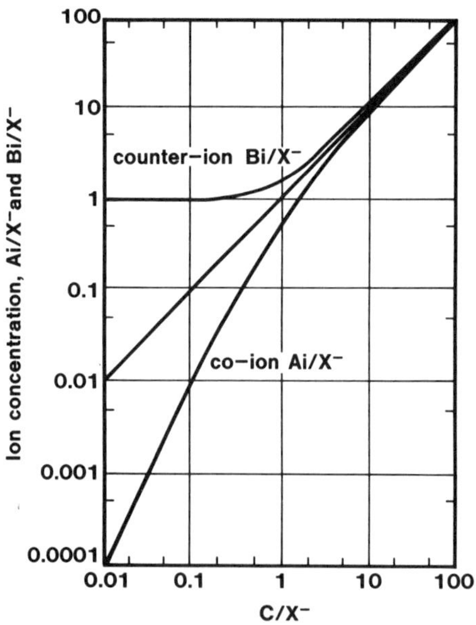

Figure 8. Concentration of counter-ions B_i and of co-ions Ai in the fixed charge phase as against the electrolyte concentration C of the outer aqueous phase.

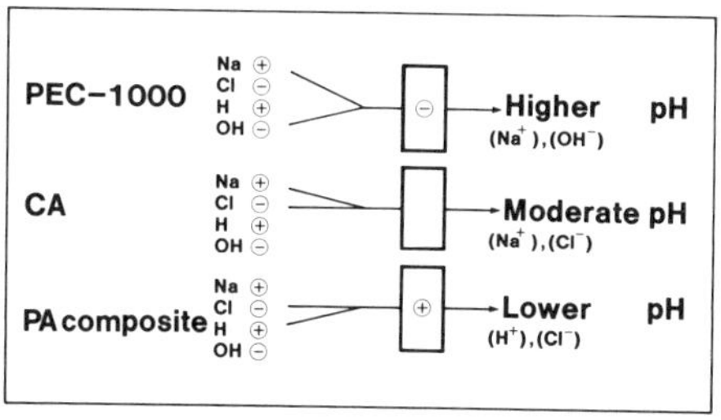

Figure 9. Permeation scheme through membranes.

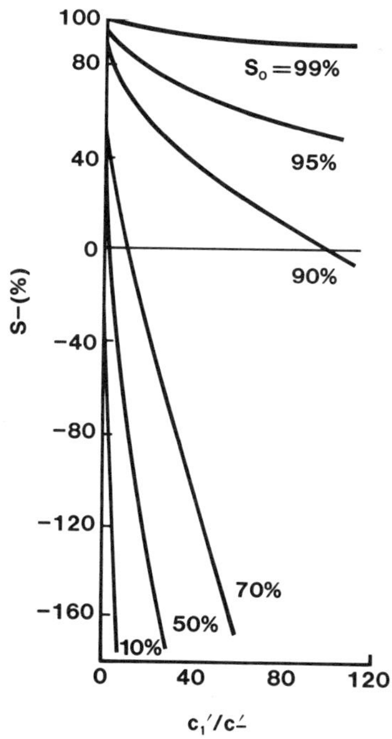

Figure 10. The rejection of co-ion as a function of the initial rejection and the ratio of membrane-impermeable ion to co-ion.

$$\begin{aligned}
&\text{as} \quad C_1'/C_-' \to 0, && S_- \to S_0 \\
&\text{as} \quad C_1'/C_-' \to \infty, && S_- \to -\infty \\
&\text{as} \quad S_0 \to 1, && S_- \to 1 \\
\text{and} \; &\text{as} \quad S_0 \to 0, && S_- \to 1 - \sqrt{(1 + C_1'/C_-')} \leqq 0.
\end{aligned} \tag{6}$$

Figure 10 shows that rejection for membrane-permeable anion (S_-) becomes worse as salt rejection in the absence of membrane impermeable anion (S_0) becomes lower at any concentration ratio of C_1'/C_-'. Though the value of rejection itself is not strictly independent on membrane materials (16), membranes which show higher S_0 value may be considered to be more perfect semipermeable membrane than those which show lower S_0 value from this theory.

PEC-1000 membrane may be considered to be the most perfect semipermeable membrane among these three kinds of membranes evaluated in this paper from the comparison of Figure 6 and Figure 10 because Mg^{2+} is considered to be almost impermeable cation and Na^+ is permeable as shown in Figure 7. The perfectness of PEC-1000 membrane is also supported by the phenomenological transport study by Pusch et al. (10).

Conclusions

Results in ion separation and transport characteristics through the membrane suggest that PEC-1000 membrane is negatively charged and it shows quite different behaviors from those of other representative RO membranes as CA membrane and PA composite membrane, which may be suggested to be neutrally charged and slightly positively charged, respectively. PEC-1000 membrane is also confirmed to be negatively charged by ion-exchange capacities and membrane potential data.

Water permeability (A value) of the membrane becomes lower in the presence of divalent metal ions, degrees of which are quantatively determined by the molar ratio to monovalent metal ions contained in the feedwater.

This membrane may be considered to be the most perfect semipermeable membrane among the membranes, reported in this paper from the ion rejection behaviors derived from "Donnan-membrane effects" through membranes.

Literature Cited

1. Satone, H. Proc. 9th Annual Conference of NWSIA. Vol.I, Session II, Washington, D.C., 1981.
2. Brandt, D.C.; Battey, R.F. Proc. 10th Annual Conference of WSIA. Vol.I, Session II, Honolulu, Hawaii, July 25-29, 1982.
3. Kurihara, M.; Kanamaru, N.; Harumiya, N.; Yoshimura, K.; Hagiwara, S. Desalination 1980, 32, 13.
4. Kurihara, M.; Harumiya, N.; Kanamaru, N.; Tonomura, T.; Nakasatomi, M. Desalination 1981, 38, 449.
5. Kurihara, M.; Nakagawa, Y.; Takeuchi, H.; Kanamaru, N.; Tonomura, T. Proc. 10th Annual Conference of WSIA. Vol. II, Session VII. Honolulu, Hawaii, July 25 - 29, 1982.
6. Kurihara, M. Membrane, 1983, 8, 97.

7. Nakagawa, Y. Workshop on Membrane Separation Technology: Application to Foods and Biochemical Processes, National Taiwan Univ., March 17 - 18, 1983.
8. Kurihara, M.; Nakagawa, Y.; Takeuchi, H.; Kanamaru, N.; Tonomura, T. Desalination 1983, 46, 101.
9. Doelle, R.A.; Kallenberg, K.H.; Heyden, W. Poster Presentations of First World Congress on Desalination and water Reuse, Florence, Italy, May 23 - 27, 1983.
10. Chen, J.Y.; Kurihara, M; Pusch W. Desalination 1983, 46, 379.
11. Kurihara, M; Watanabe, T.; Inoue, T.; U.S. Patent 4 366 062, 1982.
12. Cadotte, J.E., U.S. Patent 4 277 344, 1981.
13. Yamabe, T. et al. Nihon Kagaku Kaishi 1975, 10, 1713.
14. Lonsdale, H.K. Desalination 1973, 13, 317.
15. Hanai, T. "Maku to Ion": Kagaku Doujin, Kyoto, 1980, P171.
16. Lonsdale, H.K.; Pusch, W.; Walch, A.; J.C.S. Faraday Trans I 1975, 71, 501.

RECEIVED March 13, 1985

Hydrodynamic Properties of the Skin and the Bulk of Asymmetric Reverse-Osmosis Membranes

KAROL J. MYSELS[1], HAROLD K. LONSDALE[2], and DAVID E. WANT[3]

[1]Research Consulting, 8327 La Jolla Scenic Drive, La Jolla, CA 92037
[2]Bend Research, Inc., Bend, OR 97701-8599

The hydrodynamic resistance of a Loeb-Sourirajan type reverse-osmosis membrane can be expressed within experimental error as the sum of two independent resistances in series: one due to the skin, the other due to the bulk of the membrane. The skin resistance increases exponentially with the temperature at which the membrane is annealed but is unaffected by the operating pressure, whereas the bulk resistance is unaffected by annealing and increases linearly with the operating pressure. These conclusions are based on the analysis of water flux data obtained for samples of the same membrane annealed at six different temperatures and tested in duplicate at six operating pressures.

It is well known that upon annealing at increasing temperatures Loeb-Sourirajan type reverse-osmosis membranes increase in selectivity with a loss in water flux at constant applied pressure.(1) It is also known that their morphology is that of a very thin skin with few if any pores and a bulk whose porosity increases gradually as the distance from the skin increases,(2) and that their hydro- dynamic resistance increases with increasing applied pressure.(3) The division in the hydrodynamic resistance between the skin and the bulk and how this division depends on the applied pressure and on annealing temperature have not been thoroughly studied.(4-6) The purpose of this paper is to arrive at a self-consistent and structurally reasonable answer to these questions for fresh membranes. This turns out to be rather simple: the resistance of the skin is independent of the applied pressure but increases exponentially with annealing temperature, whereas that of the bulk is not affected significantly by annealing but increases linearly with pressure.

We consider hydraulic resistance, R, rather than its inverse, the hydraulic permeability (or membrane constant), A, of the membrane, because of the resulting simplification of all expressions. The two are defined and related by

[3]Current address unavailable.

0097-6156/85/0281-0201$06.00/0

$$J_w = A(\Delta P - \Delta\pi) = (\Delta P - \Delta\pi)/R \qquad (1)$$

where J_w is the flow of water per unit time through unit area of the membrane, ΔP the hydrostatic pressure difference, and $\Delta\pi$ the osmotic pressure difference, across it. The simplification is due to the fact that the membrane structure is layered and water has to cross these layers in series. Hence their resistances add up to the total resistance just as they do for an electric circuit. If permeabilities were used, one would have to add their inverses to obtain the inverse of the total, which is clearly an unnecessary complication.

EXPERIMENTAL

Portions of cellulose acetate reverse-osmosis membrane cast according to the formulation first proposed by Sourirajan and Govindan (7) were annealed in water for 30 minutes at 60, 70, 80, 85, and 90°C. Duplicate samples from each annealed portion were tested for water flux and salt rejection in a standard reverse-osmosis test apparatus. (8) The tests were conducted at 25±1°C with 1% NaCl feed solution over extended periods of time at applied pressures from 250 to 1500 psi in steps of 250 psi. From the measured flux and salt rejection, the membrane constant, A in Equation 1, was calculated in the conventional way and data from each sample were individually fitted by a least square treatment to a log-log relationship between membrane constant and time. (3) The 15-minute value on this line was used as the membrane constant in this study. Fifteen minutes was considered short enough to make membrane compaction and fouling negligible, yet long enough to reduce experimental uncertainties.

RESULTS AND INTERPRETATION

The resistances, R_i ($=1/A_i$), of the 72 individual samples are shown in Table I as a function of the annealing temperature, T, and the applied pressure, P. The resistances are expressed in 10^3 sec-atm/cm. (10^3 sec-atm/cm is the inverse of 0.693 gal/ft^2-day psi). Figure 1 shows the 36 averages of the duplicate samples in graphical form. If we define the average resistance of each duplicate measurement as R_a, the relative standard deviation of the duplicates, $\sqrt{\Sigma((R_i-R_a)/R_a)^2/36}$, with i extending from 1 to 72, was 8.0% and showed no systematic variation. The absolute standard deviation, $\sqrt{\Sigma(R_1-R_a)^2/36}$, was 6.0 x 10^3 sec-atm/cm and it increased methodically as R_a increased.

Inspection of Figure 1 shows that there is a rapidly accelerating increase in resistance with annealing temperature and a smaller uniform increase with operating pressure. Preliminary tests with a polynominal three dimensional least square fitting program indicated that cross terms were not significant. To test this, the data were represented by the following equation:

$$R_i = k + f_t(T) + f_p(P) \qquad (2)$$

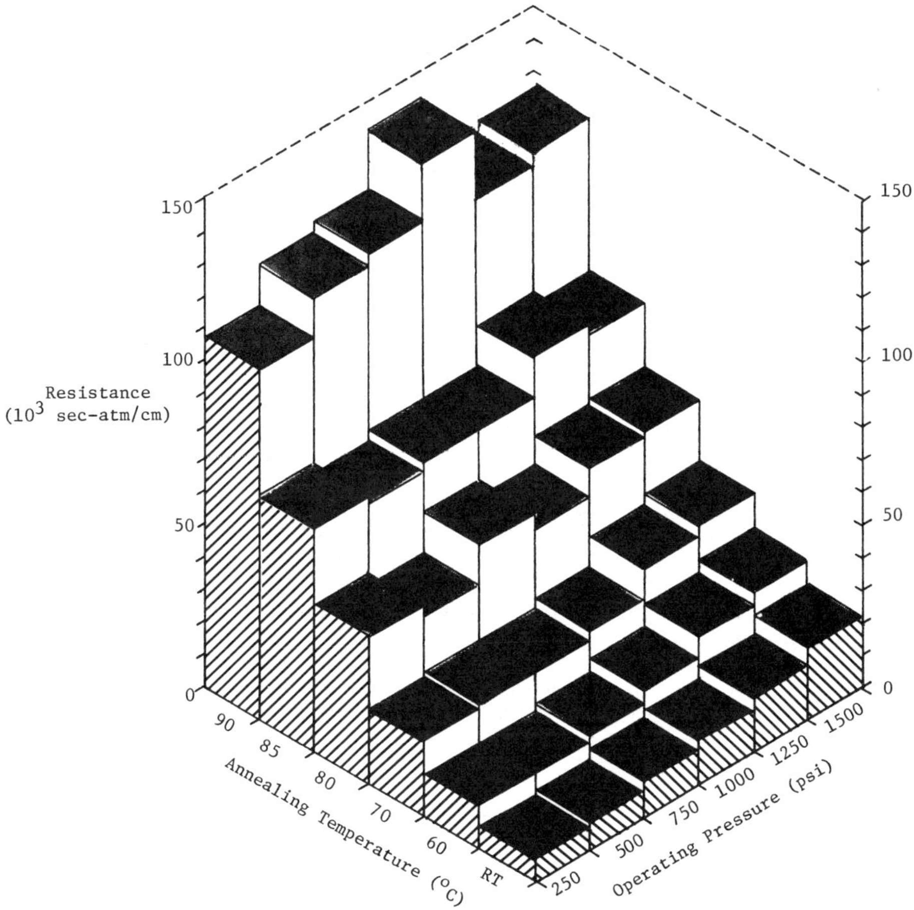

Figure 1. The average initial resistance of duplicate membrane samples as a function of the annealing temperature and operating pressure.

where f_t is function of the annealing temperature, T, only and f_p function of the operating pressure, P, only. k is an arbitrary constant whose value was selected for convenience to make both f_t (unannealed) and $f_p(250)$ zero. The values for the f functions for each of the other conditions were obtained by averaging all the R_i values corresponding to that condition and subtracting the corresponding average for f_t unannealed) or $f_p(250)$. For example, $f_p(1000)$ was taken as $(\Sigma R_{T_i,1000})/12 - (\Sigma R_{T_i,250})/12$. The results are shown in Table II which gives a set of 10 f values plus the k value from which the resistance of the membrane under all 36 conditions of annealing temperature and operating pressure can be calculated. The relative standard deviation of the results,

Table I. Initial Membrane Resistances

Operating Pressure (psi)	Resistance (10^3 sec-atm/cm)					
		Annealing Temperature (°C)				
	Room Temp.	60	70	80	85	90
250	6.8	15.1	23.2	44.6	66.9	114
	7.0	10.7	23.0	45.9	70.8	100
500	9.4	13.8	25.3	41.2	62.8	128
	8.3	12.7	23.1	42.3	70.2	110
750	11.0	13.9	26.4	52.9	66.6	132
	10.1	14.9	21.6	53.8	71.5	114
1000	13.4	19.4	29.7	50.4	69.5	129
	13.0	20.1	25.4	47.1	67.9	151
1250	17.8	25.1	38.4	56.2	79.4	134
	17.7	25.8	33.6	58.4	83.1	106
1500	21.7	31.1	41.0	57.2	76.3	128
	23.4	27.3	39.3	61.2	80.0	115

Table II. Values of the Parameters of Equation (2)

Operating Pressure (psi)	f_p	Annealing Temperature (°C)	f_t
250	0	(unannealed)	0
500	1.59	60	5.88
750	5.12	70	15.90
1000	9.06	80	37.66
1250	12.29	85	58.80
1500	15.00	90	108.68
k = 6.19			

$\sqrt{\Sigma((R_{calc}-R_i)/R_{calc})^2/72}$, was 9.9%. This is as good as could be expected in view of the fact that the deviation of the duplicate measurements is 8.0%. Thus no significant improvement can be expected from any cross-terms involving both T and P, and, as stated above, none could be found. Thus, annealing temperature and operating pressure act in series, each contributing separately and independently to the total membrane resistance.

Since we know that salt rejection in these membranes occurs in the skin and that rejection and thus skin properties are strongly affected by annealing, it is reasonable to ascribe the effect of annealing on resistance to changes in the skin and those of pressure to changes in the bulk. The nature of any such changes cannot be inferred from the present experiments, but we can conjecture that annealing slightly reduces the diameter of all capillaries, and that this has a major effect on the permeability of the skin, where there are no large pores. Conversely, annealing should have little effect on the properties of the bulk of the membrane, where there are many large pores. Similarly, the permeability of the skin should be much less influenced by compressive forces than that of the bulk, with its large pores.

Thus, we assign f_t to the skin and f_p to the bulk.

The above analysis leaves the constant k without a definite meaning since it can be assigned either to the skin or the bulk resistance. In addition this treatment of the data provides a series of distinct values rather than analytical functions for f_p and f_t. Further progress can be made by noting that f_p values lie close to a straight line as a function of P as shown in Figure 2. In fact, least-square fitting yields the equation

$$f_p = -3.76 + .0124 \; P \qquad (3)$$

No significant improvement could be obtained by using additional or fractional powers of P.

The intercept of the straight line of Figure 2 gives the resistance of an unannealed membrane at zero applied pressure as 2.43×10^3 sec-atm/cm. This is small, of the order of the experimental error for most conditions, and could be due to the skin or the bulk or both. For simplicity we shall assign it all to the skin, thus assuming that the bulk resistance is negligible at zero pressure. This assumption puts a limit on the logical rigor of our analysis but, as stated, the uncertainty thus introduced is at most of the order of experimental error. We thus replace f_t by $f_t' = f_t +$ 2.431. Combining Equations 2 and 3 this gives

$$R_i = 0.0124 \; P + f_t'(T) \qquad (4)$$

The f_t' values show a rapidly increasing curvature with temperature (see Figure 1) but their logarithms fit closely a straight line as shown in Figure 3. A least-square fit (omitting, of course, the values for the unannealed membranes) yielded the equation

$$\log_{10} f_t' = 1.295 + 0.0366 \; T \qquad (5)$$

The fit could not be significantly improved by additional or fractional powers of T.

Thus, the experimental data can be represented with only three adjustable constants as the simple sum of a linear pressure term and an exponential temperature term:

$$R = 0.0124 \; P + 10^{(-1.295 + 0.0366 \; T)} \qquad (6)$$

The former represents the resistance of the bulk, and the latter of the skin part of the membrane. We can make a rough check on the validity of our assumption that the resistance of the unannealed membrane at P=0 lies in the skin by calculating the annealing

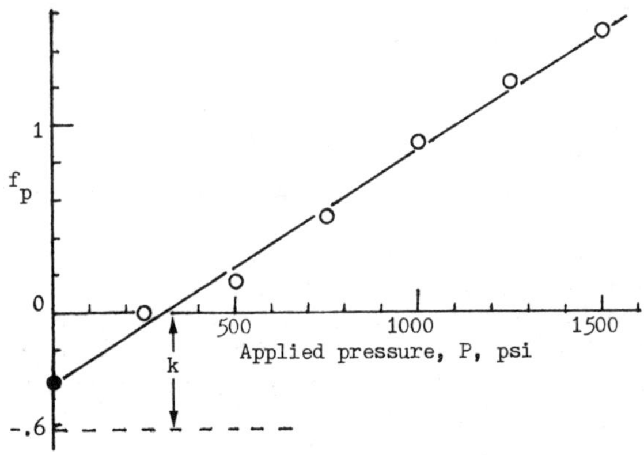

Figure 2. The contributions of pressure to the resistance of the bulk of the membrane, f_p, lie on a straight line. Extrapolation to zero pressure gives a value close to the additive constant k of Table II.

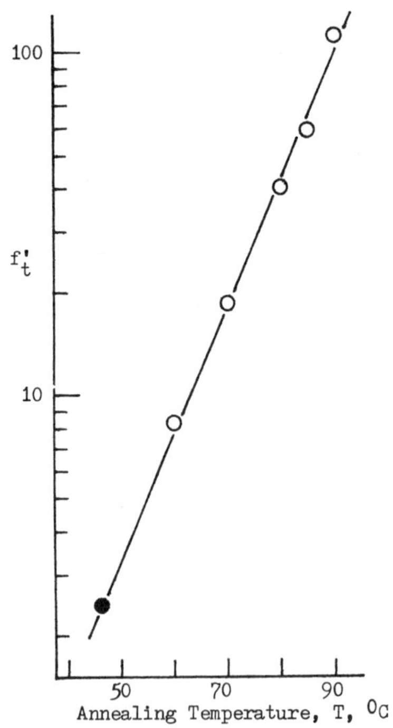

Figure 3. The contributions of temperature to the skin resistance, f'_t, lie on a straight line of a semi-logarithmic plot. The indicated effective temperature for unannealed skin is also shown.

temperature which would correspond to its resistance at zero applied pressure (2.43×10^3 sec-atm/cm) according to Equation 6. This gives 45.9°C which does not seem unreasonable. With the addition of the effective annealing temperature of 45.9°C the above equation agrees within a standard relative deviation of 6.6% with the R_a averages of the replicates, i.e., about as well as these agree with the individual points.

Figure 4 shows how the fractional initial skin resistance varies as a function of the annealing temperature and the applied pressure. It may be seen that for high annealing temperatures and low operating pressures it is the skin resistance that dominates, whereas at low annealing temperatures and high operating pressures the resistance of the compressed bulk determines the total resistance.

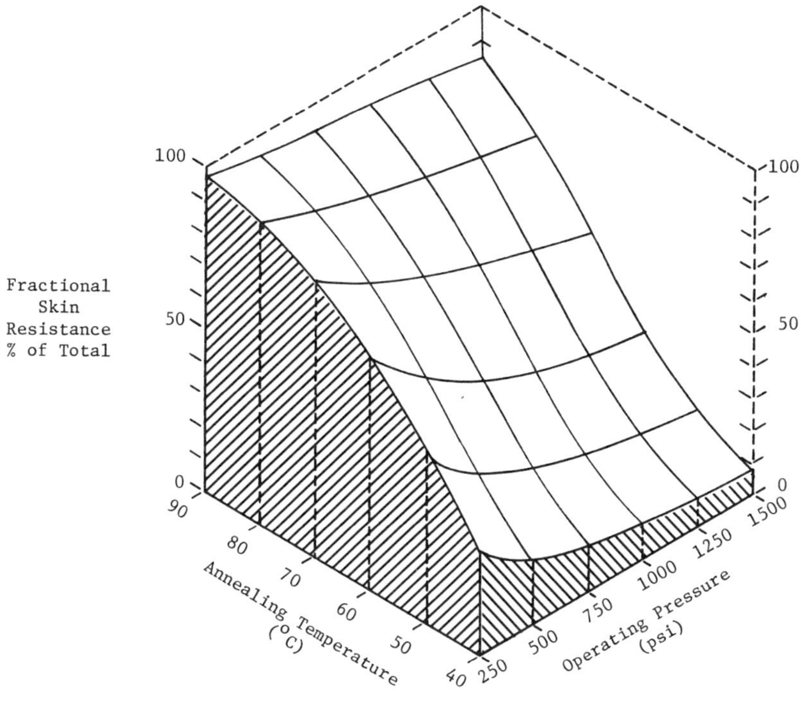

Figure 4. Skin resistance as a fraction of the total membrane resistance for different annealing temperatures and operating pressures.

Acknowledgment

This work was done in 1970-71 at the Gulf General Atomic Co., San Diego, California, with support of the Office of Saline Water, under Contract 14-30-2609.

Literature Cited

1. Loeb, S.; Sourirajan, S. Adv. Chem. Ser., 1963, 38, 117.
2. Riley, L.R.; Gardner, J.O.; Merten, U. Science, 143(1964)801.
3. Lonsdale, H.K.; Merten, U.; Riley, R.L. J. Appl. Poly. Sci., 1965, 9,1341.
4. Jagur-Grodzinski, J.; Kedem, O. Desalination, 1966, 1, 327.
5. Baayens, L.; Rosen, S.L. J. Appl. Polym. Sci. 1972, 16, 663.
6. Jonsson, G. Proc. 7th Int. Congr. Rheol., 1976, p. 298.
7. Sourirajan, S.; and Govindan, T.S. Proc. 1st Int. Symp. Water Desalination, 1968, p. 251.
8. Vos, K.D.; Hatcher, A.P.; Merten, U. Ind. Eng. Chem. Prod. Res. Dev., 1966, 5, 211.

RECEIVED February 22, 1985

16

Limiting Flux in the Ultrafiltration of Macromolecular Solutions

MAHENDRA R. DOSHI

Institute of Paper Chemistry, Appleton, WI 54912

Some of the causes for the observed flux declines in the ultrafiltration of macromolecular solutions are
(a) Osmotic pressure: the applied pressure is approximately equal to the osmotic pressure of the solution at the upstream side of the membrane.
(b) Gel formation: solute concentration at the upstream side of the membrane approaches the solubility limit.
(c) Solute-membrane interaction (membrane fouling): adsorption of solute on the membrane surface and consequent pore blockage or deposition of solute within pores, etc., are commonly termed membrane fouling.
The first two causes (a,b) are governed by boundary layer diffusion, while the third one (c) obeys a filtration equation. In some cases, both diffusion and filtration resistance may be important. A mathematical model is developed to identify causes of limiting flux in the ultrafiltration of macromolecular solutions.

An in-depth study of the ultrafiltration of macromolecular solutions was conducted by Blatt and coworkers (1) in 1970. One of their findings was that as the operating pressure is increased the solvent flux through the membrane first increases but eventually approaches a limiting value which is independent of pressure. In concurrence with Michaels (2), the formation of a gel layer at the membrane upstream surface was proposed to explain the observed pressure independent flux.

Since 1970, considerable work has been done to understand the causes of limiting flux in the ultrafiltration of macromolecular solutions. Blatt *et al.* (1) used film theory to model the gel polarized region. Subsequently, Probstein *et al.* (3-5) and Trettin and Doshi (6-8) used an integral method and more exact theories to gain insight into the limiting flux behavior.

0097-6156/85/0281-0209$06.00/0

In general, the osmotic pressure of the dilute macromolecular solutions could be negligible due to their high molecular weight. However, concentration polarization can increase solute concentration at the membrane upstream surface to such an extent that the contribution from second and higher order virial coefficient terms to osmotic pressure could be quite significant. Goldsmith (9) pointed out the importance of osmotic pressure but did not use an appropriate model to fit his data. Leung and Probstein (5) used an integral method analysis to study the effect of osmotic pressure. Vilker *et al.* (10) and Trettin and Doshi (8) developed an asymptotic solution such that the osmotic pressure of the solution at the membrane surface approaches the applied pressure. Trettin and Doshi also proposed a way to distinguish between the osmotic pressure limited case and the gel polarized region.

Recently, many researchers have observed that solute-membrane interaction could be very important in the ultrafiltration of macromolecular solutions. Solute could adsorb on the membrane and offer an additional hydraulic resistance for the solvent flow. In some cases, solute adsorption can be dominating so that diffusion can be neglected and filtration theory can be used to explain the data. For example, Dejmak (11) has shown that his unstirred ultrafiltration data could, in a large number of cases, be explained by filtration theory. Ingham *et al.* (12) have convincingly demonstrated that albumin adsorption does reduce the solvent flux through the membrane. Fane *et al.* (13-14) studied the effect of pH on protein adsorption in ultrafiltration membranes and observed that the protein adsorption was maximum at the isoelectric point (pH ≃ 5). Some interesting experiments were conducted by Reihanian *et al.* (15). They concluded that the filtration theory is adequate in interpreting their unstirred batch cell data. Howell *et al.* (16) studied flux decline due to protein adsorption and proposed a way to alleviate the problem. Many other researchers have observed the adsorption of albumin or other solutes on a membrane surface. The reader is referred to a comprehensive review by Matthiasson and Sivik (17).

In this paper a generalized mathematical model is developed to account for osmotic pressure and solute adsorption. The model developed for an unstirred batch cell reduces to the osmotic pressure limited cases when adsorption is negligible and to the filtration equation when adsorption is dominating and diffusion is negligible.

THEORETICAL DEVELOPMENT

Mathematical analysis presented here follows closely that developed by Doshi and Trettin (18). Important differences between the two analyses will be noted as the model development advances. The unsteady state diffusion equation and boundary condition for the ultrafiltration of a macromolecular solution in an unstirred batch cell are:

$$\frac{\partial C}{\partial t} + \upsilon \frac{\partial C}{\partial y} = D \frac{\partial^2 C}{\partial y^2} \tag{1}$$

$$C(o,y) = C(t,\infty) = C_o \tag{2}$$

$$C(t,o) = C_w \tag{3}$$

$$N_s = \upsilon C_w - D \frac{\partial C}{\partial y}\bigg|_{y=o} \tag{4}$$

The solute concentration at the interface of adsorbed layer is not known _a priori_ but is assumed to be constant (boundary condition, Equation 3). For the adsorption dominated process, C_w will approach the initial concentration C_o and the concentration gradient will be negligible so that the diffusion boundary layer will not develop. In this case, solvent flux will be determined by the hydraulic resistance of the adsorbed layer (Figure 1a). On the other hand if adsorption is negligible, osmotic pressure will limit the solvent flux through the membrane (Figure 1b). As shown by the analysis of Trettin and Doshi (8) and Vilker _et al._ (10) osmotic pressure of the solution on the upstream side of the membrane will approach the applied pressure. In the general case considered here both adsorption and diffusion are important. The solute concentration in the liquid phase at the interface of the adsorbed layer, C_w, will have an intermediate value between C_{wa} (osmotic pressure limited case) and C_o (adsorption limited case). The actual value of C_w is not known, and whether it will be independent of time in the asymptotic case considered here is open to question. Also, in this model we will assume that, asymptotically, permeate rate and solute adsorption flux are inversely proportional to the square root of time. More work needs to be done to understand the adsorption process under ultrafiltration conditions. The present model may have to be modified as more information on adsorption kinetics and equilibrium becomes available. For now we will accept that C_w is constant with upper and lower bounds of C_{wa} and C_o, respectively.

The coordinate y in the diffusion equation is measured from the adsorbed layer which may increase in thickness. Therefore, υ and υ_w are related by the rate at which the thickness of the adsorbed layer increases:

$$\upsilon = \upsilon_w - \frac{dS}{dt} = \upsilon_w + \frac{N_{sa}}{\rho_s(1-\varepsilon)} \tag{5}$$

where υ_w is the permeation rate with respect to the stationary membrane. The thickness of the adsorbed layer is denoted by S. The net solute flux, N_s, consists of two parts:

N_s	=	N_{sa}	+	N_{sp}	(6)
Total net solute flux		Adsorptive flux		Solute leakage through membrane	

where

$$N_{sa} = -\rho_s(1-\varepsilon)\frac{dS}{dt} \tag{7}$$

and

$$N_{sp} = \upsilon_w C_p \tag{8}$$

Finally, the phenomenological solvent flux equation completes the problem statement:

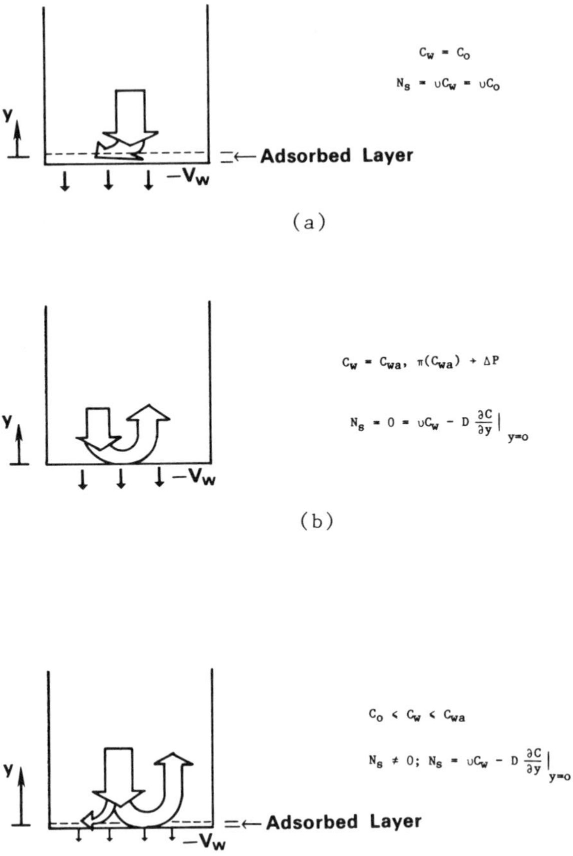

Figure 1. Schematic diagram showing causes of limiting flux in the ultrafiltration of macromolecular solution in an unstirred batch cell. $N_{sp=o}$ assumed here but accounted for in the theory. (a) Adsorption dominating. Solvent flux is independent of solute diffusion coefficient and is determined primarily by the hydraulic resistance of the adsorbed layer. (b) Adsorption is negligible. Solvent flux depends on the diffusion boundary layer thickness. Osmotic pressure of the solution at the membrane approaches the applied pressure. (c) Both adsorption and diffusion are important. The value of C_w is intermediate between C_o and C_{wa}.

$$\upsilon_w = - \frac{\Delta P - \Delta\pi}{\mu \cdot [\alpha \cdot \rho_s (1-\varepsilon) S]} \tag{9}$$

where the term in the square brackets represents the hydraulic resistance of the adsorbed layer. For the large time asymptotic solution considered here, the resistance of the membrane is neglected. This is generally reasonable, since in most cases the solvent flux for macromolecular ultrafiltration is quite low compared to that for pure solvents.

The following are the principal differences between the model presented here and the analysis of Doshi and Trettin (18):

- C_w is assumed constant in both the models, but is equal to C_g in Ref. (18), whereas its value lies between C_o and C_{wa} here;
- Solute permeation through the membrane, N_{sp}, is considered here but was neglected by Doshi and Trettin;
- The osmotic pressure effect was not important in the problem solved by Doshi and Trettin. However, the osmotic pressure is considered in this paper, Equation 9.

The large time asymptotic solution is obtained by a procedure similar to that used by Doshi and Trettin (18). The above equations are written in the dimensionless form as:

$$\frac{d^2\theta}{d\eta^2} + (2\eta + V)\frac{d\theta}{d\eta} = 0 \tag{10}$$

$$\theta(o) = \theta_w \tag{11}$$

$$\theta(\infty) = 0 \tag{12}$$

$$V = V_w + 2S^+ \tag{13}$$

$$V_w = \frac{(1-\phi)\left(1-\frac{\Delta\pi}{\Delta P}\right)}{KS^+} \tag{14}$$

$$V_w(1+\theta_w)(1-R) + \frac{2S^+}{1-\phi} = V(\theta_w+1) + \left.\frac{d\theta}{d\eta}\right|_{\eta=o} \tag{15}$$

Equation 10 and the associated boundary conditions can be used to obtain the solution in the following form:

$$\frac{\theta_w}{1+\theta_w} = \frac{\left[R.V_w - \frac{2\cdot\phi\cdot\left(1-\frac{\Delta\pi}{\Delta P}\right)}{K\cdot V_w}\right]\cdot f(V)}{1 - \frac{2\cdot\left(1-\frac{\Delta\pi}{\Delta P}\right)}{K\cdot V_w} f(V)} \tag{16}$$

where

$$f(V) = \frac{\sqrt{\pi}}{2}\cdot\exp\left(\frac{V^2}{4}\right)\cdot\operatorname{erfc}\left(\frac{V}{2}\right) \tag{17}$$

and

$$V = V_w + \frac{2\cdot(1-\phi)\cdot\left(1 - \frac{\Delta\pi}{\Delta P}\right)}{K\cdot V_w} \tag{18}$$

Let us consider some of the limiting forms of this solution.

<u>Case I</u>. Adsorption dominating and diffusion negligible.

If all of the solute convected to the membrane is adsorbed on the membrane and/or permeates through the membrane, diffusion will be negligible, as C_w approaches C_o, i.e., $\theta_w \to 0$. Then from Equation 16:

$$RV_w - \frac{2\phi\left(1 - \frac{\Delta\pi}{\Delta P}\right)}{K\cdot V_w} = 0$$

or

$$V_w{}^2 = \frac{2\phi\left(1 - \frac{\Delta\pi}{\Delta P}\right)}{R\cdot K} \tag{19}$$

which in essence is the dimensionless form of the filtration equation. In filtration, the osmotic pressure effect and solute leakage are usually not important, so that,

$$\left(1 - \frac{\Delta\pi}{\Delta P}\right) \to 1$$

and

$$R \to 1 \tag{20}$$

$$V_w{}^2 = \frac{2\phi}{K}$$

<u>Case II</u>. Adsorption negligible, diffusion dominating.

If solute adsorption on the membrane surface is negligible, the flux will be limited by osmotic pressure.

$$C_w \to C_{wa} \qquad \Delta\pi \to \Delta P \tag{21a}$$

Equation 16 simplifies to:

$$\frac{\theta_w}{1 + \theta_w} = \frac{\sqrt{\pi}}{2} R\, V_w \exp\left(\frac{V_w{}^2}{4}\right) \operatorname{erfc}\left(\frac{V_w}{2}\right) \tag{21b}$$

Equation 21 agrees with the solution for the osmotic pressure limiting case considered by Vilker <u>et al</u>. (<u>10</u>) and by Trettin and Doshi (<u>8</u>). For a given applied pressure $(\overline{\Delta P})$ and initial concentration, θ_w can be calculated from Equation 21a and dimensionless flux, V_w can be predicted from Equation 21b. Results for bovine serum albumin are presented in Figure 2. Note that V_w becomes relatively insensitive to ΔP for pressures above 600 kPa. For dilute solution and low pressure, V_w increases linearly with pressure as in filtration.

Case III. When osmotic pressure and solute leakage through the membrane are negligible,

$$\frac{\Delta\pi}{\Delta P} \rightarrow 0, \text{ and } R \rightarrow 1$$

Equations 16-18 reduce to corresponding equations obtained by Doshi and Trettin (18).

RESULTS AND DISCUSSION

The permeation rate of solvent in the ultrafiltration of macromolecular solutions in an unstirred batch cell can be predicted from Equations 16-18, provided the values of the following parameters are known: C_o, ΔP, $\pi(c)$, D, ε, μ, α, ρ_s, D_p, R, C_w. For the case of the bovine serum albumin experiments of Vilker et al. (10), C_o, ΔP, $\pi(c)$, μ, and R are known with reasonable certainty. Dimensionless permeation velocity, V_w, calculated from Equations 16-18 for different values of C_w ranging from C_o to C_{wa} is plotted in Figure 3 for pH = 7.4 and in Figure 4 for pH = 4.7. Corresponding values of the diffusion coefficient are taken from Trettin and Doshi (7). Two different values of ε and D_p, and therefore four different values of hydraulic resistance, α, are considered in these diagrams. Reihanian et al. (15) proposed $\varepsilon = 0.51$, $\rho_s = 1.2$ g/mL, and $D_p = 62$ Å for bovine serum albumin. However, Vilker et al. (10) obtained $\rho_s = 1.34$ g/mL, which will give $\varepsilon = 0.63$ if the adsorption layer is assumed to be a gel of concentration 0.59 g/mL.

Thus, it is not possible to estimate exact values of ρ_s, ε, and D_p from the literature. Selected values, shown on the figures, are considered to represent the reasonable estimate of these parameters.

Theoretical results in Figure 3 are obtained for $C_o = 0.072$ g/mL and $\Delta P = 70$ kPa. The corresponding osmotic equivalent concentration is $C_{wa} = 0.243$ g/mL. The value of V_w at $C_w = C_{wa}$ represents the osmotic pressure limiting case and is independent of the hydraulic resistance of the adsorbed layer. When $C_w = C_o$, V_w can be predicted from the filtration equation and will therefore depend on the resistance of the adsorbed layer, α. In between these two extreme cases, both adsorption layer resistance and diffusion are important. Depending on the value of α, adsorption may increase or lower the permeation rate compared to that for the osmotic pressure limiting case.

The solute adsorption affects permeation rate in two opposing directions. Adsorption can result in lower solute concentration in the liquid phase, $C_w < C_{wa}$, and as a result the driving force, $\Delta P - \Delta\pi$, for the permeation velocity will increase. However, the hydraulic resistance of the adsorbed layer will lower the flux. If a membrane is covered with a protective porous membrane to keep α low even after solute adsorption on the protective layer, it is possible to increase flux compared to that for the osmotic pressure limiting case. This may also explain why addition of colloidal particles to macromolecular solutions will increase the flux (19). The

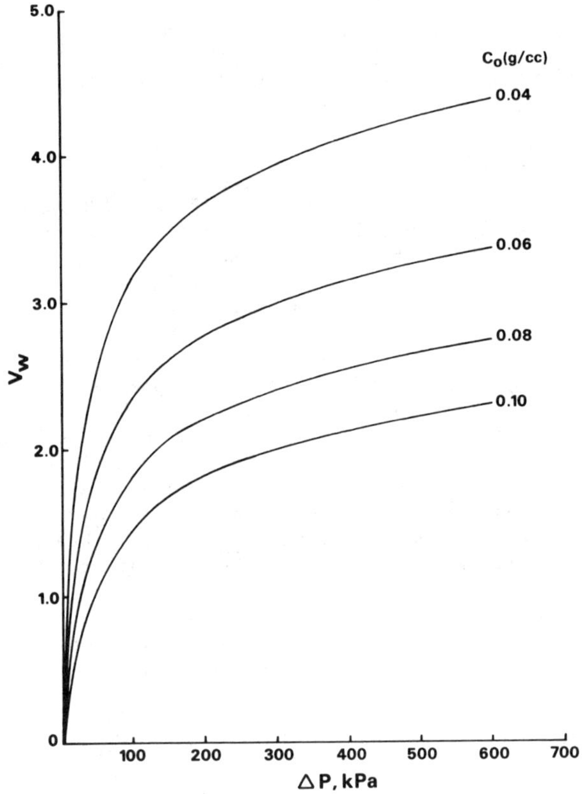

Figure 2. Dimensionless permeation velocity as a function of applied pressure for different values of the initial concentration, C_o. The solute considered is bovine serum albumin at pH 7.4.

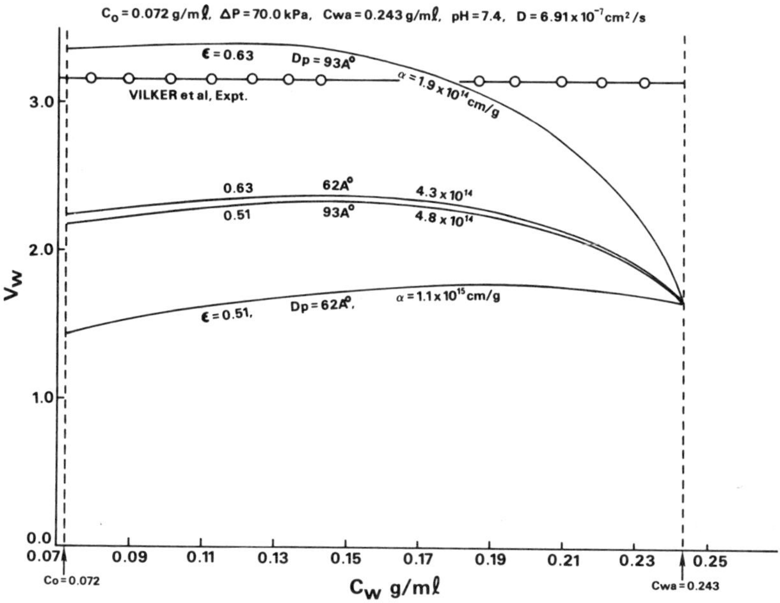

Figure 3. Dimensionless permeation velocity predicted from theory, Eq. (16)-(18) (solid lines). Vertical dotted line on the left at C_w-C_o corresponds to the filtration case (adsorption dominant), while the dotted line on the right at $C_w = C_{wa}$ corresponds to the osmotic pressure limiting case. Experimental dimensionless permeation velocity (assuming D = 6.91 x 10^{-7} cm^2/s) from the data of Vilker et al. (10) is also shown on this graph for bovine serum albumin at pH 7.4.

concept of protecting cover has also been advanced by Belfort *et al.* (20,21).

In Figures 3 and 4, experimental values of V_w reported by Vilker *et al.* (10) are also shown. It appears from these figures that adsorption of albumin on a cellulose acetate membrane was probably significant at pH 4.7 (close to the isoelectric point) rather than at pH 7.4. But Vilker's concentration profile data indicate that C_w is close to C_{wa} in both the cases. More experimental data are needed to test the theory. Specifically, independent measurements of adsorbed layer porosity and resistance will be quite useful. Also one would like to know more about the nature of the adsorption process under ultrafiltration conditions and the equilibrium solute concentration in the liquid phase.

Dimensionless permeation velocity, V_w, is shown as a function of the applied pressure in Figure 5 for the following limiting cases:

(a) Osmotic pressure limiting, $C_w = C_{wa}$

(b) Adsorption limiting (filtration theory applies): two values of α are considered to get the estimate on lower and upper bound.

(c) Gel formation, $C_w = C_g$. (V_w is independent of pressure in this case).

These theoretical results are obtained for bovine serum albumin at pH = 7.4 and C_o = 0.08 g/mL. Note that it is difficult to distinguish between the lower bound of filtration theory result and the osmotic pressure limiting case for operating pressure below 300 kPa. Hence at C_o = 0.08 g/mL of bovine serum albumin, experiments in an unstirred batch cell should be conducted at pressures above 300 to 400 kPa so as to be able to identify the cause of limiting flux. For other values of C_o, pressures above which experiments should be conducted can be represented by a straight line as shown in Figure 6. Experimental conditions should be in the upper portion of the line so that the difference between the filtration theory prediction and the osmotic pressure limiting case is greater than 25%.

CONCLUSION

A generalized theoretical model is developed to account for the adsorption of solute and diffusion of solute in the ultrafiltration of macromolecular solutions in an unstirred batch cell. The theory reduces to the osmotic pressure limiting case when adsorption is negligible and to the filtration equation when adsorption is dominating. In the intermediate case when both adsorption and diffusion are important, C_w (solute concentration in the liquid at the interface of the adsorbed layer) and hydraulic resistance of the adsorbed layer will determine the value of the solvent flux. It is interesting to note that solute adsorption can result in higher flux compared to that for the osmotic pressure limiting case with no adsorption, provided the hydraulic resistance of the adsorbed layer is not too high. Finally, we have shown that experiments conducted at low ΔP and C_o are not adequate to identify the cause or causes of limiting flux in the ultrafiltration of macromolecular solutions in an unstirred batch cell. For example, experiments with BSA at pH 7.4 should be conducted under the following guidelines:

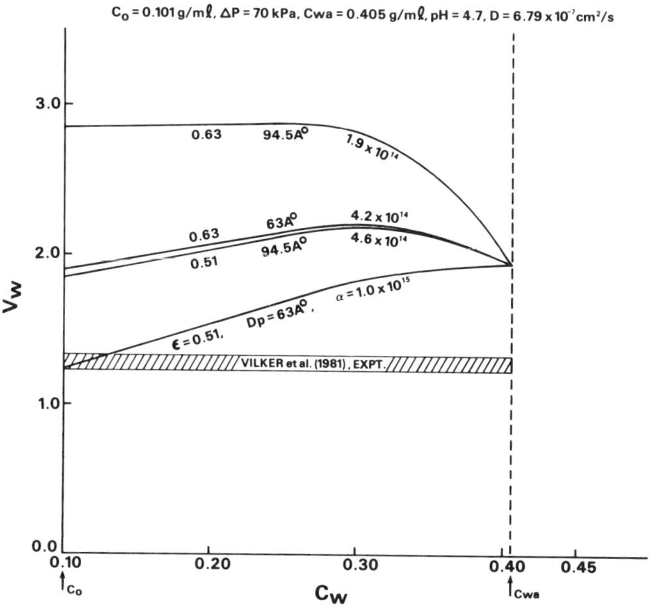

Figure 4. Dimensionless permeation velocity predicted from theory, Eq. (16)-(18) (solid lines). Vertical dotted line on the left at C_w-C_o corresponds to the filtration case (adsorption dominant), while the dotted line on the right at $C_w = C_{wa}$ corresponds to the osmotic pressure limiting case. Experimental dimensionless permeation velocity (assuming D = 6.79 x 10^{-7} cm^2/s) from the data of Vilker et al. (10) is also shown on this graph for bovine serum albumin at pH 4.7.

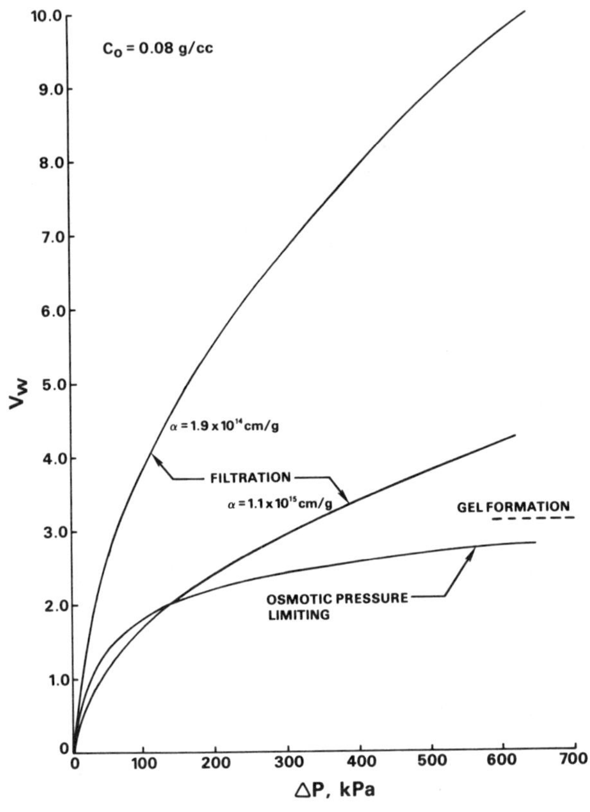

Figure 5. Limiting flux in the ultrafiltration of bovine serum albumin (pH 7.4) in an unstirred batch cell. Flux could be limited by adsorption (filtration equation applies) or osmotic pressure or gel formation. Initial concentration, C_o = 0.08 g/cc.

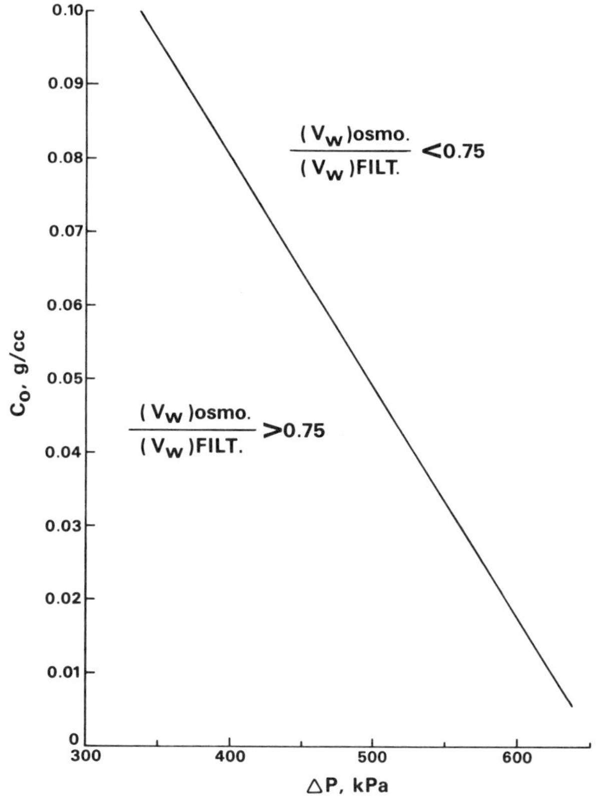

Figure 6. Threshold value of operating pressure above which experiments should be conducted to distinguish between osmotic pressure limited and adsorption limited cases.

for $0.01 \leq C_o \leq 0.15$ g/mL, ΔP (kPa) $\geq$ 655-3170 (Figure 6)
and for $C_o \geq 0.15$ g/mL, ΔP (kPa) $\geq 10\ \pi_o$

Nomenclature

C = solute concentration, g/mL
C_g = gel concentration, g/mL
C_o = initial solute concentration, g/mL
C_p = solute concentration in permeate, g/mL
C_w = solute concentration at y = o, g/mL
C_{wa} = asymptotic solute concentration at y = o so that its osmotic pressure equals the applied pressure, g/mL
D = diffusion coefficient, cm^2/s
D_p = particle diameter, Å or cm
K = $\alpha.\mu.D.C_o/\Delta P$, dimensionless
N_s = $N_{sa} + N_{sp}$ = net solute flux to the membrane, g/cm^2 s
N_{sa} = adsorptive flux, g/cm^2 s
N_{sp} = solute leakage through the membrane, g/cm^2 s
ΔP = applied pressure drop, g/cm s^2 or kPa
S = thickness of the adsorbed layer, cm
S^+ = $S/\sqrt{4Dt}$
R = $1 - C_p/C_w$, membrane rejection constant
t = time, s
υ = transverse velocity in the cell w.r.t. the coordinate y, cm/s
υ_w = permeate velocity w.r.t. stationary membrane, cm/s
V = $-\ \upsilon\ \sqrt{4t/D}$
V_w = $-\ \upsilon_w\ \sqrt{4t/D}$
y = transverse distance measured from the adsorbed layer, cm

Greek Letters

α = specific resistance of the adsorbed layer, cm/g
ε = adsorbed layer porosity
η = $y/\sqrt{4Dt}$
θ = $(C-C_o)/C_o$
θ_w = $(C_w-C_o)/C_o$
π = osmotic pressure of the solution, g/cm s^2 or kPa
$\Delta\pi$ = $\pi_w - \pi_p$
π_w = $\pi\ (C_w)$
π_p = $\pi\ (C_p)$
μ = solvent viscosity, poise
ϕ = $1 - C_o/[\rho_s(1-\varepsilon)]$
ρ_s = solute density, g/mL

Literature Cited

1. Blatt, W. F.; Dravid, A.; Michaels, A. S.; Nelsen, L. "Solute polarization and cake formation in membrane ultrafiltration: Causes, consequences, and control techniques" in Membrane Science and Technology, J. E. Flinn, ed., p. 47, Plenum Press, NY, 1970.
2. Michaels, A. S., Chem. Eng. Prog. 1968, 64, 31.

3. Shen, J. S.; Probstein, R. F., Ind. Eng. Chem. Fundam. 1977, 16 (4), 459.
4. Probstein, R. F.; Shen, J. S.; Leung, W. F., Desalination 1978, 24, 1.
5. Leung, W. F.; Probstein, R. F., Ind. Eng. Chem. Fundam., 1979, 18 (3), 274.
6. Trettin, D. R.; Doshi, M. R., Chem. Eng. Commun., 1980, 4, 507.
7. Trettin, D. R.; Doshi, M. R., Ind. Eng. Chem. Fundam., 1980, 19, 189.
8. Trettin, D. R.; Doshi, M. R. "Pressure independent ultrafiltration - is it gel limited or osmotic pressure limited?" in Synthetic Membranes, A. F. Turbak (ed.), ACS Symp. Series No. 154, Vol. II, p. 373 (1981).
9. Goldsmith, R. L., Ind. Eng. Chem. Fundam. 1971, 10, 113.
10. Vilker, V. L.; Colton, C. K.; Smith, K. A., AIChE J. 1981, 27 (4) 637.
11. Dejmek, P. "Concentration polarization in ultrafiltration of macromolecules." Ph.D. Thesis, Lund Institute of Technology, Lund, Sweden, 1975.
12. Ingham, K. C.; Busby, T. F.; Sahlestrom, Y.; Castino, F. "Separation of macromolecules by ultrafiltration: Influence of protein adsorption, protein-protein interactions, and concentration polarization," in Ultrafiltration Membranes and Applications, A. R. Cooper, ed., p. 141, Plenum Press, NY, 1980.
13. Fane, A. G.; Fell, C. J. D.; Suki, A., J. Memb. Sci. 1983, 16, 195.
14. Fane, A. G.; Fell, C. J. D.; Waters, A. G., J. Memb. Sci. 1983, 16, 211.
15. Reihanian, H.; Robertson, C. R.; Michaels, A. S., J. Memb. Sci. 1983, 16, 237.
16. Howell, J.; Velicangil, O. "Protein ultrafiltration: Theory of membrane fouling and its treatment with immobilized proteases," in Ultrafiltration Membranes and Applications, A. R. Cooper, ed., p. 217, Plenum Press, NY, 1980.
17. Matthiasson, E.; Sivik, B., Desalination 1980, 35, 59.
18. Doshi, M. R.; Trettin, D. R., Ind. Eng. Chem. Fundam. 1981, 20, 221.
19. Bixler, H. J.; Rappe, G. C., U.S. Patent 3 541 006, Nov., 1970.
20. Belfort, G.; Marx, B., Desalination 1979, 28, 13.
21. Green, G.; Belfort, G., Desalination 1980, 35, 129.

RECEIVED February 22, 1985

17

Mineral Ultrafiltration Membranes in Industry

DANIÈLE GERSTER[1] and RENÉ VEYRE[2]

[1]Comission de l'Energie Atomique, 29 Rue de la Fédération, 75015 Paris, France
[2]SFEC, BP 33, 84500 Bollene, France

The technology and components developed for gas phase separation and, more specifically, for isotopic enrichment of uranium by gaseous diffusion, have been adapted to separations in liquid media by SFEC, with the technical backing of the French Atomic Energy Commission. SFEC manufactures "Carbosep" the first mineral membranes in the world commercially available. The industrial use of the advanced membranes is discussed.

The concept of "mineral membranes" and their specific properties and characteristics are summarized, and the corresponding industrial modules and ultrafiltration units are then described. Many industrial facilities based on Carbosep membranes are in operation in France and other European countries, treating a wide variety of compounds.

This paper deals in particular with:

- the food industries: milk, whey, cheese, fruit juices, alcoholic beverages, vinegar, eggs etc.,
- the biotechnology industries: membrane fermentation reactors, enzyme reactors etc.

In these two fields, the ability of Carbosep units to withstand successive steam sterilizations unaffected is most welcome.

Apart from France, only a small number of countries, such as the USA, Japan, and more recently Holland, have tried to develop entirely mineral separation membranes. The most widely used materials are porous nickel, sintered stainless steel, carbon, alumina and zirconia. A few composite mineral/organic membranes have also been developed, of the zirconia/polyacrylate or zirconia/polyamide type. As a rule, the applications planned are related to hyperfiltration for mixed membranes, and microfiltration for ceramic membranes.

Until very recently, however, the use of these mineral membranes expanded much less than that of organic membranes, and only rarely left the laboratories. On the other hand, the mineral ultrafiltration membranes manufactured and marketed under the Carbosep trade name by Société de fabrication d'Eléments Catalytiques (SFEC), a CEA subsidiary, are well established in Europe today,

0097–6156/85/0281–0225$06.00/0

particularly in the agri-foodstuffs and biotechnology fields. The different industrial references compiled in Europe since 1981 now foreshadows the worldwide commercial development of the process.

Before presenting the process (and its applications), it should be pointed out that it results from the merger of American and French technological know-how in the area of mineral microporous substances initially developed for gas separation. The Carbosep process, which stems from the licence issued by Union Carbide to SFEC, has been developed and adapted by the company since 1979 mainly for agri-foodstuffs and biotechnology applications.

The Carbosep Ultrafiltration Process

Membranes. The microporous mineral membranes exhibit a composite morphology. The composite character is obtained by the superposition of several homogeneous microporous media:

- a sintered carbon support,
- an ultrafiltering layer of metallic oxide, usually zirconia.

The geometry selected is tubular. The tube is 1200 mm long, has an outside diameter of 10 mm, and is 2 mm thick. Its main characteristics are given in the Table I, Table II and Table II.

Table I. Separation Characteristics of Carbosep Membranes

CUT OFF	
M4 membrane	20,000 daltons
M1 membrane	50,000 daltons
M6 membrane	2,000,000 daltons
WATER FLUX (25°C, 4 bars)	180-600 $l/m^2 \cdot h$

Table II. Mechanical Properties of Carbosep Membranes

BURST PRESSURE	60 bars
OPERATING PRESSURE	15 bars
CRUSH STRENGTH	30 kgf/20 mm
YOUNG MODULUS	1000

Table III. Operating Conditions of Carbosep Membranes

PRESSURE	15 bars
pH	0 to 14
TEMPERATURE	
Membrane	300°C
System	150°C
STERILIZATION	
Oxidant	YES
Steam	YES

The performance of these membranes offers outstanding industrial reliability. They can withstand the most severe stresses engendered by hot process fluids, and those of chemical and especially steam sterilization. Moreover, they accept the highest concentrations and viscosities hitherto reached in ultrafiltration. A number of original properties merit emphasis:

- possibility of back-washing,
- dry storage
- long life time: some installations have been operating for over three years without any decrease in performance.

Modules. The tubular membranes are assembled in parallel bundles in stainless steel housings. The steel grade is adapted to the specifications of the agri-foodstuffs and biotechnology industries. The type of assembly adopted allows the unit replacement of the membranes. The diffusing areas of the modules range from 0.02 to 5.73 m^2 (see Table IV).

Table IV. Carbosep Modules

REFERENCE (number of tubes)	DIFFUSING AREA m^2
S 1	0.02
S 7	0.16
S 37	0.84
S 151	3.43
S 252	5.73

Systems. The industrial systems are of the "multistage" type, operating in steady state conditions with a recirculation loop. The equipment is gradually approaching standardization, from laboratory units to industrial units with several hundred square meters of diffusing area. SFEC now has a total of 30 mobile pilot units performing in situ tests required for the design of industrial units.

Industrial Applications

The main industrial applications developed since 1981 concern different product lines of the agri-foodstuffs and biotechnology industries.

Dairy industry. Due to the dimensions of milk proteins, the dairy industry is the privileged field of ultrafiltration applications. Since its commercial introduction in Europe, SFEC has established 25 installations which have not yet required membrane replacement.

The different processes developed in the dairy field are the following:

- Milk protein standardization: which serves to correct seasonal variations in milk protein content and to raise their average content, which tends to fall in certain areas. This process helps to achieve a higher cheese yield and to improve operating reliability.

This process is applied to various cheese technologies, such as soft, blue-veined and pressed cheeses, and also to different goat's milk base technologies, particularly the freezing deferment process.

- Extraction of serum proteins from whey: ultrafiltration represents the first upgrading of this by-product of the cheese industry. The protein concentrates are characterized by a protein/dry extract ratio of 0.4 to 0.9, and, after spray drying, they display very high nutritious value. They are intended for infant diets (maternized milks) and therapy diet (hyperprotein foods) as well as cured meat productions. The industrial installations are of the multistage type, and operation is fully automated. Constant protein content is guaranteed by refractometric or flow rate regulation. Daily cleaning is performed at 85°C in alkaline medium. From the hydrodynamic standpoint, the turbulence at every point of the installation yields protein concentrates of very satisfactory bacteriological quality. The permeate or deproteinated lactosed juice is either concentrated to 32% by evaporation for sale to pig breeders, or used as a substrate for protein production from single cell organisms.
- Production of fresh cream cheeses by the ultrafiltration of curdled milk: this application is currently revolutionizing the French and German cheese industry, because it helps to produce fresh cream cheeses such as Petits Suisses, low-salt cheeses, and a variety of cream cheeses with a much higher cheese yield than obtained by centrifugation and thermo-centrifugation.

Fruit juice clarification. In 1983, Carbosep mineral membranes witnessed further development with the construction of three industrial plants for apple juice clarification. A comparison of the conventional process with the ultrafiltration process immediately shows that:

- the conventional batch process is vastly simplified by the elimination of operations such as cooling, fining, tank cleaning, and filtration through diatomaceous earth and cellulose plates.
- passage from a batch process to a continuous process facilitates the operating conditions.
- equipment: fewer tanks, elimination of the centrifugal separator and conventional filtration units.
- products: less enzymation, elimination of gelatin, bentonite, kieselguhr, and cellulose plates.
- manpower: reduction in transfer operations, possibility of bottling 2 hours after startup of the ultrafiltration unit compared to 30 hours in a conventional process.
- material yield: evaluated as not less than 97%, against approximately 90 to 93% in conventional processing, depending on location.

Despite closely comparable compositions, two specific characteristics may be noted.

- Ultrafiltered juices preserve apple tannins, whereas they are partly complexed by gelatin fining in conventional processing.

- Ultrafiltered juices are more deeply colored than conventional juices, as the fining operation causes a loss in color. Moreover, taste tests are more favorable to juices clarified by ultrafiltration.

Cross flow microfiltration of fermented alcoholic beverage. Since 1983, SFEC and CEA have launched a research and development program to step up the Carbosep process for the clarification of fermented alcoholic beverages. France's three leading beverages are:

- wine 50 to 80.10^6 hl/year
- beer 20 to 25.10^6 hl/year
- cider 15 to 20.10^6 hl/year

The advantage of the Carbosep process for this product line is: passage from a batch process to a continuous process and simplification of the operation of the conventional filtration system.

The various constraints pertaining to wines, both qualitative and cost related, led us to the concept of cross flow microfiltration. A mineral microfiltration membrane was specially developed for this application and for the microfiltration of cider and beer (M6 membrane recently marketed in Europe).

Eggs. Carbosep systems have witnessed original development in France for the concentration of whole eggs and egg whites, due to their capacity to convey viscous liquids with minimum mechanical shear of the product. Moreover, the lysozyme contained in the egg is now subjected to extraction/concentration.

Biotechnologies. Mineral membranes offer a new tool for biotechnologies. Micro-organism separation, purification and concentration are ideally performed by Carbosep ultrafiltration units. The "product" yield is nearly 100%. Here also, Carbosep ultrafiltration, which serves to treat highly viscous liquids, due to its hydrodynamics, competes with centrifugation equipment.

These units also allow the coupling of a bioreactor or a biofermentor with the ultrafiltration device. The ultrafiltration system forms an integral part of the fermentor. This coupling offers the following main functions:

- joint sterilization of the fermentor and the ultrafiltration system,
- continuous, controlled elimination by the ultrafilter of metabolites or of the sterile upgradable phase,
- homogenization of the biomass by recirculation between the fermentor and the ultrafilter,
- if necessary, purification of the biomass by continuous washing without exposure to air,
- concentration of the residual biomass, whose value may be raised by the ultrafiltration system to levels of several hundred g/l,
- sterile drainage and recovery.

On the whole, bioreaction and biofermentation yields are boosted. Several ultrafiltration systems coupled with fermentors or reactors, are operational.

They can be used for the concentration of micro-organisms, the synthesis of amino and enzyme compounds, the extraction of biomolecules produced by fermentation, hydrolysis reactions of highly viscous complex media, the synthesis of organic acids resulting from a bioreaction, such as vinegar, the extraction of yeasts and the synthesis of different type of ferments, the sterile concentration of viruses and the extraction of specific toxins.

The capacity to operate in a totally sterile environment also endows these new membranes with potential applications, such as the preparation of thermally fragile biochemical solutions, by "cold" sterilization by passage on the membrane wall. This property is of particular interest to the pharmaceutical industry.

Conclusion

The Carbosep ultrafiltration process, characterized by the use of so-called 3rd generation mineral membranes offers a high performance tool for the agri-foodstuffs and biotechnology industries. After three years of industrial and commercial expansion, its operating reliability has enabled the process to penetrate the industries discussed above successfully in Europe, and is now aimed at North America.

Yet, the inherent virtue of every new technology is to require continued research and development activities to hold on to markets.

To achieve this, each link in the "membrane separation" chain (membranes, modules, systems, applications) is the subject of sophisticated research, and development in our laboratories and our engineering design departments, or jointly with other French research institutions and industrial firms.

RECEIVED February 22, 1985

18

Membrane Bioreactors for High-Performance Fermentations

MUNIR CHERYAN and MOHAMED A. MEHAIA

University of Illinois, Urbana, IL 61801

Membrane bioreactors provide an opportunity to vastly improve the productivity of fermentations by (a) converting batch processes into continuous processes, (b) maintaining very high cell concentrations in the system, (c) allowing high dilution (i.e., flow) rates with complete substrate conversion, and (d) removing inhibitory end-products continuously. The Membrane Recycle Fermentor, where the membrane module forms a semi-closed loop with a conventional fermentation vessel, appears to give much better performance than the Hollow Fiber bioreactors, where the microbial cells are loaded onto the shell-side and the feed is pumped through the tube side. Further advantages of this system are that the product stream is free of cells and particulates and capital expenditure for fermentation equipment can be dramatically reduced.

Microbial conversion processes (fermentations) are commonly used in the production of foods, pharmaceuticals, organic chemicals, liquid fuels and other biological products. The traditional, and still the most common, method of fermentation is the use of "free" cultures of microbial cells in a batch reactor. Batch fermentors have several problems, including

(a) their inherent inefficiency due to their start-up and shut-down nature,

(b) large capital costs for equipment, due to the low productivity,

(c) batch-to-batch variation in the product,

(d) the need to separate out the microbial cells at the end of the fermentation, and

(e) the long times needed for the fermentation, sometimes measured in days.

The low productivity of batch fermentations (expressed as amount of product produced per unit volume per unit time) is also due to low concentration of the biocatalyst (i.e., the microbial cells), end-product inhibition and substrate depletion. There is now great

0097-6156/85/0281-0231$06.00/0

interest in high-performance fermentations, especially with the new breed of genetically-engineered microorganisms. Thus any high-performance fermentation system should have the following characteristics built into it:

-- It should be a continuous process
-- It should operate at high dilution rates (i.e., high flow rate)
-- High cell densities should be maintained at all times
-- Inhibitory end-products should be removed continuously from the system

The advantages of conducting fermentations in a continuous mode instead of a batch mode include reduction in capital cost, better process control and usually improved productivity; in the case of ethanol production, it is usually about three times better. The simple continuous stirred tank reactor (CSTR) has one major disadvantage. Dilution rate cannot exceed the maximum growth rate of the microorganism, or there will be cell "washout". Recycling the cells, typically with a centrifugal separator to separate and recycle the cells, results in much higher productivity. Although modern centrifuges are fairly efficient for cell separation, they are expensive to buy, to operate and maintain. An obvious choice for replacement of centrifuges and mechanical settlers would be membrane separation devices.

One way to overcome these disadvantages and to achieve the efficiency of a continuous process has been to "immobilize" these biocatalysts. However, immobilization in the conventional manner, e.g., by fixing on to solid supports or entrapping the cells in a gel matrix, creates its own problems, such as diffusional restrictions and steric hindrance, loss in activity upon immobilization, possibility of contaminating the culture during immobilization and the expense of the immobilization step.

An alternate approach involves the use of synthetic semi-permeable membranes of the appropriate chemical nature and physical configuration to either trap or confine the biocatalyst within the bioreactor, or to continuously separate the biocatalyst from the reaction mixture and recycle it back to the main reaction vessel for further reaction. These membrane bioreactors take advantage of the size differences that exist between the biocatalyst (whether an enzyme or microbial cell) and the product.

The idea of using "semipermeable" membranes with microorganisms actually dates back to 1896 when Metchnikoff et. al. (1) showed the existence of diffusable cholera toxin in cultures of _Cholera vibrios_ contained in collodion sacs. In the 1960s and 70s, Gerhardt and coworkers performed several pioneering experiments on _in vitro_ dialysis culture systems for a variety of applications (2-4). Although they showed that dialysis culture was superior to the non-dialysis process, it may not be practical on an industrial scale since the reaction rate will be limited by the rate at which substrate and product can diffuse through the membrane. Pressure-activated membrane processes, such as ultrafiltration and microfiltration should be more efficient. Two approaches have been used: the membrane recycle bioreactor, operated essentially as a continuous stirred-tank reactor (CSTR), and the hollow fiber bioreactor, operated essentially as a plug-flow reactor.

Membrane Recycle Fermentor

In this configuration, the main reaction vessel in which the fermentation occurs is coupled in a semi closed-loop configuration to a membrane module (Figure 1). The membrane module is serving its intended purpose, i.e., as a separation device, to separate the microbial cells from the product and recycle the cells back to the fermentation vessel. In operation, the reaction vessel is initially filled with the culture with the required cell concentration, and feed pumped into the vessel at a controlled rate. The vessel contents are pumped through the membrane module and recycled back to the reaction vessel. The permeate, containing product molecules, and unconverted substrate molecules that are small enough to pass through the pores of the membrane, will be removed at the same volumetric flow rate as the feed rate, to keep the total reaction volume constant. The membrane should be selected to retain the biocatalyst while allowing free passage of product molecules. Considering the size differences between microbial cells and most products of fermentation, microfilters or large-pore ultrafilters can be used, in the appropriate module configuration.

Compared to enzymatic conversions, little work has been done with fermentations in membrane bioreactors (5). Until recently, recycle of microorganisms was rarely practised with the notable exception of activated sludge processes for wastewater treatment, where recycling is essential for process stability and high performance (18). To the authors best knowledge, Budd and Okey (6) were the first to study this concept, primarily for sewage treatment. This work is apparently the basis for the MARS process (Membrane Anaerobic Reactor System) marketed by Dorr-Oliver (Figure 2). Others (7,8) also suggested using the membrane recycle fermentor (MRF) for conducting enzymatic and microbial conversion processes. Little hard experimental data was provided until about 1970, when Wang et. al., (9) studied the production of extracellular proteases during continuous cultivation of *Clostridium histolyticum* (Figure 3). The cell concentration and the product (enzyme) concentration were higher in the continuous membrane system than in the batch system. In this particular case, since neither the cells nor the enzyme were permeable, the only possible explanation for their higher yields is the removal of toxic metabolites during fermentation.

Ethanol Fermentation

The most attention has been given to the production of ethanol by fermentation. Margaritis and Wilke (10,11) and Haroldsen and Rosen (12) suggested using "Rotofermentors", a schematic of which is shown in Figure 4. The membrane is fixed to a rotating module, which apparently helped to minimize fouling and concentration polarization due to the high turbulence at the rotating membrane surface. Productivity of 36.5 g/L/hr was obtained at cell concentrations of 51 g/L. The rotofermentors, however, appear to be too complex and expensive to be practical, and a much simpler system has been developed at the University of Illinois, Urbana by the authors, (13-15) and at the University of New South Wales, Australia (16). Figure 5 is a schematic of the University of Illinois process. Al-

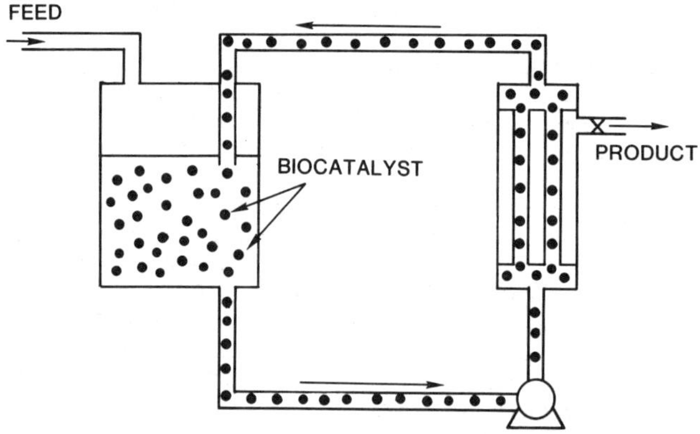

Figure 1. Schematic of membrane recycle bioreactor. The drawing is not drawn to scale. The membrane unit on the right is much smaller in volume than the fermentation vessel on the left.

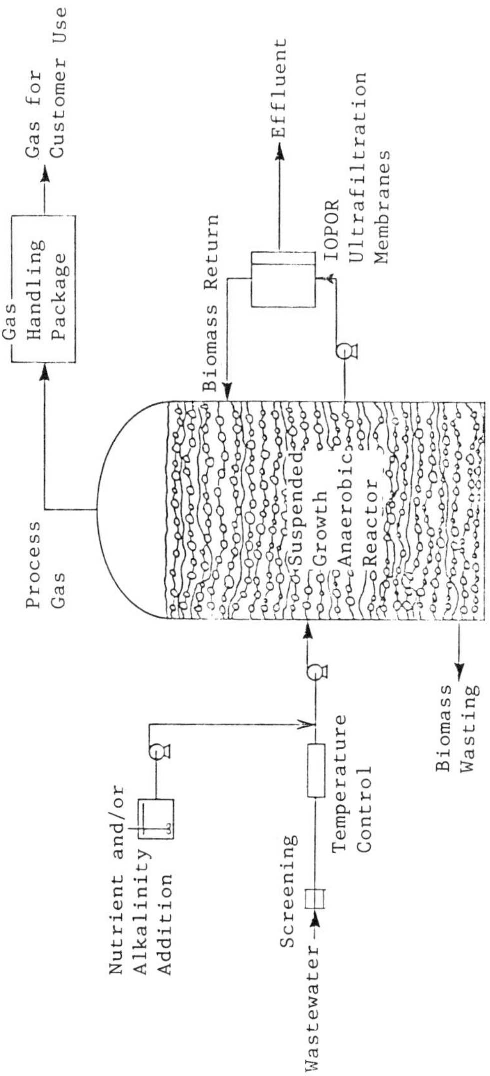

Figure 2. Process schematic of the Dorr-Oliver MARS process for anaerobic digestion of wastewater. (Reproduced with permission. Copyright Dorr-Oliver.)

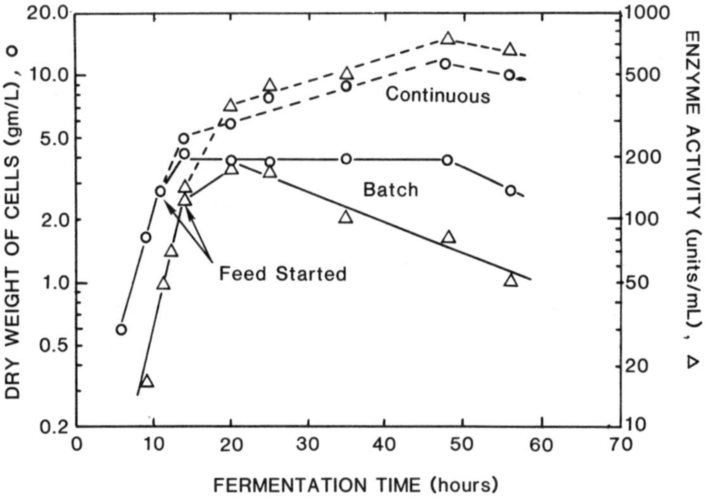

Figure 3. Comparison of batch and continuous membrane fermentor for the culture of Clostridium histolyticum. Solid lines are batch data, and broken lines are membrane fermentor data. (Reproduced from Ref. 8. Copyright 1972, American Chemical Society.)

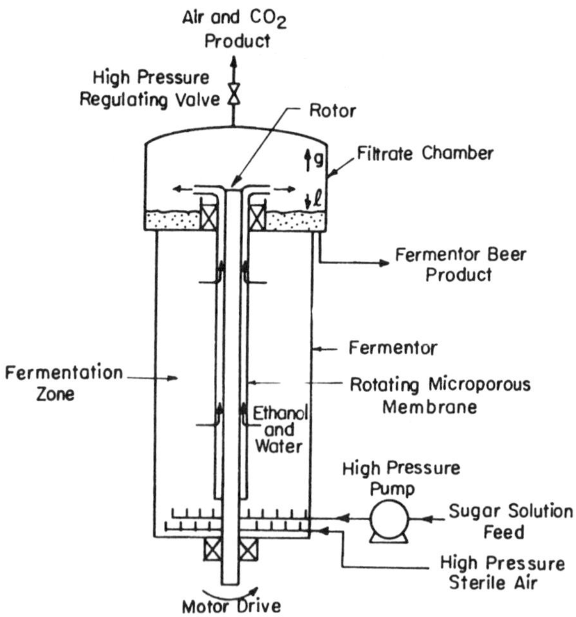

Figure 4. Schematic of Rotofermentor. (Reproduced with permission from Ref. 35. Copyright 1981, Springer.)

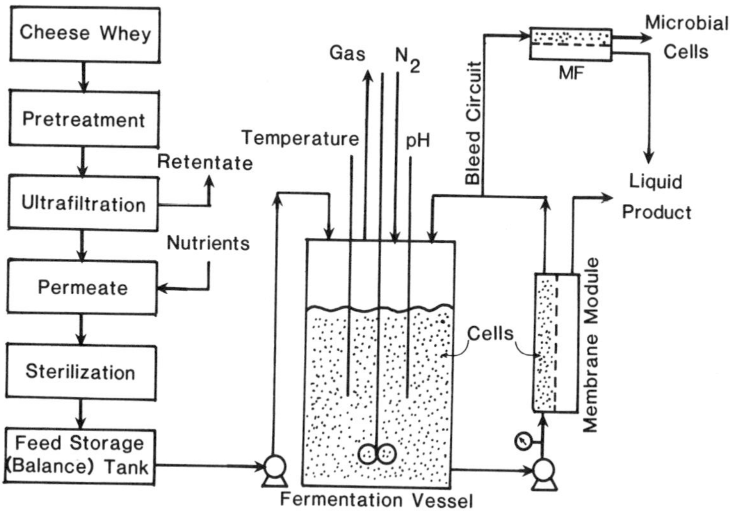

Figure 5. Process schematic for conversion of cheese whey into fermentation products with a membrane recycle fermentor.

though the general process scheme can be applied to almost any fermentation process, the fermentation of cheese whey to ethanol will be used as a model to show the applicability of the membrane recycle concept.

Cheese whey is a by-product of the cheese manufacturing industry. Over 40 billion kg are produced annually, over half in the U.S. alone, of which about half is not utilized in any way. This not only creates a potential pollution problem, but is also a waste of good quality protein and a renewable, easily fermentable carbohydrate source. The scheme in Figure 5 shows a method for utilizing both major components of cheese whey. The cheese whey, containing about 5.6% total solids (4.5% lactose, 0.4-0.6% true protein) undergoes the necessary pretreatment prior to ultrafiltration, (such as pasteurization, clarification and/or microfiltration), to maximize UF performance (17). The retentate from the UF unit, after diafiltration if necessary to improve the protein content of the retained solids, can be marketed as a highly functional and nutritious protein ingredient.

The permeate contains a solution of lactose, non-protein nitrogen and dissolved salts. After addition of the appropriate additives/nutrients to maintain optimum growth of the microorganisms, it is then concentrated by reverse osmosis (if necessary) to the appropriate sugar concentration. The feed is then sterilized by heat or by microfiltration through a 0.22-micron cross-flow module, before being stored in a sterile balance tank. This prepared feed is then pumped into the MRF, which contains a highly concentrated suspension of the yeast, *Kluyveromyces fragilis*. The entire fermentation vessel contents are continuously pumped through the appropriate membrane module and the membrane operating parameters adjusted to give the required flux. The MRF is operated as a CSTR, as determined by residence time distribution studies. Table I is a comparison of the performance of the MRF with conventional systems for lactose-to-ethanol fermentations.

Table II shows a comparison of ethanol productivities of different bioreactors for the fermentation of glucose to ethanol. The apparent superiority of the membrane recycle bioreactor is obvious. Similar process schemes have been used for the production of ethanol from glucose using the yeast *Saccharomyces cerevisiae* and the bacteria *Zymomonas mobilis* by our research group. Figure 6 shows the productivity as a function of dilution rate and cell concentration. Increasing the cell concentration allows a higher dilution rate (i.e., throughput rate) for the same conversions. These productivities are 20-60 times higher than traditional batch processes and better than immobilized whole cell reactors. In comparing the data in Table II, it should be borne in mind that the organism *Zymomonas mobilis* has a higher growth rate and is reputed to be a better ethanol producer than the yeasts we studied. Rogers *et al.* (16) used a Millipore Pellicon flat-sheet microfiltration system for cell recycle while we used hollow (fat) fibers of Romicon and Amicon for our studies.

Other Fermentations

Bull and Young (19) have recently described partial cell recycle using tubular ceramic filter elements as the recycle device for the

Table I. Comparison of Bioreactors for Conversion of Lactose to Ethanol Using *K. fragilis*

System	Lactose Concentration (g/L)	Ethanol Concentration (g/L)	Lactose Utilization (%)	Productivity (g/L/hr)	Reference
BATCH	50	25	99	2.7	13
	150	75	99	3.0	13
	45(WP)	22	97	1.9	14
CONTINUOUS CULTURE	50	16	64	1.5	21
MEMBRANE RECYCLE	50	20	80	118	13
	50	25	99	46	13
	150	40	64	240	13
	150	71	99	75	13
	45(WP)	19	97	45	14
	100	47	88	7.1	27
	120	58	89	2.9	27
	120	47	81	7.1	27
HOLLOW FIBERS	50	24	97	34	28
	45(WP)	21	97	30	28

WP = whey permeate

Table II. Comparison of Bioreactors for Conversion of Glucose to Ethanol

System	Sugar Concentration (g/L)	Microorganism	Ethanol Concentration (g/L)	Utilization of Sugar(%)	Productivity (g/L/hr)	Reference
BATCH	100	*S. cerevisiae*	49	99	2.5	29
	200	*S. cerevisiae*	98	98	4.0	15
	250	*Z. mobilis*	102	80	3.4	16
CONTINUOUS CULTURE	160	*S. cerevisiae*	31	38	4.1	30
	100	*Z. mobilis*	49	80	9.0	16
IMMOBILIZED WHOLE CELL	197	*S. cerevisiae*	71	74	25	31
	100	*Z. mobilis*	44	97	29	32
	150	*Z. mobilis*	74	98	57	33
MEMBRANE RECYCLE	100	*Z. mobilis*	44	90	120	16
	100	*S. cerevisiae*	49	99	100	15
	200	*S. cerevisiae*	65	65	130	15
HOLLOW FIBER	89	*S. cerevisiae*	12	27	26	34
	100	*S. cerevisiae*	40	85	10	29

conversion of D-sorbitol to L-sorbose by *Gluconobacter oxydans* subspecies *suboxydans* 1916B and for the conversion of glucose to 2-keto-gluconic acid by *Serratia marcescens* NRRL B-486. Cell recycle improved productivity of the former fermentation from 48 g sorbose per liter per hour to 76 g/L/hr, with only a partial cell recycle of 0.49. The productivity doubled for the latter fermentation. In our opinion, however, both fermentations were operated far from the optimum conditions and much greater improvements could have been obtained if the fermentation and the membrane system had been optimized viz-a-viz their operating parameters.

The production of lactic acid in membrane recycle fermentors has been studied by Vick Roy *et al.* (20). Their data is shown in Table III. Similar improvements in lactic acid productivity using different microorganisms and different substrates have been obtained by our own research group (21). Minier *et al.* (22) have reported an improvement in acetone-butanol fermentation productivity using cell recycle with Carbosep tubular membranes.

Hollow Fiber Fermentor

Hollow fiber bioreactors, where the microbial cells are trapped within the shell-side of the cartridge and the feed is pumped in through the tube side (Figure 7), have not fared as well as membrane recycle fermentors (Tables I and II). Either productivities have been much lower than with membrane recycle fermentors, or the substrate utilization has been poor. For example, substrate conversions of only 3-4% were obtained for the production of lactic acid in hollow fiber bioreactors by Vick Roy *et al.*, (23) and our group (21).

The relatively poor performance is due to several practical problems that arise during the operation of hollow fiber fermentors. Because the biocatalyst (the microbial cells) is separated from the substrate and product stream by a physical barrier (the membrane), the rate-limiting step becomes the diffusion of substrate into, and the diffusion of the product out of, the shell-side of the hollow fiber cartridge. In our experiments, visual inspection of the shell-side indicated that most of the microbial cells were attached to the

Table III. Conversion of Glucose (45 g/L) to Lactic Acid by *Lactobacillus delbreuckei* in a Membrane Recycle Fermentor

Dilution Rate (hr-1)	Cell Concentration (g/L)	Lactic Acid (g/L)	Productivity (g/L/hr)
0.26	2-5	24	6.2
0.54	5-25	33	17.8
1.14	25-38	34	38.7
2.16	38-54	35	76.0

Source: Adapted from Ref. 20.
Note: For comparison, batch fermentors generally have cell concentration of 7-11 g/L and productivity of 1-3 g/L/hr.

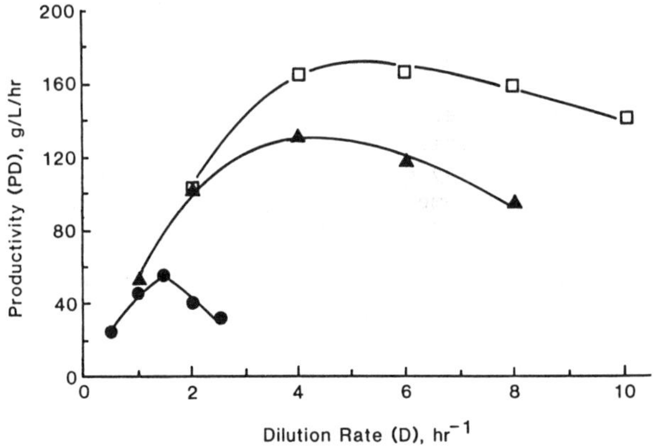

Figure 6. Effect of dilution rate and cell concentration on ethanol productivity. The feed was glucose (100 g/L) and the microorganism was S. cerevisiae. Cell concentration (x): ●, 14 g/L; ▲, 60 g/L; and □, 100 g/L.

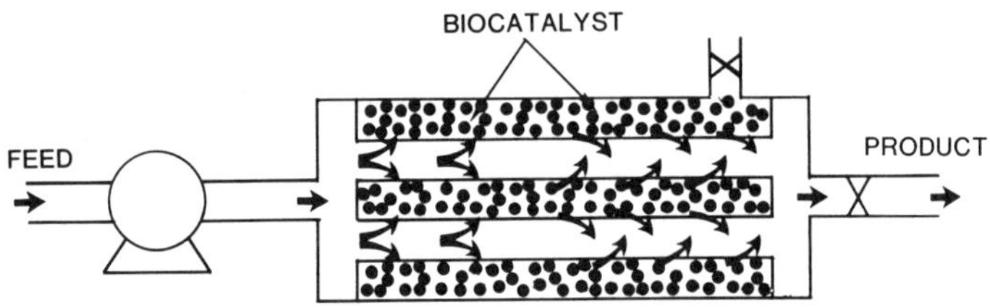

Figure 7. Schematic of hollow fiber bioreactor. The biocatalyst is packed into the shellside and feed pumped through the tubes.

fibers or were localized to an annular volume just around the fibers. Thus the cells far from the fibers were probably starved of nutrients and dying, and so a means must be found of removing these dead cells continuously. If not, the growing mass of cells may even rupture the relatively sensitive fibers. One of the biggest problems is to purge the gas that is formed in the shell-side as a by-product of the fermentation. Michaels and Robertson et. al., have recently been awarded patents for applications of the hollow fiber concept (24,25); it appears to be most useful for reactions involving non-growing cells.

Conclusions

Membrane bioreactors provide an opportunity to vastly improve the performance and productivity of fermentations. Specific advantages of membrane recycle fermentors include:

- The inherently more efficient continuous process can be used
- High cell densities can be obtained, exceeding 100 g/L of dry weight of cells, if necessary. This usually implies cell concentrations of 10^{11}-10^{12} per mL, close to theoretical loadings. The upper limit to cell concentrations is the physical properties of the fermentation mixture, which will affect the pumpability, polarization and fouling problems with the membrane module.
- High dilution rates can be used, as a direct result of the high cell concentrations. For ethanol fermentation, dilurates of 1-2 hr^{-1} can be used with complete substrate conversion, provided the substrate concentration is less than 15% w/v and the cell concentration is above 90 g/L. Higher substrate concentrations will lead to high ethanol concentrations and consequent lower fermentation rates, which will necessitate slowing down the flow rate to obtain complete conversion. A considerable amount of optimization work remains to be done to determine kinetic parameters in the high product concentration operating regions.
- Control of the environment is easy with the membrane recycle fermentor. If the membrane module is properly designed, the bulk of the fermentation or reaction should occur in the reaction vessel. Thus control of temperature, pH, dissolved oxygen, etc., can be done with the usual methods as currently practised. However, we suspect that hardware design aspects such as heat transfer and mass transfer will be the limiting factors in these high-rate fermentations, rather than the microbiology or the membrane separations *per se*. For example, membrane bioreactors used for ethanol production will be producing at least 50 times as much heat and gas per unit volume of bioreactor than conventional fermentors, which will necessitate drastic redesign of the hardware.
- Membrane recycle bioreactors can process both soluble and insoluble substrates, unlike the hollow fiber and most immobilized whole cell bioreactors.
- The product stream from membrane bioreactors is free from microbial cells and other particulate material, which should reduce downstream processing costs.

-- There are no gas hold-up problems as with packed-bed, immobilized whole cell reactors or hollow fiber reactors.
-- This concept is probably the easiest to incorporate into an existing plant. At the very least the membrane modules can replace conventional centrifuges used for separating microbial cells from the fermentation broth which are expensive to buy, operate and maintain.

Some potential limitations should also be mentioned: the membrane recycle bioreactor is operating essentially as a continuous stirred tank reactor (CSTR). The outlet (permeate) stream will have the same concentration as the reaction mixture, which means the system will be ideally operating with a high product concentration at all times. This means a low fermentation rate, since most fermentations are severely product-inhibited, sometimes at very low product concentrations (for ethanol, it is usually 12-15% v/v). The long-term effects of continuous contact of microbial cells with high product concentrations has not yet been evaluated.

In addition, as already alluded to, the key factor in successful long-term operation of a membrane recycle bioreactor is the performance of the membrane module. Several studies undertaken by our group (21) have indicated that the flux and "fouling" is controlled not only by the expected operating parameters, such as pressure, temperature, flow velocity and feed concentration, but also by specific interactions between the membrane and feed-stream components. Thus it is important not only to have the appropriate physical configuration of the module, but the chemical nature of the membrane should be such so as to minimize these interactions. Fermentation broths are highly complex mixtures of macromolecules, salts, vitamins, etc., and it is important to understand and minimize their interactions with the membrane in order to minimize membrane-related problems during long-term operation. Some membranes completely inactivate some enzymes by mere contact (5, 26).

Finally, sterilizability of the membranes is also critical to its application in continuous bioreactors. Ideally, the membrane module should be repeatedly steam-sterilizable so as to fit in with the routine steam sterilization cycles of the fermentors and associated apparatus. Unfortunately, the membranes that have the desired properties as far as flux and fouling are concerned tend to have very poor temperature stability and vice versa. Perhaps a new generation of "biocompatible" membranes need to be developed for these specific applications.

Literature Cited

1. Schultz, J. S.; Gerhardt, P. Bacteriol. Rev. 1969, 33, 1.
2. Abbott, B. J.; Gerhardt, P. Biotechnol. Bioeng. 1970, 12, 577.
3. Abbott, B. J.; Gerhardt, P. Biotechnol. Bioeng. 1970, 12, 601.
4. Steiber, R. W.; Gerhardt, P. Biotechnol. Bioeng. 1981, 12, 535.
5. Cheryan, M.; Mehaia, M. A. In "Membrane Separations in Biotechnology"; McGregor, W. C., Ed.; Marcel-Dekker, NY, 1985.
6. Budd, W. E.; Okey, R. W. U.S. Patent 3 472 765, 1969.

7. Michaels, A. S. In "Progress in Purification and Separation"; Perry, E. S., Ed; Interscience, New York, 1968. p. 297.
8. Porter, M. C.; Michaels, A. S. Chem. Tech. 1972, 2, 5.
9. Wang, D. I. C.; Sinskey, A. J.; Butterworth, T. A. In "Membrane Science and Technology", Flinn, E. D., Ed.; Plenun, New York, 1970.
10. Margaritis, A.; Wilke, C. R. Biotechnol. Bioeng. 1978, 20, 709.
11. Margaritis, A.; Wilke, C. R. Biotechnol. Bioeng. 1978, 20, 727.
12. Haroldsen, A.; Rosen, C. G. Eur. J. Appl. Microbiol. Biotechnol. 1982, 14, 216.
13. Cheryan, M.; Mehaia, M. A. Biotechnol. Lett. 1983, 5, 519.
14. Mehaia, M. A.;Cheryan, M.; Argoudelis, C. A. Cult. Dairy Prod. J., 1984, in press.
15. Cheryan, M.; Mehaia, M. A. Process Biochem. 1984, 19(Dec).,
16. Rogers, P. L.; Lee, K. J. ; Tribe, D. E. Process Biochem. 1980, 15, 7.
17. Kuo, K. P.; Cheryan, M. J. Food Sci. 1983, 48, 113.
18. Hamer, G. Biotechnol. Bioeng. 1982, 24, 511.
19. Bull, D. N.; Young, M. D. Biotechnol. Bioeng. 1981, 23, 373.
20. Vick Roy, T. B.; Mandel, D. K.; Dea, D. K.; Blanch, H. W.; Wilke, C. R. Biotechnol. Lett. 1983, 10, 665.
21. Cheryan, M. Unpublished results.
22. Minier, M.; Ferras, E.; Goma, G.; Soucaille, P. 7th International Biotechnology Symposium, New Delhi, India, 1984.
23. Vick Roy, T. B.; Blanch, H. W.; Wilke, C. R. Biotechnol. Lett. 1982, 4, 483.
24. Michaels, A. S.; Robertson, C. R.; Cohen, S. N.; Inloes, D. S.; Smith, W. J. U.S. Patent 4 442 206, 1984.
25. Michaels, A. S.; Robertson, C. R.; Cohen, S. N. U.S. Patent 4 440 853, 1984.
26. Kohlwey, D. K.; Cheryan, M. Enzyme Microb. Technol. 1981, 3, 64.
27. Janssens, J. H.; Bernard, A.; Bailey, R. B. Biotechnol. Bioeng. 1984, 26, 1.
28. Mehaia, M. A.; Cheryan, M. Enzyme Microbial Technol. 1984, 6, 117.
29. Mehaia, M. A.; Cheryan, M. Applied Microbiol. Biotechnol. 1984, 20, 100.
30. Ghose, T. K.; Tyagi, R. D. Biotechnol. Bioeng. 1979, 21, 1387.
31. Ghose, T. K.; Bandyopadhyay, K. K. Biotechnol. Bioeng. 1980, 22, 1489.
32. Maragritis, A.; Bajpai, P. K., Wallace, J. B. Biotechnol. Lett. 1981, 3, 616.
33. Klein, J.; Kressdorf, B. Biotechnol. Lett. 1983, 5, 497.
34. Inloes, D. S.; Taylor, D. P.; Cohen, S. N.; Michaels, A. S.; Robertson, C. R. Appl. Environ. Microbiol. 1983, 46, 264.
35. Maiorella,B; Wilke,C.R.;Blanch,H.W. Adv.Biochem.Engr. 1981, 20, 43.

RECEIVED April 5, 1985

19

Single-Stage Seawater Desalting with Thin-Film Composite Membrane Elements

W. G. LIGHT, Z. B. TAYLOR, and A. B. RIEDINGER

Fluid Systems Division, UOP Inc., San Diego, CA 92131

To simplify system design and reduce capital and operating costs, Fluid Systems has developed spiral-wound, reverse osmosis membrane elements for producing potable water from seawater using a single-stage system. The elements contain a high salt rejection, thin-film composite (TFC) membrane and are fabricated in a close-coupled configuration to maximize productivity. The performance of commercial-size elements has been monitored at a seawater pilot plant since September 1983. The single-stage desalting capability of the elements (rejection > 99.5%) is demonstrated at pH 4-6.5 and for the duration of testing in terms of constant membrane salt passage. Single-stage elements display a high, stable productivity comparable to first-stage elements in two-stage systems. Productivity stability is explained by a new flux decline model.

Many reverse osmosis (RO) systems which have been used to produce potable water from seawater are either of a two-stage design or a partial two-stage design. For these systems, all or part of the permeate from the first stage unit is processed through a second unit. This has been necessary because the solute rejection of previously available membrane elements has not been high enough nor stable enough to produce an acceptable permeate quality at an economical recovery for a reasonable element life. A single-stage system is preferred over a two-stage system because it is simpler, not requiring interstage pumps nor the

0097–6156/85/0281–0247$06.00/0

pressure and flow controls for the second stage. Consistent with the simpler design, capital and operating costs can be lower for a single-stage than for a two-stage system.

For producing potable water, typically the most difficult criterion to meet for the permeate for a seawater desalting system is to satisfy the World Health Organization (WHO) Drinking Water Quality Guideline (1) for chloride ion concentration of < 250 mg/L, rather than the guideline for total dissolved solids (TDS) concentration of < 1,000 mg/L. To do this from seawater having a chloride ion concentration of 19.8 g/L and using an RO system operated at 30% recovery, requires a membrane chloride ion rejection of about 99%. To ensure acceptable permeate quality for the duration of membrane element life, it is often presumed that an initial membrane rejection of about 99.5% is required to compensate for salt passage and flux changing in time. Once the required rejection is achieved, the costs for a single-stage system are primarily determined by the element productivity and flux stability.

The objective of the work presented here is to demonstrate the single-stage desalting capability and high, stable productivity of spiral-wound RO elements containing a high-rejection, thin-film composite (TFC) membrane. As the long-term performance characteristics of such elements have not been well documented in the literature, the technical approach taken has been to monitor performance since September 1983 at a seawater pilot plant facility. Data are presented here of element productivity and salt rejection as functions of elapsed operating time, net pressure, permeate recovery, and feedwater concentration and pH. Based on the data obtained, procedures are proposed for predicting membrane flux and rejection over the course of element life.

Materials and Methods

Seawater Characterization. The seawater test facility is located at a coastal site in Chula Vista, several miles south of San Diego. The source of water is a 120-ft deep seawell on the bayfront. The concentrations of well water components are routinely monitored in support of a National Pollution Discharge Elimination System permit and to ensure constant feedwater quality. Individual chemical species analyses are performed using standard techniques (2).

The raw well water has virtually the same composition as seawater but with each species at a lower concentration. To maintain a constant system feedwater TDS concentration of about 36 g/L, the RO system is operated in a feed-and-bleed mode with part of the concentrate being recycled and blended with well water to give the desired system feedwater concentration. The fraction of feedwater that is recycled concentrate (x) can be calculated in terms of the fractional recovery per pass (k), the fractional membrane rejection (R), and the ratio of RO feedwater concentration to raw wellwater concentration (y) using the following equation:

$$x = \frac{(y - 1)(1 - k)^R}{y - (1 - k)^R} \qquad (1)$$

The compositions of the well water and RO feedwater as compared to standard seawater are given in Table I. It can be seen that the feedwater composition is comparable to that for seawater with a low standard deviation over the course of eight months. The only notable exception to typical seawater is the relatively high calcium and sulfate concentrations. Even so, the feed does not become saturated in $CaSO_4$ until about 60% recovery.

Table I. Chemical Analysis for Reverse Osmosis System Feedwater

	Concentrations (mg/L)			
Component Species	Well Water	Well Water Standard Deviation	Standard Seawater (3)	Feedwater for RO System
TDS	16,700	826	36,000	36,600
Calcium	340	35	420	760
Magnesium	650	46	1,330	1,220
Strontium	3	-	13	6
Sodium	5,500	436	11,030	11,200
Potassium	150	25	380	310
Bicarbonate	330	56	146	65
Sulfate	1,300	198	2,770	3,700
Chloride	8,550	423	19,800	19,400
Silica	17	3.6	-	28

<u>Membrane Element Assembly</u>. Spiral-wound RO elements (UOP Inc., Fluid Systems Division) having a chloride ion rejection of > 99.5% were employed in the pilot unit. The single-stage (SS) elements (6-in diameter, 40-in long) are constructed using a TFC membrane and

are designed for operation with feedwater at temperatures < 50° C and pH 3-11. The principal test assembly was a 6-in diameter fiberglass reinforced plastic (FRP) pressure housing (UOP Inc., Fluid Systems Division) containing six elements in series (4). The test assembly was designed so that permeates from the three lead-end elements and the three downstream elements were segregated and could be collected separately.

Reverse Osmosis Pilot Plant. The RO test facility is shown schematically in Figure 1. Well water (23.1 ± 0.3 ° C, pH 7.3) is pumped from the well, blended with concentrate recycle water, and for long-term tests adjusted to pH 5.8 ± 0.3 using sulfuric acid. A 25-μm pore size cartridge filter is employed to protect the high-pressure, positive-displacement pump which delivers feedwater to the unit at 800 psig and 17.5 gpm/housing. To maintain typical seawater concentrations while operating at about 30% recovery, 9 gpm concentrate is blended with 8.5 gpm wellwater to produce the system feedwater. The remaining concentrate (3 gpm) and total permeate flows (5.5 gpm) are discharged to drain after being adjusted to seawater pH. No precipitation inhibitor is used. Also, for stable long-term operation, the TFC single-stage elements do not require removal of dissolved oxygen using sodium bisulfite or a vacuum deaeration system.

Short-term tests were performed to determine performance as a function of feedwater TDS concentration (18-50 g/L), pH (pH 3.5-8), net pressure (200-500 psi), and recovery (20-40%). Long-term experiments were carried out to measure flux and salt passage changes in time. The test unit has been in operation since September 7, 1983.

Data Analysis. Optimum system performance is achieved by selecting a system design and operating conditions and an element design that maximize element productivity for a given net pressure (P_{net}) and that meet the single-stage desalting requirements for the duration of element life. In assessing experimental data, it is often convenient to quantify these concepts in terms of mass transfer coefficients for the solvent or water (A) and the solute or salt (B) (5). The coefficient A is the membrane water flux divided by its driving force, P_{net} (the hydraulic applied pressure minus the osmotic pressure). The term B as used here is the chloride ion flux divided by its driving force, transmembrane chloride ion concentration gradient (ΔC). Solute accumulation at the membrane-liquid interface is accounted for in ΔC by multiplying the bulk feed concentration (C_F) by a boundary layer correction factor. This is an exponential function of the ratio of the element productivity (Q) to the average feed flowrate.

Values for A and B were calculated from experimental measurements of Q for elements having a given membrane area (S) and measurements of C_F and permeate chloride ion concentration (C_P) using the following equations:

$$A = \frac{Q}{S\ P_{net}} \tag{2}$$

$$B = \frac{Q\ C_P}{S\ \Delta C} \tag{3}$$

In the calculation of P_{net}, osmotic pressure was estimated as the sum of the individual ion contributions calculated using the van't Hoff equation.

As it has become common practice to report system performance characteristics in terms of Q and element chloride ion rejection (R) normalized to a typical P_{net} of 400 psi, data are given here for these variables as well as A and B. Normalized values for Q (represented as Q_n) were calculated from:

$$Q_n = Q\ \frac{400}{P_{net}} \tag{4}$$

Values for R were calculated in terms of the linear average bulk feed chloride ion concentration ($\bar{C}$) and a normalized permeate concentration (C_{pn}) according to:

$$R = \frac{\bar{C} - C_{pn}}{\bar{C}} \tag{5}$$

where

$$C_{pn} = C_p\ \frac{P_{net}}{400} \tag{6}$$

Values of R can also be calculated in terms of A and B, the net pressure, and the ratio of transmembrane concentration gradient to feed concentration, as shown below:

$$R = 1 - \frac{B\ \Delta C}{A\ P_{net}\ \bar{C}} \tag{7}$$

Results

Short-term tests were performed that confirm the form of Equations 2-4 for ranges of feedwater TDS concentration (18-50 g/L), permeate recovery (20-40%), and applied pressure (600-800 psi). The interdependence of these variables was utilized by combining data to demonstrate that Q is directly proportional to P_{net} for P_{net} in the range 200-500 psi, while A and B are independent of P_{net}. For example, when P_{net} was increased by a factor of two from 200 to 400 psi, Q also increased by a factor of two from approximately 1,000 to 2,000 gal/day. Results also show that at 30% recovery and 800 psig applied pressure, permeate meets WHO guidelines when using feedwater having high TDS (40 g/L) and chloride ion (22 g/L) concentrations.

Effects of pH on Performance. Figure 2 is a plot of membrane chloride ion rejection as a function of feedwater pH for normalized data obtained after 146 days of operation. Maximum rejection is achieved at about pH 5. Although rejection is pH dependent, the dependency is modest, with the average rejection being > 99.5% for feedwater pH 4-6.5.

The actual pH that the internal membrane is exposed to is less than the feedwater pH, as shown in Figure 3, a plot of permeate pH as a function of feedwater pH. These data show that the feedwater pH range of 4 to 6.5 cited above corresponds to a permeate pH range of about 3 to 4.5. This difference in pH is due to a Donnan exclusion effect and to the high rejection of bicarbonate ion and the low rejection of carbon dioxide in feedwater.

Element Productivity Data. A logarithmic plot of A and Q_n as functions of elapsed operating time is given in Figure 4. An important trend shown here is that rather than A and Q_n declining from the onset of system operation, a delay period (t_1) of approximately 20 days is observed during which A and Q_n actually increase according to:

$$A = A_o \left[\frac{t}{t_o} \right]^{0.05} \qquad (8)$$

where A_o = 2.32×10^{-5} cm^3/cm^2-sec-atm
t = time, days
t_o = 1 day

The average value of A during t_1 is 2.6×10^{-5} cm^3/cm^2-sec-atm, which corresponds to Q_n of 2,200 gal/day. These values are equal to nominal values for first-stage elements for two-stage systems (e.g., Model 1501, UOP Inc., Fluid Systems Division).

After the 20 day delay period, A and Q_n gradually decrease with a flux decline slope of -0.06. The equation for A as a function of t for t > 20 days is the following:

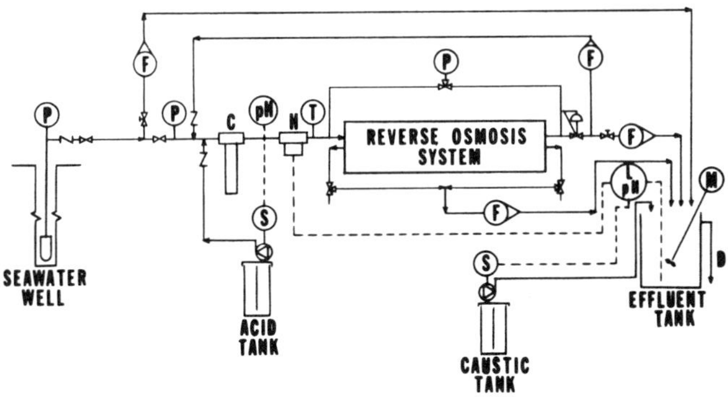

Figure 1. Flow schematic for reverse osmosis test facility. Key: C, cartridge prefilter, 25 μm pore size; F, flow indicator; H, high-pressure pump; L, level indicator/pump shutdown; M, mixer; P, pressure indicator; pH, pH indicator; S, speed controller; and T, temperature indicator.

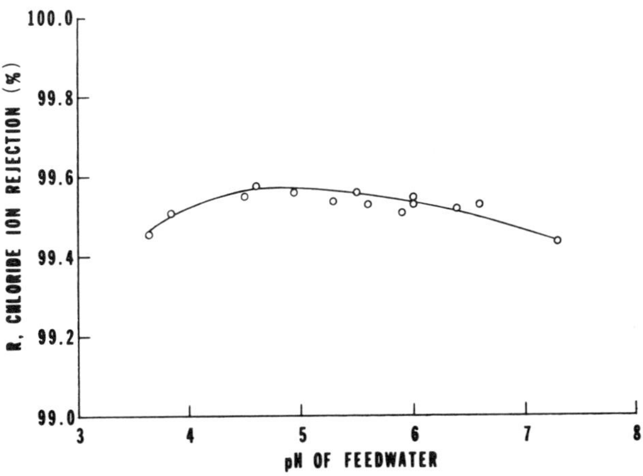

Figure 2. Effect of feedwater pH on TFC membrane chloride ion rejection for data normalized to 400 psi net after 146 days of operation.

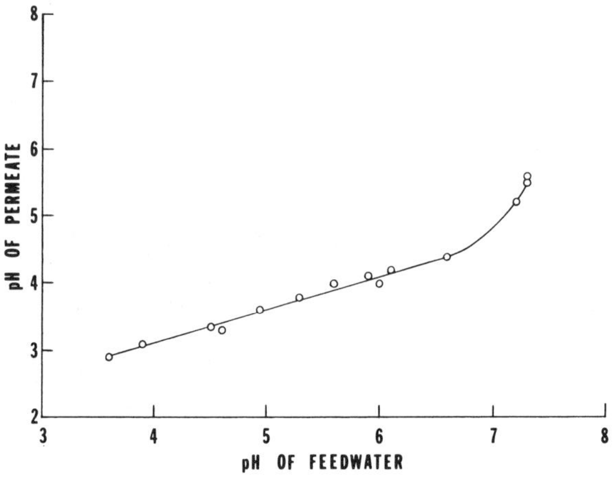

Figure 3. Permeate pH as a function of feedwater pH showing the effect of Donnan exclusion and selective membrane passage of CO_2 relative to HCO_3^- in reducing the pH of permeate.

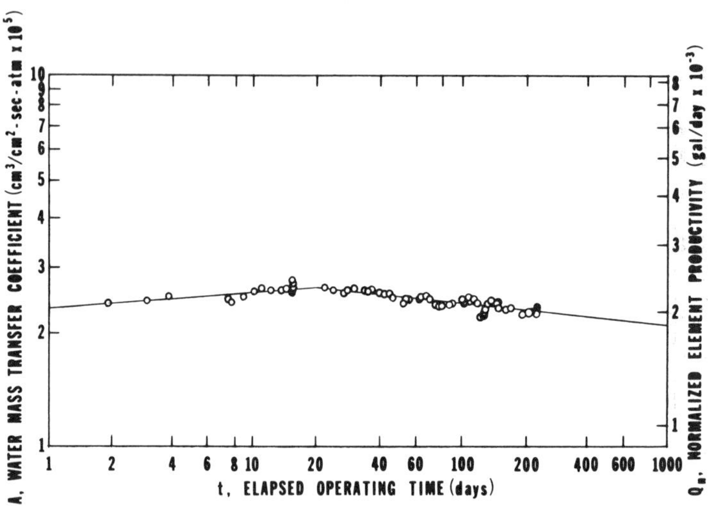

Figure 4. Membrane water mass transfer coefficient and normalized element productivity as functions of elasped operating time showing the delay period of 20 days followed by a flux decline with a slope of -0.06.

$$A = A_1 \left[\frac{t}{t_1} \right]^{-0.06} \quad (9)$$

where $A_1 = 2.7 \times 10^{-5}$ cm^3/cm^2-sec-atm
$t_1 = 20$ days

Using this equation and extrapolating to five years of operation indicates a decline to 79% of the delay period average values for A (to 2.06×10^{-5} cm^3/cm^2-sec-atm) and Q_n (to 1,780 gal/day).

To determine whether fouling had occurred, the three lead-end elements were cleaned after 100 days of operation using UOP Solution Y (citric acid solution at pH 3) and Solution Z (Borax, EDTA, trisodium phosphate solution at pH 9.5) for 45 minutes each at 35 °C. Results did not demonstrate that any flux-related fouling had occurred.

Membrane Salt Passage Data. Figure 5 is a plot of B as a function of time. A statistical analysis of eight months of data indicates that there is no significant correlation between B and time (r = 0.21), as determined by comparing the best fit least squares regression line slope and a slope of zero. It is thus concluded that B is independent of time, the mean value of B being $0.280 \pm 0.027 \times 10^{-5}$ cm/sec. The fact that salt passage is constant over this time period indicates that there has been no degradation of the membrane mechanism responsible for retaining salt.

Although B is constant in time, R decreases in proportion to the decrease in A, as indicated by Equation 7. This effect is shown in Figure 6, a plot of R as a function of elapsed operating time. These data show a decline in element rejection of from about 99.6% over the first 20 days to 99.55% at 240 days corresponding to a decline in Q_n and A of 11%. For this situation where B, ΔC, $\bar{C}$, and P_{net} are constant, R at time t can be calculated as a function of a reference rejection (R_2) and A values for time t (A) and a reference time (A_2) according to:

$$R = 1 - (1 - R_2) \frac{A_2}{A} \quad (10)$$

Normalized element rejection data at 30% system recovery are given in Table II for individual components in feedwater after 219 days of operation. These data show that multivalent ions are very highly reected, resulting in permeate containing essentially only sodium and chloride ions, which confirms that the most rigorous and meaningful measure of rejection for seawater is chloride ion rejection.

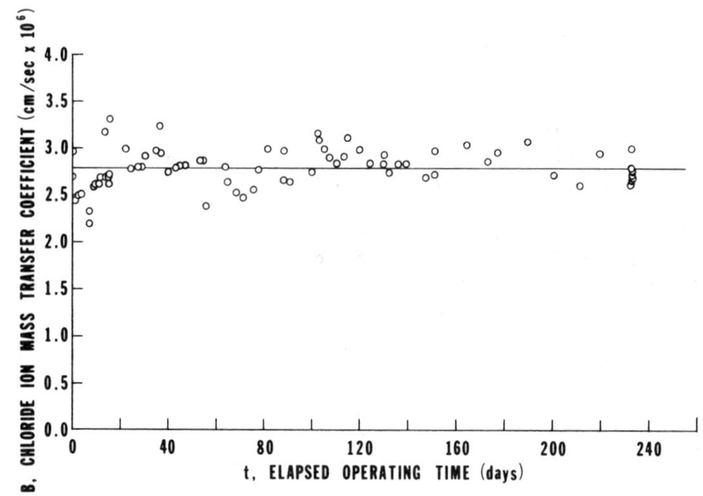

Figure 5. Constant membrane chloride ion mass transfer coefficient over the course of 8 months of operation (r = 0.21).

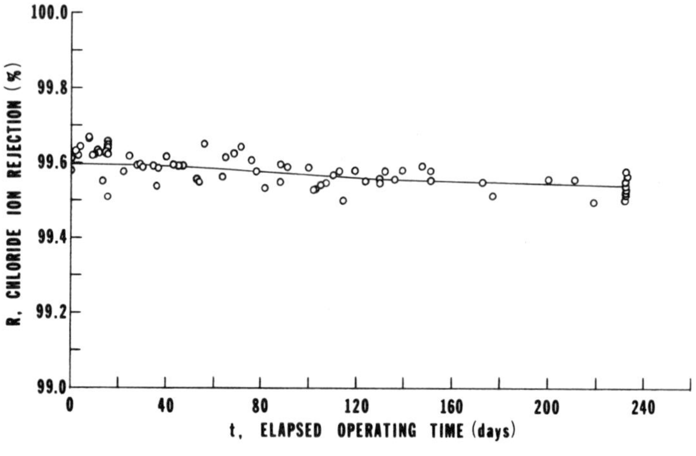

Figure 6. Membrane chloride ion rejection as a function of elapsed operating time showing the effect of flux decline on rejection.

Table II. Average Element Rejection Data Normalized to 400 psi net at 30% Recovery for Individual Components in Feedwater After 219 Days of Operation

Component Species	Concentrations (mg/L) Feedwater	Concentrate	Permeate	Element Rejection (%)
TDS	36,400	52,100	168	99.62
Calcium	700	1,200	<.5	>99.95
Magnesium	1,120	1,900	0.8	99.95
Sodium	10,300	16,440	60	99.55
Potassium	323	515	1.6	99.62
Sulfate	3,400	4,900	<.5	>99.99
Chloride	18,186	26,446	109	99.51
Bicarbonate	178	283	-	-
Silica	28	39	0.8	97.61

Discussion

For 8 months of constant P_{net} operation using single-stage TFC elements with seawell water, a statistical evaluation of salt passage data indicates that B is constant in time and that a decline in element rejection is due to a decline in A in time. The membrane chloride ion mass transfer coefficient remained constant, even after cleaning, without requiring a rejection restorative treatment such as poly(vinyl methyl ether) or tannic acid. Were this trend of constant B to continue, an R value of 99.5% would decrease to about 99.3% at 5 years, as calculated using Equation 10, which is well above the value of 99% required to meet the WHO guideline for chlorides of 250 mg/L.

The new model for flux decline presented here includes a delay period of 20 days during which element productivity increases and after which productivity declines logarithmically with a slope of -0.06 at 800 psig applied pressure at 23.1°C. The delay period could be a result of several effects. There may be a grace period during which there is a relaxation in the resistances in series that together act to impede flow, such as the membrane itself, support fabric, and permeate carrier. There also may be transient effects that act to facilitate permeate flow and that are compensated for by increases in the resistances to permeate flow in the element. The transient effects could relate to membrane hydration or swelling, the low pH of permeate (Figure 3) or other phenomena.

Previous flux decline models predict an initial, rapid drop in flux followed by an extended, moderate decline, which on a logarithmic plot appears as a straight line having a constant, negative slope from initial time. The origin of these models is the measurement of flux decline for cellulosic membrane (6) for which the rate of flux decline is directly proportional to the rate of increase in resistance to permeate flow due to membrane compaction (densification by internal pore structure collapse and pore narrowing). Recent work has expanded this model to account for compaction related flux decline as a function of temperature and pressure as well as operating time (7).

Unlike previous models, the flux decline for non-cellulosic TFC membrane elements presented here appears not to be dictated by membrane compaction. In an effort to deduce the mechanism responsible for flux decline, an analytical approach was devised to determine whether flux decline could be accounted for by increases in resistance to flow (H) in permeate carrier material. By equating Equation 9 and the standard equation for A as a function of H (8) which is commonly used for selecting permeate carriers (9), an equation was derived for H as a function of time for $t > 20$ days:

$$H = at^{b} \tag{11}$$

where values of a (0.50) and b (0.219) have been calculated at 23 °C for an applied pressure of 800 psi. The validity of Equation 11 and the effect of temperature and pressure on the values of a and b are being evaluated empirically. To minimize the effect of permeate carrier resistance, efforts are also being expended to select permeate carrier materials with high permeability and low compressibility and to design elements with appropriate leaf lengths and permeate channel configurations.

New, single-stage elements are commercially available in a variety of sizes in a "close-coupled" configuration. Compared to previous designs, the "close-coupled" design permits elements to be loaded in pressure housings abutting one another, thus increasing the membrane area per housing by about 10% and minimizing the risk of elements disconnecting in service. For large seawater systems, Model 2021 SS elements (8-in diameter, 40-in long) are preferred. As for all single-stage elements, the membrane contained in the 2021 SS elements displays the same flux and rejection characteristics as the elements tested here. Being 8-in diameter "close-coupled" elements, however, they have a higher productivity, a nominal initial value of 4,000 gal/day.

Conclusion

Spiral-wound RO elements containing Fluid Systems' TFC membrane are capable of producing potable water from seawater in a single-stage system for a broad range of operating conditions and with little change in performance in time. Rather than incurring an initial membrane flux decline, the single-stage elements maintain a high productivity for the initial 20 days of operation which is followed by a gradual exponential decline. The high, stable membrane flux for single-stage elements is comparable to that for first-stage elements in two-stage systems, indicating substantially lower costs for single-stage systems. For the seawell water processed for 8 months, no increase in salt passage was observed, and changes in rejection are accounted for in terms of a new model for flux decline. Performance projections indicate single-stage desalting capability well beyond five years of operation.

Nomenclature

A = solvent or water mass transfer coefficient, cm^3/cm^2-sec-atm

A_i = value of A at a given time, where: i = 0 (time zero), 1 (time t), or 2 (reference time t), cm^3/cm^2-sec-atm

B = chloride ion mass transfer coefficient, cm/sec

$\bar{C}$ = linear average bulk feedwater chloride ion concentration, g/L

C_F = bulk feedwater chloride ion concentration, g/L

C_P = permeate chloride ion concentration, g/L

C_{pn} = normalized permeate chloride ion concentration, g/L

ΔC = transmembrane chloride ion concentration gradient, g/L

H = permeate channel resistance to flow, atm-sec/cm^3

k = permeate recovery, system productivity divided by feed flowrate, %

P_{net} = net pressure, applied hydraulic pressure minus the feedwater osmotic pressure, psi or atm

Q = element permeate flowrate or productivity, gal/day or cm^3/sec

Q_n = element productivity normalized to P_{net} of 400 psi, gal/day or cm^3/sec

r = statistical correlation coefficient, where a value of zero indicates no correlation

R = membrane chloride ion rejection, %

R_o = membrane chloride ion rejection at time zero, %

R_2 = membrane chloride ion rejection at a reference time, %

S = membrane area in an element, ft^2 or cm^2

t = time, days

t_o = 1 day
t_1 = duration of the flux decline delay period, days
x = fraction of feedwater that is recycled concentrate
y = ratio of wellwater concentration to feedwater concentration

Literature Cited

1. "Guidelines for Drinking-Water Quality, Vol. 1, Recommendations," World Health Organization, Geneva, Switzerland, 1984.
2. "Standard Methods for the Examination of Water and Wastewater," American Public Health Association, Washington, D.C., 1981, 15th ed.
3. "1981 Annual Book of ASTM Standards, Part 31 (Water)," American Society for Testing and Materials, Philadelphia, 1981.
4. Chu, H.C.; Ewanowich, W.; Light, W.G. Water Supply Improvement Association Conference, Orlando, Florida, 1984.
5. Riley, R.L.; Hightower, G.; Lyons, C.R. In "Reverse Osmosis Membrane Research"; Lonsdale, H.K.; Podall, H.E., Eds.; Plenum Publishing Corp.: New York, 1972; pp. 437-456.
6. Sourirajan, S.; Govindan, T.S. Proc. First Desalination Symp., Washington, D.C., 1965.
7. Dale, M.C.; Okos, M.R. Ind. Eng. Chem. Prod. Res. Dev. 1983, 22, 452-6.
8. Miller, C.; Spatz, D.D. "Research and Development on Low Pressure RO Membranes and Design and Construction of a Small Unit for Purification of Brackish Water," Office of Saline Water, Contract 14-01-001-939, 1967.
9. Kremen, S.S. In "Reverse Osmosis and Synthetic Membranes"; Sourirajan, S., Ed.; National Research Council of Canada: Ottawa, 1977; Chap. 17, pp. 371-385.

RECEIVED February 22, 1985

Production of Boiler-Feed Quality Water from Bitumen–Heavy Oil–Oil-in-Water Emulsions by Ultrafiltration

B. A. FARNAND, S. COULOMBE, and H. SAWATZKY

Energy Research Laboratories, Energy, Mines, and Resources Canada, Canada Center for Mineral and Energy Technology, Ottawa, Ontario, Canada K1A 0G1

In order to reduce the large water demand of enhanced bitumen/heavy oil recovery operations, the use of porous membranes in a water recycle process using wellhead o/w emulsions were destabilized by this process, which would aid in further processing to make pipeline quality crude oil. Polar material in the membrane permeate was considered to be the stabilizing agent whose removal caused the emulsion's destabilization. Selective removal of stabilizing agents with subsequent destabilization of produced bitumen/heavy oil o/w emulsions has not been previously reported and attempts were made to characterize the polar material. The major difficulties with the use of membranes in this process are severe fouling of the membrane caused by the bitumen and operation at wellhead temperatures.

The amount of water required for enhanced oil recovery (EOR) of heavy oil and bitumen with steam floods and other water dependent fluid methods is so large that it may become the restricting factor for development of some projects (1). Also, methods of EOR and in situ recovery produce stable wellhead oil-in-water (O/W) emulsions which resist conventional treatment and can be processed only at great expense by chemicals or energy intensive treatment units (1-6). Even with existing specialized steam generation facilities the production of steam from produced water is difficult and accounts for only a small amount of EOR and in situ steam supply (3, 7-9).

Previous work in our laboratory has demonstrated the possibility of using porous membranes as a physical method of breaking bitumen/water emulsions in a batch process to produce high quality water (10). This report is an extension of the previous work to a quasi-continuous membrane process with analysis of the permeate for boiler feed quality parameters. The emulsion used to demonstrate this was a wellhead bitumen/water/mineral emulsion of approximately

0097–6156/85/0281–0261$06.00/0
Published 1985, American Chemical Society

4% bitumen supplied by Texaco Canada Resources Ltd., from a steam flood in situ pilot plant at Fort McMurray, Alberta. This wellhead emulsion is currently treated by adding chemical demulsification agents at elevated temperatures (3). Further processing is concerned with producing low water content bitumen with little emphasis on the product water quality.

The use of porous membranes for breaking O/W emulsions is not unique; trade literature and scientific literature are available (11-14). However, no reports exist on the treatment of bitumen/water emulsions with porous membranes. This could be a result of the unusual properties of the bitumen/water emulsions related to the presence of clay, other minerals, water, and dissolved species including salts. The effect of clay on stabilizing O/W emulsions has been demonstrated by Gewers with Athabasca bitumen and he has indicated that this effect is reduced at pH 7 to 8 (6). Broughman reports the density of bitumen and water overlaps at 34°C and 115°C, with water slightly denser between these two temperatures (15). He also reports that this difference is maximized from 65°C to 80°C. With this information, operating conditions for Athabasca bitumen/water/mineral separations should be at pH 7 to 8 and temperatures from 65°C to 80°C. Although the pH range is acceptable for the membrane used in this work, the temperature range is the extreme of usual membrane practice. The emulsion as supplied was pH 7.

Parameters describing the quality of boiler feed water for several types of steam generation have been summarized by Fulton as shown in Table I (9). Although the Thermosludge Generator has lower water quality standards, it also has higher capital and operating costs than a once-through steam generator. For this reason, the once-through generator's standards were considered a test of any water recycle process. Other water standards have been included in Table I for comparison.

Experimental

Unlike the previous work (10), a continuous system for the evaluation of membrane performance was assembled as shown in Fig. 1. When operating with the emulsion, there was considerble difficulty in controlling the operating pressure, which was apparently caused by free bitumen "slugging" through the back pressure control valve. This caused rapid but small variations in the operating pressure which were visually averaged to determine the system's operating pressure. As well, the emulsion was opaque and the rotameter shown in Fig. 1 was not used in the emulsion experiments. For the elevated temperature experiments, no attempt was made to insulate the system from the surroundings, and heat was only applied to the reservoir. Accordingly, the reservoir's temperature was 75°C and the liquid's temperature at the end of the series of cells was 45°C. The latter was considered more significant because of the large amount of tubing and fittings between the reservoir and the first membrane cell.

The experiments were conducted in the following order. First, a pure water permeation (PWP) experiment was performed to determine the PWP rate for each membrane, followed by the replacement of the pure water with KI solution. The permeation rate (PR) of the KI solution experiment was determined for each membrane, and the concentration of KI in the permeate and the reservoir was measured. A function describing the separation is defined as:

Table I. Water Quality Objectives for Steam Generation and Other Uses (9)

Parameter	Once-through steam generator[d]	Thermosludge generator[a]	Deep Well disposal[b]	Irrigation	Cooling water make-up
Total hardness, mg/L as $CaCO_3$	0.5	100-2000	-	-	c
Silica, mg/L	50	10-250	-	-	c
Oil, mg/L	1	50-1000	20	5	1
TDS, mg/L	8000	1000-50000	-	1000	-
Suspended solids, mg/L	0	10-500	5	-	d
Dissolved O_2, mg/L	0.04	-	-	-	-
Iron, mg/L	0.05	-	-	-	

a These are vendor recommendations. Some of these values have been exceeded in actual practice(9).

b Water should be chemically compatible with formation waters.

c The product of Mg as $CaCO_3 \times SiO_2$ should be less than 35,000 mg/L.

d Total suspended solids in the recirculating water should be less than 200 mg/L.

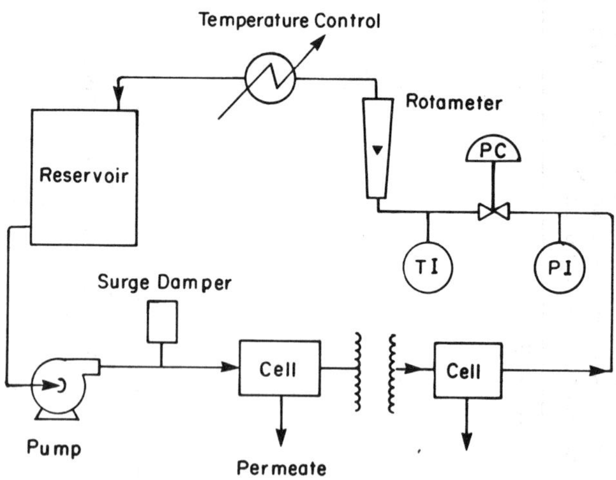

Figure 1. Continuous system for the evaluation of membrane performance.

$$f = \frac{\text{reservoir concentration - permeate concentration}}{\text{reservoir concentration}} \quad (1)$$

where f is the fraction of separation and is expressed as a percent.

The second step consisted of replacing the KI solution with the bitumen/water/mineral emulsion. The PR was determined, and the permeate was analyzed for total dissolved solids (TDS), metals (Mg, Ca, Fe), silica, and pH. The feed solution consisted of 2 L of emulsion, and was not adjusted throughout this step. The third step consisted of replacing the emulsion in the reservoir with a new sample and performing the experiment as in the second case, except the temperature was increased to 75°C in the reservoir and 45°C for the membrane cells.

All of these experiments were performed at 1800 kPa operating pressure and with a circulation rate of 0.5 L/min. The pressure drop throughout the cells at this flowrate was less than 30 kPa, and constant operating pressure was assumed.

Results and Discussion

The characterization of the membranes is given in Table II with typical KI solution experimental results. The degree of separation is considered inversely proportional to the pore size of the membranes. This assigns the cellulose acetate (CA) membrane the smallest pore size, and the nitrocellulose (NC) or the polytetrafluorethylene (PTFE) membranes the largest pore sizes. It should be noted that heat treatment usually reduces the pore size of polymeric membranes. A wide range of pore sizes was made available for this study in order to evaluate membrane facility for the removal of dissolved species from the permeate. The penalty for small pore sizes is reduced permeation rate, although this is mitigated by the "population" of pores which is considered a function of the membrane fabrication method, membrane material morphology, and molecular weight.

In order to compare the variation of permeation rates with time of the different membranes, the (PR) value of Table II for each membrane was chosen to represent the 100% permeation rate. Subsequent permeation rates for each membrane during the emulsion experiments are expressed as a percentage of the (PR) value of Table II. These are presented in Fig. 2 for both the 25°C and 45°C operating temperatures. This figure demonstrates the large decrease of permeation rate that occurs when the bitumen/water/mineral emulsion is broken on membrane surfaces. The exception to this is the regenerated cellulose (RC) membrane permeation rates. While a slight decrease of permeation rate was observed for the two RC membranes during the 25°C experiment, they did not suffer the drastic decreases observed for the other membranes. This ability of cellulose membranes to resist organic component fouling has been reported elsewhere (16). However, operating at 45°C caused a large decrease in permeation rate for all membranes. Because of this, this temperature is not desirable for the long term use of the membrane types studied.

In general, the appearance of the permeates was clear with varying degrees of amber colour, and the intensity was approximately inversely proportional to the KI solution. There was no detectable odour, and no evidence of sediment even on long standing, although there was evidence of long-term bacterial activity. As well, the

Table II. Membrane Performance at 1800 kPa and 24 °C with Dilute KI Solution

MEMBRANE NO.:	1	2	3	4	5	6
Material[a]	NC	PTFE	RC	NC	RC	CA
Pore size, m × 10^{10}	100	200	50	50	50	-[b]
Supplier	Sartorius	S and S[c]	S and S	S and S	S and S	CANMET[d]
Heat treatment						
°C × 10 min	80°C	80°C	-[e]	90°C	80°C	-[e]
(PWP), avg. kg/m^2/day	1.208	529	98	936	93	735
(PR), kg/m^2/day	1.202	475	102	955	96	722
f, %[g]	7.3	7.3	12	12	20	32

[a]NC = Nitrocellulose; PTFE = Polytetrafluorethylene;
RC = Regenerated Cellulose; CA = Cellulose Acetate.

[b] = undetermined

[c]S and S = Schleicher and Schuell

[d]CANMET = Fabricated in-house

[e]No heat treatment was applied

[g]Calculated separation (by eq 1) for a feed of 2050 ppm of KI

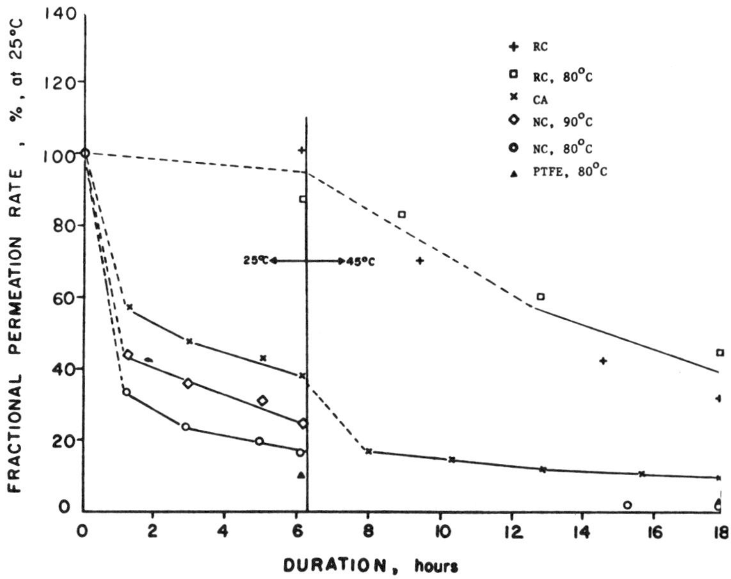

Figure 2. Variation of permeation rate with run duration.

permeate supported a large foam formation when it was shaken. More specific analyses are shown in Table III and the limits for single pass boiler feed water of Table I are included for comparison. In all cases, TDS passes the limits set in Table I. However, the PTFE membrane exhibited unusual two-phase permeation for the 45°C experiment, and demulsified bitumen could be observed in the permeate as 2 to 3 mm diam droplets. This phenomena could be the result of the large pore size of the membrane combined with the strong preferential sorption of the bitumen on the PTFE surface.

To confirm that bitumen did not pass through to the permeate (the PTFE case excluded), attempts were made to extract the permeate with toluene. There was no evidence of the transfer of coloured material from the aqueous to the organic phase, nor was there any visible residue left by the evaporation of the toluene suggesting that the material is highly polar. To further evaluate the composition of the various liquid streams, infrared analyses of the aqueous samples for the cellulose acetate case was investigated by their evaporation onto polycrystalline zinc sulphide plates (IRTRAN, Perkin-Elmer) at 50°C with subsequent drying at 105°C in a vacuum oven for one half hour.

The infra red spectra of the bitumen from the emulsion as received and of evaporation residue of the permeate are shown in Figure 3. The former is characteristic of bitumen while the latter bears little resemblance to the former. Thus, the material in the permeate which appears to be the surfactant that stabilized the emulsion is very different to the bitumen. Acidification of the permeate to pH2 caused the formation of precipitates. Studies on characterizations of this material are in progress.

An unexpected effect of the membrane process was the separation of the emulsion in the reservoir to form two layers: bitumen rich and water rich. The bitumen-rich layer floated as scum on the surface. This layer was retrieved and analyzed for water content by Dean and Stark analysis. The bitumen content was 23% at 25°C and 35% at 45°C. Since this effect was unexpected, the retrieval of the bitumen-rich layer may have incorporated water-rich emulsion into it. No attempt was made to measure the rate of accumulation of this layer. However, the instability of the remaining water-rich emulsion was evident as it continued to break down at room temperature to a very dark bitumen-rich layer on the bottom and sandy/muddy brown water layer with visible large particles. The original emulsion as received did not segregate to the same extent.

On this basis, it is possible that the polar material observed in the permeate was the stabilizing agent for the bitumen/water/mineral emulsion, and its removal would cause this segregation. This effect has been described by Gewers (6), where the increase in stability of Athabasca bitumen is enhanced by the presence of asphaltenes and other polar material, as observed by infrared analyses.

Analysis for the concentration of selected metals in the permeates is reported in Table III. Low concentrations of these metals were observed with the exception of Si, although the CA membrane met the requirements of Table I and most of the others were close to this requirement. For Fe, the detection limit for the ion inductively coupled plasma (ICP) analysis is 0.118 ppm, which is greater than the limit given in Table I. Although the exact concentration of Fe for these permeates was not measured, it is at least in the same range as

Table III. Analysis of Permeate for the Membranes of Table II

MEMBRANE NO.:	1	2	3	4	5	6	Limit from Table I	Original emulsion
Material/ heat treatment	NC/ 80°C	PTFE/ 80°C	RC	NC/ 90°C	RC/ 80°C	CA	-	
24°C Run								
TDS, ppm[a]	635	740	506	738	661	371	8,000	55,000
Ca, ppm[b]	0.07	-	0.07	0.07	0.07	0.07	-	1.8[e]
Mg, ppm[c]	0.15	-	0.15	0.15	0.15	0.15	-	0.45[e]
Fe, ppm[d]	0.12	-	0.12	0.12	0.12	0.12	0.05	4.7[e]
SiO_2, ppm	63.3	-	59.8	63.2	56.6	39.4	50	157[e]
Suspended solids	no	no	no	no	no	no	0 mg/L	yes
Suspended oil	no	no	no	no	no	no	1.0mg/L	yes
pH	7	7	7	7	7	7	-	7
Total hardness as $CaCO_3$, ppm	0.18	-	0.18	0.18	0.18	0.18	0.5	4.6
TOC, ppm of C	45.6	76.8	81.4	69.0	65.0	48.3	-	4%[f]
45°C Run								
TDS, ppm[a]	431	NA	771	-	755	173	8,000	55,000
Ca, ppm[b]	0.07	1.62	0.07	-	0.07	0.07	-	1.8[e]
Mg, ppm[c]	0.15	0.52	0.15	-	0.17	0.15	-	0.45[e]
Fe, ppm[d]	0.12	2.75	0.12	-	0.12	0.12	0.05	4.7[e]
SiO_2, ppm	59.6	85.3	54.9	-	58.2	26.4	50	157[e]
Suspended Solids	no	-	no	-	no	no	0 mg/L	yes
Suspended oil	no	visible	no	-	-	no	1.0 mg/L	yes
pH	7	7	7	-	7	7	-	7

Continued on next page.

Table III. Continued

MEMBRANE NO.:	1	2	3	4	5	6	Limit from Table I	Original Emulsion
Material/ heat treatment	NC/ 80 °C	PTFE/ 80 °C	RC	NC/ 90 °C	RC/ 80 °C	CA		
Total hardness as $CaCO_3$, ppm	0.18	4.1	0.18	-	0.18	0.18	0.5	4.6
TOC, ppm of C	50.1	-	63.9	-	61.3	35.5	-	4%[f]

a TDS was calculated as residue at 90°C

b The minimum detectable limit for Ca is estimated to be 0.07 ppm. The corresponding limit for hardness as $CaCO_3$ is 0.175 ppm.

c The minimum detectable limit for Mg is estimated to be 0.152 ppm.

d The minimum detectable limit for Fe is estimated to be 0.118 ppm.

e Analysis by ICP on fused ash.

f Approximately 4% bitumen.

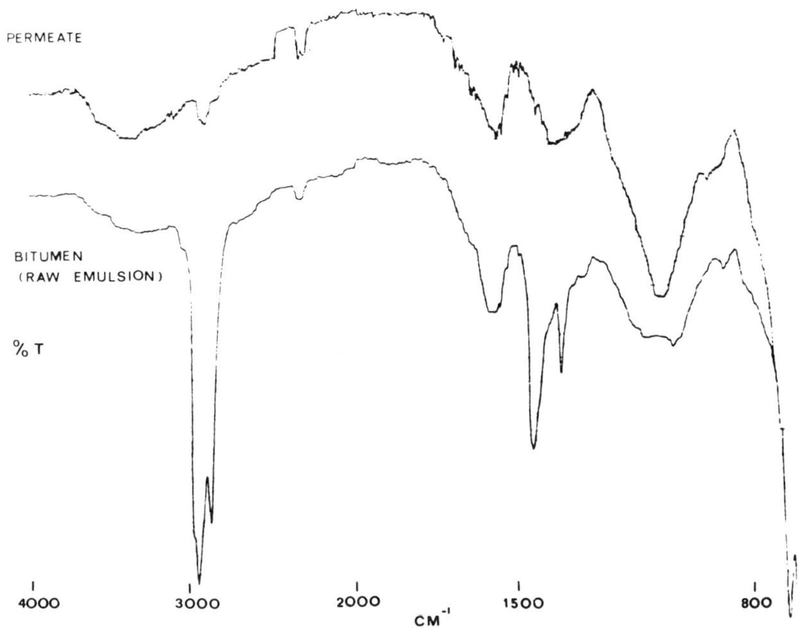

Figure 3. Infrared spectra of original emulsion and the CA membrane permeate (25°C experiment).

the limit in Table I; it is assumed that further refinement of the process will give satisfactory levels.

Conclusions

The use of porous membranes can be considered for the production of boiler feed quality water from bitumen/water/mineral emulsion obtained from the Texaco Athabasca site. The use of a continuous membrane process is possible as demonstrated by this work. An associated difficulty caused by the emulsion is membrane fouling, although the RC membranes do not foul to the same extent.

Acknowledgments

The authors wish to acknowledge the assistance of A. Martineau, CANMET for the ICP analysis, M. Curtin, Waterloo Coop student for the TDS analysis, H. Barber for the infrared analyses, T. Matsuura, NRC Canada for the total organic carbon analyses, and the technical assistance of C. Nader. Also, the cooperation of Texaco Canada Resources Ltd. is gratefullly acknowledged.

Literature Cited

1. Nagendran, J., Hrudey, S.E., 7th National Conference on Energy and the Environment, Phoenix, Nov. 30 to Dec. 3, 1980, p. 66.
2. VanHam, N.J., Behie, L.A. Svrcek, W Y., Can J Chem Eng, 61:541; 1983.
3. Young, K.B., Private Communication, 1983.
4. McCoy, D.R., Gipson, R.M., Young, K.B., U.S. Patent No. 4,321,146, Mar. 23, 1982.
5. Farnand, J.R., NRC Canada Report, Chem. Div. No. C1090-815, Dec. 16, 1981.
6. Gewers, C.W.S., J Can Pet Technol, 7:85; (1968).
7. Cannon, L., Private Communication, 1983.
8. Langley, M., Private Communication, 1983.
9. Fulton, C.W., CANMET Report "Steam generation with low-grade fuels and produced water for in-situ oil recovery", CH2M Hill Canada Ltd., Oct. 1982.
10. Farnand, B.A., Poirier, M.A., CANMET Report ERP/ERL 83-15(CF), May 1983.
11. Spatz, D.D. In "Synthetic Membranes", Vol II; Turbak, A., Ed.; ACS SYMPOSIUM SERIES No. 154, American Chemical Society: Washington, D.C., 1981; p. 221.
12. Quemeneur, F., Schlumpf, J.P., Entropie 16:22; 1980.
13. Dhwan, G.K., in "Waste treatment and utilization", Young, M.M. and Farqhuar, G.J. Editors, p. 189, 1976.
14. Sourirajan, S. In "Synthetic Membranes", Vol. 1; Turbak, A., Ed.; ACS SYMPOSIUM SERIES No. 153, American Chemical Society: Washington, D.C., 1981; p. 11.
15. Boughman, G.L. "Synthetic fuels data handbook", 2nd Ed. Cameron Engineers Inc., p. 274; 1978.
16. Farnand, B.A., PhD Thesis, University of Ottawa, Canada 1983.

RECEIVED March 26, 1985

21

Treatment of Paper-Plant Wastewater by Ultrafiltration

A Case History

H. OKAMOTO[1], M. MIZUHARA[2], Y. NUMATA[2], K. NAKAGOME[3], T. OCHIUMI[3], and T. KURODA[3]

[1]Taio Paper Industrial Company, Ltd., 2-60, Kamiya-cho, Iyomishima, Ehime, Japan
[2]Mitsubishi Kakoki Kaishi, Ltd., 4-28, 1-Chome, Mita, Minato-ku, Tokyo, Japan
[3]Nitto Electric Industrial Company, Ltd., 1-2, 1-Chome, Shimohozumi, Ibaraki, Osaka, Japan

The design and operation of the world's largest tubular membrane ultrafiltration unit at a Japanese Kraft paper plant is described. The unit treats the plant's KP-El effluent stream. By reducing the effluent load to the plant's activated sludge system, it has brought the plant into compliance with environmental regulations. The unit treats more than one million gallons of feed per day to produce a concentrate stream of 52,000 gallons per day that contains 79% of the COD content of the feed. An average membrane flux of 57 gallons/ft^2/day has been maintained.

The Taio Pulp Company has recently installed the world's largest tubular membrane ultrafiltration unit at its paper plant on Shikoku Island, Japan. The plant was designed by Mitsubishi Kakor Kaisha, Ltd. Company and uses Nitto Electric Industrial Company's tubular membranes and modules.

The Taio paper plant produces approximately 3,000 tons of paper per day. During the production process, approximately 220,000 tons (58 million gallons) of effluent water are also produced. In the past, all of the plant's effluent streams were pooled and treated sequentially by conventional sedimentation, activated sludge, coagulation precipitation, and sand filtration. However, recently the government's environmental discharge regulations were changed. The new regulations require a discharge of pH of 5.8 to 8.0, suspended solids (SS) of less than 38 ppm, and chemical oxygen demand (COD) of less than 80 ppm. The old treatment processes had difficulty in meeting these discharge limits, particularly during the summer months when high temperatures reduced the efficiency of the activated sludge processes. Membrane processes have been suggested as a solution to this type of problem by a number of workers, (1-3) and thus a membrane ultrafiltration system was developed and installed.

To reduce the load on the activated sludge plant, an ultrafiltration unit was installed to treat one of the most polluted streams from the plant, the KP-El stream from the Kraft pulping processes.

0097-6156/85/0281-0273$06.00/0

This stream has a COD of 1250 to 1900 ppm and a volume of 0.8 to 1.0 million gallons per day. It was anticipated that a reduction of the COD of this stream would bring the plant into compliance.

Pilot Plant Results

The composition of the initial KP-El feed solution is shown in Table I. The solution contains a total of 0.2 to 0.3% dissolved and suspended solids; this solution must be concentrated tenfold to 2 or 3%. Because of the highly fouling nature of the feed solution, only tubular ultrafiltration membranes appeared to be suitable for this application. The high temperature and pH of the feed also mandated the use of chemcially resistant membranes. Thus, Nitto's 3508 polysulfone membranes were chosen. These membranes have a nominal molecular weight cutoff of 8000 and can be used at temperatures of up to 60 degrees centigrade.

Table I. Composition of Taio Paper KP-E1 Effluent Stream after Treatment with the Nitto NTU-3508 Membrane

	Feed	Concentrate	Permeate
Temperature (°C)	45-55	45-55	45-55
COD (ppm)	1,250-1,900	18,000-32,000	300-700
pH	10-11.5	10-11.5	10-11
SS (ppm)	20-110	300-1,800	-

Figure 1 shows gel permeation traces of the feed, permeate, and concentrate from the 3508 membrane. The feed solution contains four peaks labeled A, B, C, and D in this figure. Experience has shown that substances represented by B are highly aromatic and are thus particularly difficult for the activated sludge plant to digest. Ulfiltration through the 3508 membrane produces a permeate that is essentially free of the substances represented by A and B peaks, which are retained in the concentrate solution.

The effect of UF treatment followed by activated sludge and coagulation on the product water quality of the system is shown in Table II, compared to normal activated sludge coagulation treatment. As this table shows, the conventional process is rather inefficient at removing COD, and only 51% of the total COD is removed. However, ultrafiltration followed by conventional treatment removes 97% of the COD. This improvement is obtained by the ultrafiltration membrane. In addition, the low molecular weight solutes that permeate the membrane are more easily digested in the activated sludge process than are the larger molecules retained by the membrane. As a result, the efficiency of the activated sludge process also increases.

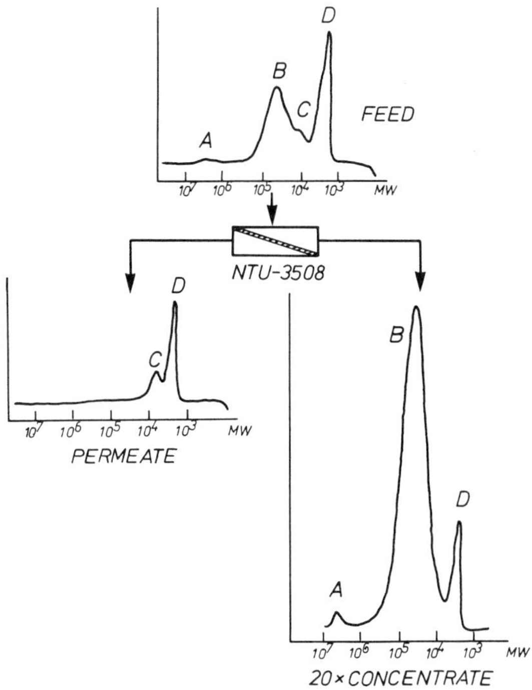

Figure 1. Gel permeation chromatography plots obtained on the KP-E1 feed, permeate, and concentrate (20x) solutions after ultrafiltration with Nitto polysulfone NTU- membrane.

Table II. COD Removal from the KP-E1 Stream by Conventional Treatment with and Without Ultraflitration

Treatment Step	Total COD Removed (%) Conventional Treatment	UF + Conventional Treatment
Ultrafiltration	--	83
Activated Sludge	40	13
Coagulation Plus Sedimentation	11	1
Total	51	97

Design of the Plant

Figure 2 shows a plot of water flux vs. concentration factor for the KP-E1 feed solution. As this figure shows, the flux through the membrane falls substantially as the feed solution becomes more concentrated. Several design options are possible in this situation. These are illustrated in Figure 3. The most efficient design is the single-pass system in which the solution is pumped through a number of modules in series. The flux obtained is the average value of the flux vs. concentration factor curve shown in Figure 2. This method of operation is commonly used in reverse osmosis plants, where 5 to 20% of the feed solution is removed as the solution passes through the module. Therefore, only 5 to 10 modules in series are normally required to produce the appropriate concentration. In ultrafiltration systems, however, only 0.1 to 1% of the feed solution is removed per pass through a module. Thus, an impossibly large number of modules would be required to achieve the desired concentration.

Many small ultrafiltration systems therefore operate in the one-stage feed-and-bleed mode shown in Figure 3b. These systems only contain a few modules, and, thus, the feed solution in the loop is allowed to build up until it reaches the required final concentration. This is an inefficient mode of operation because the membrane is always in contact with the most concentrated solution.

A final mode of operation is a multi-stage feed-and-bleed system, as shown in Figure 3c. In this system, the concentration is allowed to build up to the maximum concentration in stages. The Taio paper plant is a very large installation, and it was economical to consider this type of multi-stage design.

The system installed consists of two lines of six stages. Five stages of each line are in use at any one time while the sixth stage is being cleaned or maintained. The design of each individual stage is shown in Figure 4. Each stage consists of seven banks of membranes, each bank consisting of seven or eight modules in a series. Thus, each stage contains 49 to 56 modules. The design of the individual modules is shown in Figure 5. Each module consists of 18 ½-inch tubular membranes manifolded inside a stainless steel tube and end pieces. The surface area of each module is 247 sq. ft.

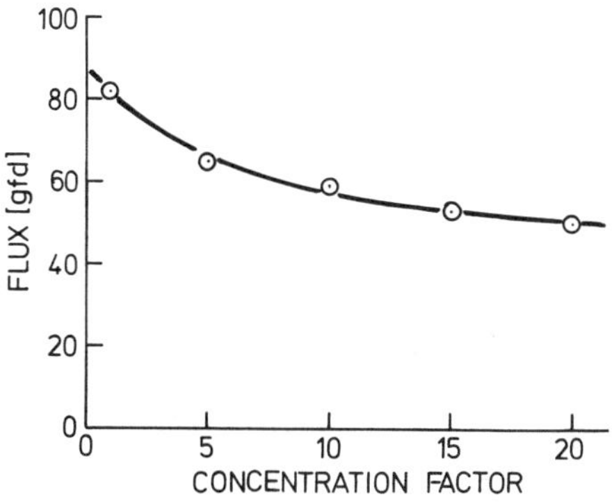

Figure 2. Flux vs. concentration factor for NTU-3508 membranes operating on KP-E1 feed solutions. The average operating pressure was 115 psi.

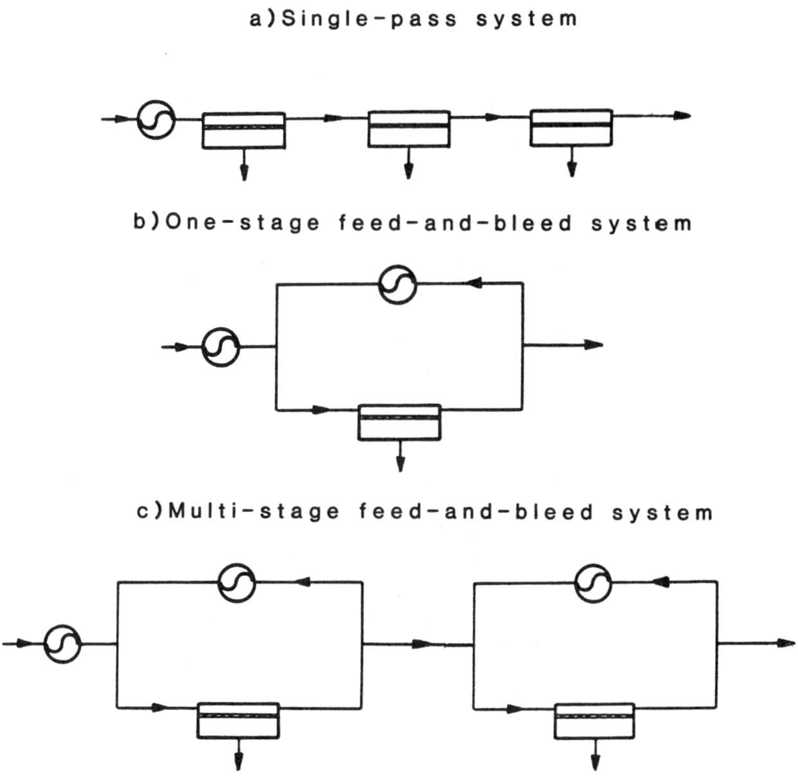

Figure 3. System design options.

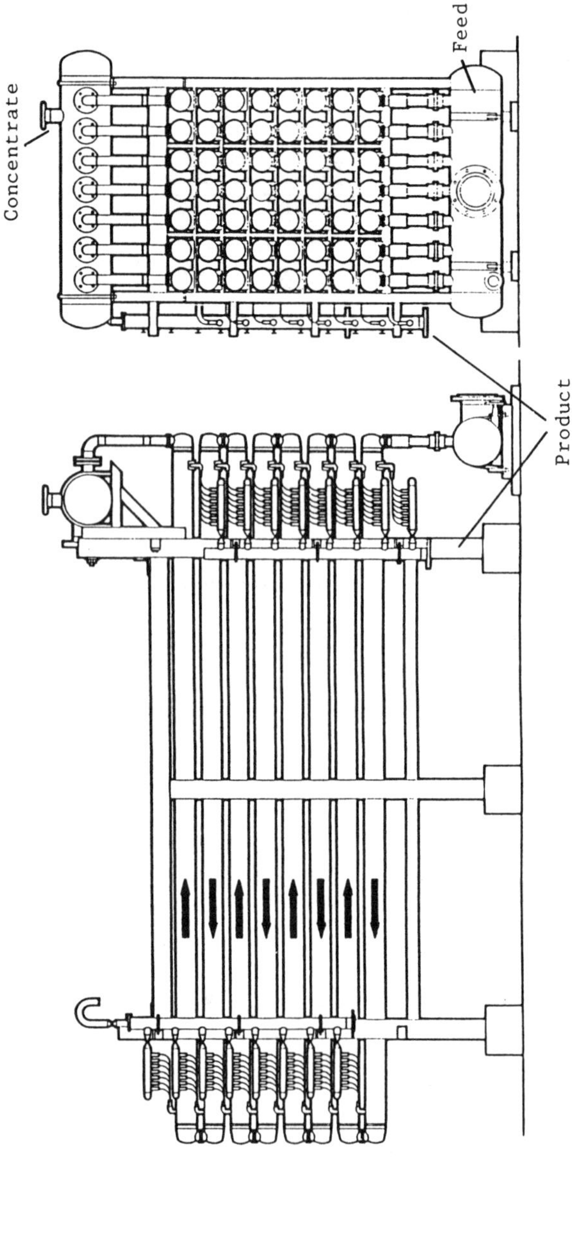

Figure 4. Design of a single 56-module stage at the Taio plant.

The layout of the entire system is shown in Figure 6. The feed solution is first screened to remove wood fibers and very large particulates. A 370 gpm pump is then used to pressurize the solution to 90 psi. Several 650 gpm pumps are used to recirculate the solution through each stage. The UF permeate is collected and sent to the treatment facility, while the UF concentrate is sent to the incinerator. A portion of the permeate is sent to a holding tank where it is mixed with NaOH and detergent solution and used for periodic cleaning of the membrane.

Operating History

The Taio plant has now (May, 1982) been in full operation for two years. Very few problems have been experienced. The average performance of the plant vs. the design performance is shown in Table III. Because the COD concentration in the feed solution has proved to be considerably higher than expected, the average flux of the unit has been somewhat lower than expected. However, an average flux of 57 gallons/ft^2/day (gfd) is being maintained, and no problems with membrane lifetime are expected. The COD removal has remained good and, in fact, is slightly above the design value. Some plugging of the feed solution screen by pulp fibers was experienced, but periodic cleaning and backwashing of the screen eliminated this problem.

Using this plant, the Taio paper company has been able to maintain the composition of its effluent streams well within the discharge limits of environmental regulations.

Table III. Design and Actual Performance of the Taio Plant

		Design	Actual
Feed	(gfd)	1,050,000	980,000
Concentrate	(gfd)	53,000	52,000
Volumetric Concentration Factor		20	16.5
COD Feed	(ppm)	1,250	1,900
COD Concentrate	(ppm)	18,500	24,700
COD Removal	(%)	75	79
Total COD Removal by Plant	(tons/day)	3.75	5.5
Average Flux	(gfd)	63	57

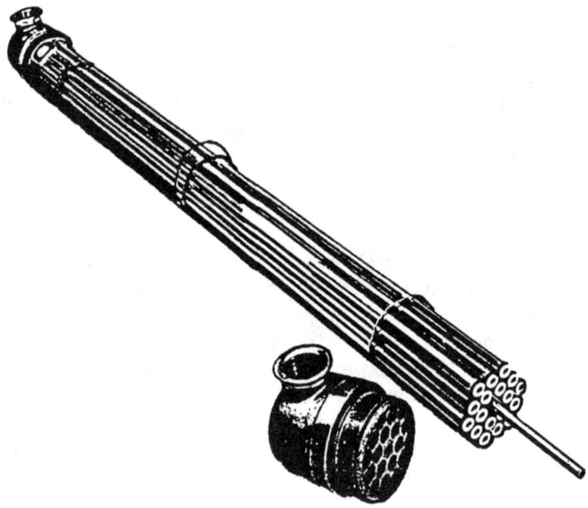

Figure 5. Design of a cut-away UF module of the type used at the Taio plant.

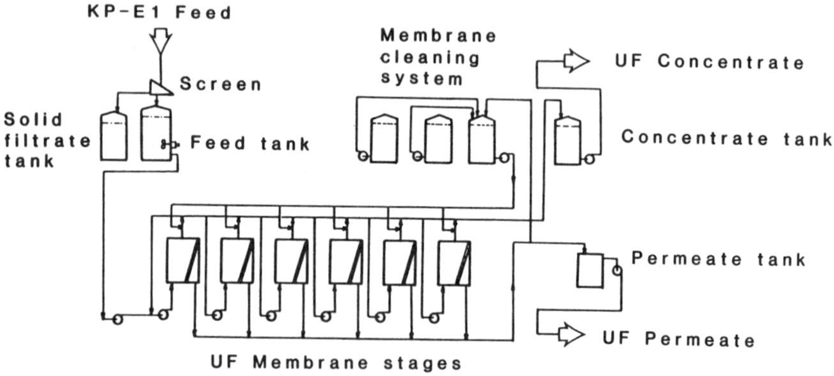

Figure 6. Flow diagram of the Taio UF plant.

Literature Cited

1. Wiley, A.J.; Scharpf, K.; Bansal, I.; Arps, D. Tappi 1972, 55, 1671.
2. Leitner, G.F. Tappi 1972, 55, 258.
3. Ahlgren, R.M. In "Industrial Processing with Membranes"; Lacey, R.E.; Loeb, S., Eds.; Wiley-Interscience, New York, 1972.

RECEIVED February 22, 1985

22

Concentration and Recovery of ε-Caprolactam from the Process Waste Stream

Y. NAKAGAWA[1], Y. NOGUCHI[2], M. KURIHARA[3], N. KANAMARU[1], and T. TONOMURA[1]

[1]Technical Development Department, Toray Industries, Inc., 3-Chome, Sonoyama, Otsu, Shiga 520, Japan
[2]Technical Department, Nagoya Plant, Toray Industries, Inc., 9-1 Ooe-cho, Minato-ku, Nagoya, Aichi 455-91, Japan
[3]Pioneering Research and Development Laboratories, Toray Industries, Inc., 3-Chome, Sonoyama, Otsu, Shiga 520, Japan

PEC-1000 membrane exhibits excellent RO performance in the field of separation for water-soluble organic compounds with low-molecular-weight as well as in the seawater desalination. After the basic test and the field test, a commercial plant for concentration and recovery of ε-caprolactam has been on continuous operation for two years. RO system is constructed as so-called "Christmas Tree" and 250 t/d of feedwater containing less than 0.1 % of ε-caprolactam is concentrated to 5 % and returned to the "Nylon-6" production process through evaporation process. It was most important to take care of fouling for designing and operating the RO systems since the feedwater contains considerable amount of suspended materials. Flux decline is controlled by periodical cleaning with anionic detergent.

On the concentration and recovery of valuable materials from their aqueous solutions, reverse osmosis (RO) technology has been expected to be employed for the purposes not only of saving energy to recover it but also of environmental pollution problems. Moreover, it is considered to be able to separate organic compounds of remarkably lower molecular-weight than with ultrafiltration (UF). However, the practical applications of this technology by means of conventional RO membranes, so far, have been limited due to the insufficiently low solute rejection of low molecular-weight organic compounds such

0097-6156/85/0281-0283$06.00/0

as ethanol or acetic acid. Therefore, it has also been necessary to develop a membrane of very high performance to overcome this limitation in this work.

Frang et al. reported that the separation of low-molecular-weight organic compounds including various functional groups in dilute aqueous solution (0.1 % by weight were not so effective as that of inorganic salts for all the membranes tested (1). Since then, many newly developed RO membranes have been examined for the selective separation performance of valuable materials in aqueous solutions by RO technology.

PEC-1000 membrane has excellent RO performance (especially in solute rejection), which may lead to wider applications in both fields of seawater desalination and concentration and recovery of valuable materials (2-6). In this paper, basic examination and application to RO system design and operation for the concentration and recovery of ε-caprolactam are reported as one of the successful examples for the latter case.

Steps toward Planning RO Systems for Recovery of Valuable Materials

Before planning RO systems for the concentration and recovery of valuable materials, following examinations and estimations are to be employed.

1st step. Basic separation performance of the objective solute.
2nd step. Effects of operating conditions on RO performance.
3rd step. Durability test.
4th step. Investigation on pretreatment and/or cleaning.
5th step. Designing of RO system after feasibility investigation.

Basic RO Performance of PEC-1000 Membrane

From the comparative RO performance data of PEC-1000 membrane and several conventional RO membranes to the solutes such as ε-caprolactam, dimethylformamide and dimethylsulfoxide, PEC-1000 membrane is known to be by far superior to NS-200 membrane as well as other conventional membranes (2,5). Pusch et al. examined the RO performance to benzyl alcohol and 1,4-dioxane as well as sodium chloride with various kinds of RO membranes (7) and reported the excellent rejection by PEC-1000 membrane. Further, Ohya et al. reported the possibility for concentration of ethanol (8,9), ethyleneglycol (10) and other solutes by RO technology using PEC-1000 membrane. Similarly the performance data of PEC-1000 membrane and those in the literatures (11-13) of the other commercialized membranes were compared and those of PEC-1000 membrane showed high solute rejections even to various kinds of water-soluble organic compounds of low-molecular-weight such as ethanol, phenol, acetic acid, and ε-caprolactam (4-6). Basic RO performance of PEC-1000 membrane for various kinds of organic compounds in aqueous solutions are tabulated in Table I. The membrane rejects quite sufficiently many of organic solutes and it exhibits its potential application to a variety of fields where the application of conventional RO membranes are recognized as quite impractical.

Based on the basic performance of PEC-1000 membrane for various solutes in aqueous solution, long-term field tests were conducted on the PEC-1000 membrane elements for ε-caprolatam, acetic acid and ethylene glycol solutions (3,4). In all cases, the RO performance remained stable for about one year without the deterioration of membrane performance. PEC-1000 membrane is not resistant to oxidative substances, where dissolved oxygen (DO) as well as chlorine affects the membrane (3). Therefore, it is very important to remove such oxidative substances from a viewpoint of durability of the membrane. Chlorine and DO can be easily removed by reducing reagents such as sodium bi-sulfite (SBS) or sodium sulfite. DO can be also removed by physical method such as vacuum deaeration. The level of DO in the latter two cases (acetic acid and ethylene glycol) is controlled below 1 ppm not by SBS dosing but by bubbling and sealing with nitrogen gas.

Background of RO Plant Planning for ε-Caprolactam Concentration

As shown in Table I, ε-caprolactam, which is the raw material or monomer of "Nylon-6", is separated or rejected over 99.9% by PEC-1000 membrane. Therefore, the concentration and recovery of ε-caprolactam by RO technology using PEC-1000 membrane elements was expected to be practical. There is a limit of concentration to be concentrated from the viewpoint of osmotic pressure which, for example, is estimated to be about 20 atms at 10% by weight of ε-caprolactam solution at 35°C. Consequently, it is necessary to apply RO technology in combination with other separation processes such as evaporation to obtain pure ε-caprolactam. There are several kinds of aqueous ε-caprolactam solutions originating from the processes of "Nylon-6" production as follows:

A. Water extracts from nylon tips or nylon fibers.
B. Distillates from the evaporator for the recovery of ε-caprolactam from the water extracts described in case A.
C. Distillates from the evaporator for the production of ε-caprolactam.
D. Waste streams from other processes for the production of ε-caprolactam.

The solute, ε-caprolactam in case A is concentrated by

Table I. Basic RO Performance of PEC-1000 Membrane for Various Organic Solutes in Aqueous Solutions (Evaluated at 56 Kg/cm^2, 25 °C)

Functional Group	Solute	(Molecular Weight)	Performance: Rejection (%) *1)	Performance: Flux ($m^3/m^2 \cdot day$)	Operating conditions: Concentration (%)	Operating conditions: pH
Alcohol	Methanol	(32)	41	0.38	5	6.9
	Ethanol	(46)	97 *4)	0.23	6	6.9
	iso-Propanol	(60)	99.5	0.32	5	6.5
	n-Butanol	(74)	99.4 *2)	0.26	4	7.0
	Ethylene glycol	(62)	94 *2)	0.36	5	6.8
	Propylene glycol	(76)	99.4 *2)	0.12	10	7.0
	Glycerine	(92)	99.8 *2)	0.38	5	7.0
	Phenol	(94)	99.0 *2)	0.24	1	5.2
Carboxylic acid	Acetic acid	(60)	86 *3)	0.38	6	2.6
	Propionic acid	(74)	98 *2)	0.31	10	3.4
	Oxalic acid	(90)	99.1 *2)	0.69	0.5	1.8
Aldehyde	Formaldehyde	(30)	66 *2)	0.21	6	5.3
Ketone	Acetone	(58)	97	0.29	4	6.7
	Methyl ethyl ketone	(72)	98	0.21	4	6.6
Ester	Ethyl acetate	(88)	99.2	0.18	4	6.8
	n-Butyl acetate	(116)	99.6	0.38	1	6.9
Ether	Tetrahydrofuran	(72)	99.8	0.28	5	6.7
	1, 4-Dioxane	(88)	>99.9 *2)	0.40	5	6.6
Amine	Ethylene diamine	(60)	99.5 *2)	0.08	5	12.1
	Aniline	(93)	95 *2)	0.11	1	8.3
	n-Butylamine hydrochloride	(109)	>99.9 *2)	0.13	1	7.0
Amide	Urea	(60)	85 *2)	0.56	1	6.9
	N, N-dimethylformamide	(73)	98	0.32	5	6.5
	N, N-dimethylacetamide	(87)	99.6	0.30	5	6.5
	ε-Caprolactam	(113)	>99.9 *2)	0.38	5	6.5
Sulfoxide	Dimethylsulfoxide	(78)	99.6	0.34	5	6.4

*1) Gas Chromatography Analysis *2) Total Organic Carbon Analysis *3) Titration Analysis *4) Refractometric Analysis

evaporation, because concentration of ε-caprolactam in the process is rather high (5% by weight) and its solution contains the oligomeric material which can be separated with difficulty. ε-Caprolactam in cases C and D is recovered or decomposed by activated sludge method (ASM) according to the quantity and the concentration of ε-caprolactam in the water to be treated. The concentration and recovery of ε-caprolactam in case B is the object in this paper.

Preliminary Result and Discussion

Examinations with Reagent ε-Caprolactam and Actual Process Waste Stream. Ohya et al. examined the relationship between RO performance and ε-caprolactam concentration in relation to the variation of operating pressure with reagent ε-caprolactam solutions (14, 15).

Aqueous ε-caprolactam solution composed of reagent or actual process water was evaluated by a membrane in this order in the closed loop system in the open air. Figure 1 shows the test results in which ε-caprolactam contents are kept at 0.15% by weight. The latter solution contains very small amounts of other components which do not affect the osmotic pressure of the solution. In spite of almost the same osmotic pressure exhibited by both solutions, permeate flow rate (flux) was rapidly decreased in the case of actual process water and it was about the half of the initial value after 50 hours operation. The feedwater quality or fouling index was evaluated by MF value, which means the time in second to pass through the membrane filter (0.45 μm) at a reduced pressure. The procedure and conditions for MF value measurement are illustrated in Figure 2. Changes of MF value are shown in Figure 3 in this series of experiments, which gradually increased in relation to the decline of flux. However, during the evaluation with ε-caprolactam reagent, both flux decline and increase of MF values are slight, therefore, some species in the actual process water must affect the feedwater quality and the RO performance. The actual process water contains a little organic materials such as oil, disperser, surface active agent and inorganic materials such as sulfates or phosphates. These materials are considered to be sources of nourishment of microorganisms. The difference of MF value between the feedwater using ε-caprolactam reagent and the actual process water may be explained by the presence of microorganisms in the latter case.

Improvement of Water Quality (MF Value) of the Feedwater to be Treated. As described above, the remarkable increase of MF values may be due to the growth of microorganisms. Table II shows the variation of MF values in relation to the elapsed time. When the actual

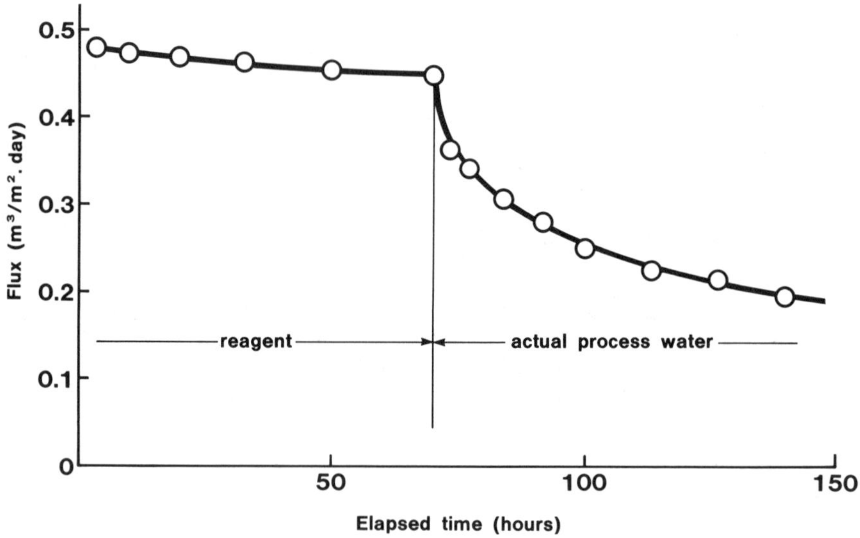

Figure 1. Flux decline curves using ϵ-caprolactam reagent or actual process water with PEC-1000 membrane. Operating conditions: 56 Kg/cm^2; 25 °C; 0.15% solution.

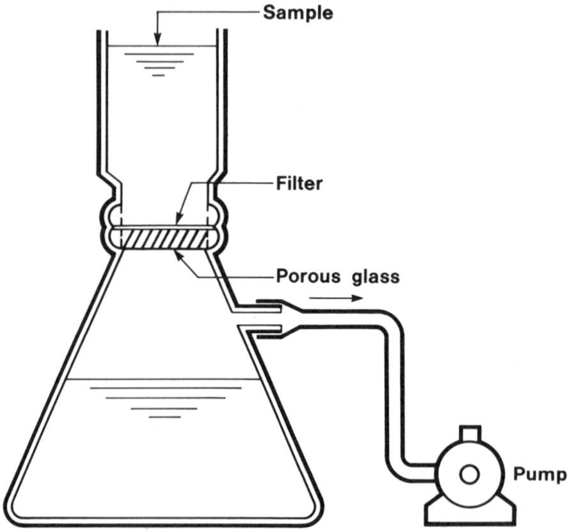

Figure 2. Measuring procedure of MF value. MF value means the time (second) required to filtrate all the sample. Measurement conditions: filter (Millipore HAWP 047 00), 0.45 μm: effective membrane area, 17.3 cm^2; pressure, -50mmHg; temperature, 25 °C; and water sample, 100 ml.

process water was stored in the open air, MF values became higher. On the other hand, when the same actual process water was stored in the nitrogen gas atmosphere and DO values were kept below 1.0 ppm, MF values remained at the initial value. However, the fouling index value of the actual process water was excessive and worse than that recommended for RO process such as seawater desalination.

Table II. Variations of Water Quality (MF values) of Dilute ε-Caprolactam Solutions Preserved in the Open Air and in the Nitrogen Gas Atmosphere

Elapsed time (hours)	Ratio of MF values to that at the beginning	
	In the open air	In the N_2 gas atmosphere
0	1.00	1.00
50	7.44	1.17
100	21.95	1.39
350	100.00	1.39

Figure 4 shows the field test data using PEC-1000 membrane elements with the actual process waste stream. In this figure, the difference of the variation of flux in relation to the elapsed time between two cases using the actual process water are exhibited, one of which was examined in the open air and the other of which was examined in the atmosphere of nitrogen gas. There is a big difference in the curves.

Effect of the Operating Pressure. As shown in Figure 4, flux at first rapidly and then rather gradually declined. As the operating pressure is one of the effective factors to increase flux, it was increased from 40 Kg/cm^2 to 70 Kg/cm^2. Flux instantly increased. However, the flux rapidly declined to a similar value before operating pressure was increased. Therefore, in the RO process in which the water is not so clean and may involve suspended materials, high pressure operation produces only higher operating cost and there exists optimum operating pressure in the lower range.

Cleaning of the Membrane. As shown in Figure 4, flux gradually declined and there may exist a certain asymptotic value in relation to the feedwater quality. At any rate, the feedwater quality is by far worse than the limitation of normal RO operation. It is important to improve the declined flux by membrane cleaning.

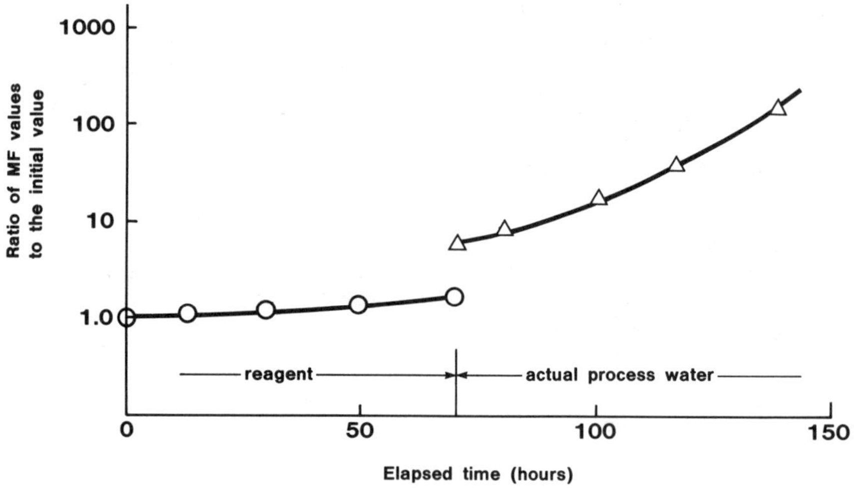

Figure 3. Variations of MF value in the case of ϵ-caprolactam reagent and actual process water.

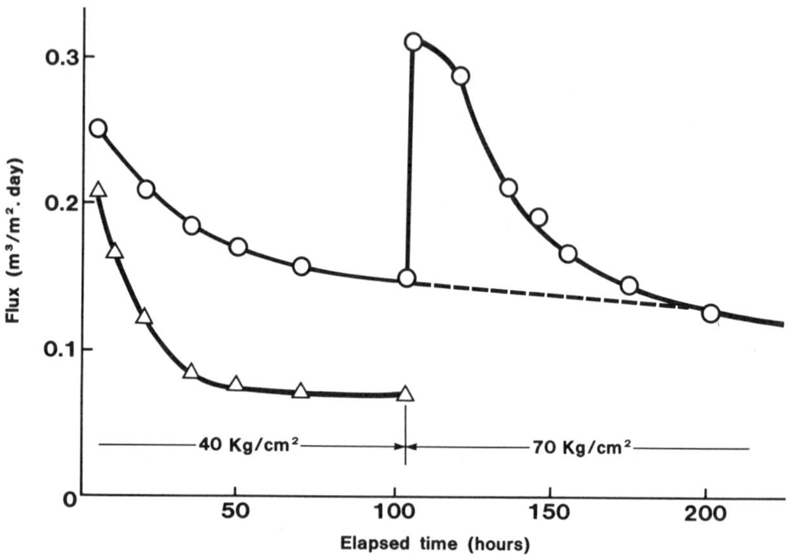

Figure 4. Field test results of ϵ-caprolactam rejection using actual process water on flux variation with PEC-1000 membrane element. Key: O, in the N_2 gas atmosphere; and Δ, in the open air. Operating conditions: 5% ϵ-caprolactam solution; 35 ^{O}C.

Since components of the foulants on the membrane surface are presumed to be organic materials such as ε-caprolactam oligomers, microorganisms and their secretions and oils, several kinds of cleaning agents were examined. The basic cleaning test results using membranes fouled with the actual process water were illustrated in Table III. The performance after the cleaning with that before was compared by the degrees of improvement in flux by using aqueous 3.5 % sodium chloride solution. Using methanol as cleaning reagent was most effective, but it was denied in this case from more practical point of view as there exists a problem in explosion. As a matter of fact, anionic surface active detergent such as dodecyl sodium sulfate (DSS) was considered to be better. Ammonium citrate, which is an effective cleaning agent for metal hydroxides, displayed no effect, which proves that the primary origin of foulants is not metal hydroxides. Rejection of ε-caprolactam was over 99.9% even after cleaning and was not damaged by any cleaning agent.

Table III. Membrane Cleaning Effects of Cleaning Agents on Fouled Membranes by Using Actual Process Water

(Evaluated at 3.5% NaCl-56 Kg/cm^2-25°C, pH 6.5)

Cleaning agents	Flux ($m^3/m^2 \cdot day$)	Flux ratio to the initial value (%)
None	0.08	22
1/100 N NaOH solution	0.010	28
10% ε-Caprolactam solution	0.010	28
2% Ammonium citrate solution	0.012	33
0.5% BIZ solution	0.32	89
0.5% DSS solution	0.32	89
Methanol	0.36	100

(Fouled membranes were soaked into cleaning solutions for 2 days)

As the second step to develop cleaning procedure, the effect of the cleaning time was examined as shown in Table IV with 0.5 % by weight of DSS using the membranes obtained from the membrane element operated for one month with the actual process water. The data shown in Table IV exhibit that three hours are enough for cleaning in this case.

Table IV. Effect of Cleaning Periods by DSS Solution on Membrane Elements Fouled by Actual Process Water

(Evaluated at 3.5% NaCl-56Kg/cm^2-25°C, pH 6.5)

	Accumulated cleaning period (hours)	Flux ($m^3/m^2 \cdot day$)	Flux ratio to the initial value (%)
Before cleaning	0	0.11	40
After first cleaning	3	0.24	85
After second cleaning	6	0.24	85
After third cleaning	9	0.24	85

Cleaning conditions

DSS concentration	0.5% by weight
Temperature	30°C
Pressure	5 Kg/cm^2
pH	6.5
Brine flow rate	40 l/min

Estimation of Membrane Life. RO systems using PEC-1000 membrane elements, as is known from seawater desalination studies, require removal of DO as well as chlorine (3). However, in this case, SBS dosing for removal of DO was not permitted from a viewpoint of the influence to the whole processes for "Nylon-6" production. Therefore, it was important to estimate the membrane life in the presence of DO for economic evaluation. Figure 5 shows the relationship between the rejection of ε-caprolactam and that of sodium chloride with various kinds of deteriorated PEC-1000 membranes, which were intentionally deteriorated by oxidation. This figure shows that the rejection of ε-caprolactam is high enough as long as the rejection of sodium chloride reaches 60-70%. From another experiment which examined the relaltionship between the rejection variations of sodium chloride and DO values, it was concluded that membrane life for the concentration of ε-caprolactam from 0.1 to 5 % lasts one and a half years even at 35°C whenever DO in the feedwater is maintained below 1 ppm.

RO Plant Operation

Process Flow Diagram and RO Plant Size. Process flow diagram of ε-caprolactam including the RO process is shown in Figure 6. Dilute aqueous solution of ε-caprolactam (below 0.1 %) obtained as distillate from the top of the evaporator is carried to the RO process. The solute ε-caprolactam is concentrated to 5 % and returned to the same evaporator, and the permeate is treated as waste water by ASM. The concentrated pure

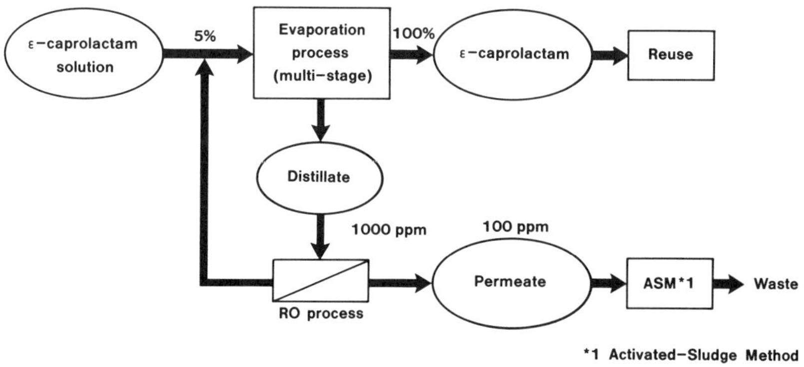

Figure 5. Relationship between ϵ-caprolactam rejection and NaCl rejection with various RO performance of PEC-1000 membranes. Evaluation conditions: 56 Kg/cm^2; 25 ^{o}C; pH 6.5; 3.5% NaCl solution; 5% ϵ-caprolactam solution.

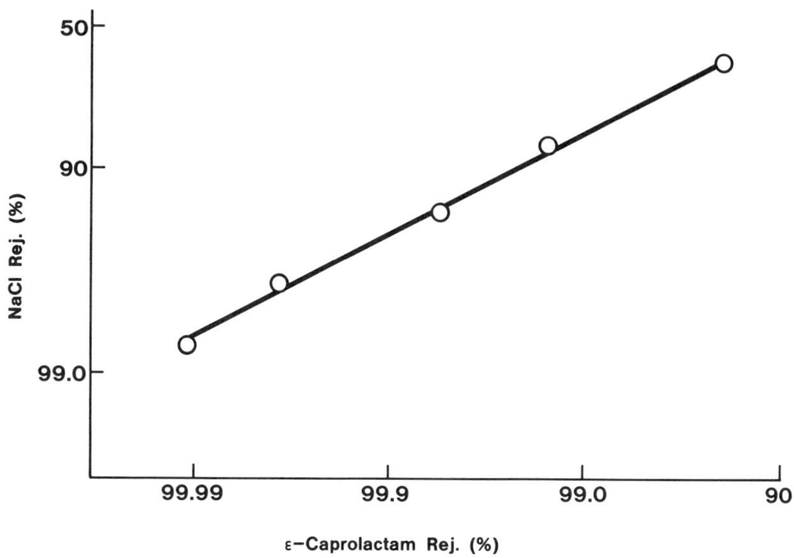

Figure 6. Process flow scheme for concentration and recovery of ε-caprolactam by using combination of RO process and evaporation process.

-caprolactam which is obtained from the bottom of the evaporator, is returned to the process for "Nylon-6" production. The distillate from the evaporator is cooled to 35°C in a heat exchanger and DO in the distillate is kept below 1 ppm by carring inert (oxygen free) gas into the cooler. The RO process for the recovery and concentration of ε-caprolactam from below 0.1 to 5 % by weight is illustrated in Figure 7. The RO system is composed of a three bank system, consisting of eleven, three and one module. Each module consists of four 8" elements (SP-120) in series. Dimension of the RO plant are 5.0 m(L) x 1.9 m (W) x 2.5 m(H) (See Figure 8). Feedwater at 254 m^3/day is treated by this RO system and 4 m^3/day of the concentrate is returned to the evaporator as described above. Therefore, the water recovery ratio is over 98%. As the water recovery ratio is very high and the water quality of the feedwater is very bad and contains suspended solids, double safety filter systems (10 µm and 5 µm) are adopted. Membrane cleaning is carried out by using the low pressure pumps for the ordinary operation with raw feedwater DO in the feedwater is maintained below 1 ppm by sealing with inert gas, therefore SBS for removing DO is not required.

Operation Results. The RO plant has been continuously operated at 40-45 Kg/cm^2 and at 35-40°C for two years. Permeate productivity varies depending on feedwater quality, levels off to a certain value in about one week, and remains at the constant value for about two or three weeks. Therefore, cleaning with aqueous DSS solution is adopted at intervals of two to four weeks, while flushing with the feedwater is carried out once a week. Safety filters of 10 µm and 5 µm are changed three times and twice per month, respectively. The rejection of ε-caprolactam was over 99 % and the concentration of ε-caprolactam in the permeate was kept sufficiently low for two years.

Conclusions

An actual plant for concentration and recovery of ε-caprolactam by RO technology was constructed and has been successfuly operated for two years using PEC-1000 membrane elements. While planning and designing RO process for the concentration and recovery of valuable materials, it is important to examine RO performance with actual process water and it is to be determined from the viewpoint of total cost whether complete pretreatment system is necessary or not. When complete pretreatment system is not adopted, however, it is necessary to establish cleaning procedures, which may be the most important and the most time consuming.

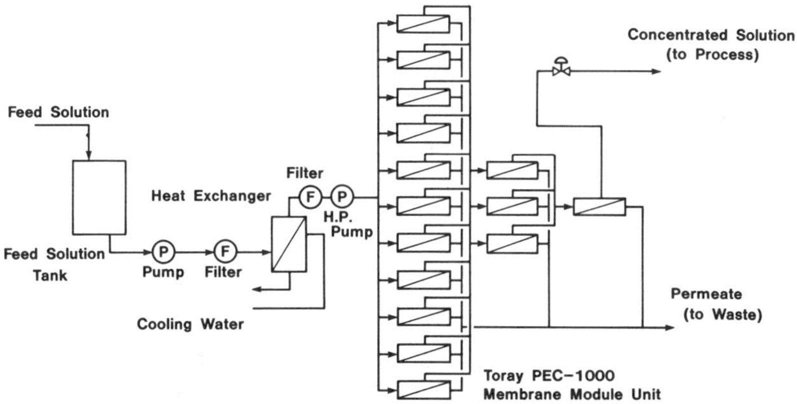

Figure 7. Flow diagram of RO plant for concentration and recovery of ε-caprolactam using PEC-1000 membrane modules.

Figure 8. Photograph of RO plant for concentration of dilute ε-caprolactam solution using PEC-1000 membrane modules.

Literature Cited

1. Fang, H.H.P.; Chien, E.S.K. Environ. Sci. Tech. 1976, 10, 364.
2. Kurihara, M.; Kanamaru, N.; Harumiya, N.; Yoshimura, K.; Hagiwara, S. Desalination 1980, 32, 13.
3. Kurihara, M.; Harumiya, N.; Kanamaru, N.; Tonomura, T.; Nakasatomi, M. Desalination 1981, 38, 449.
4. Kurihara, M.; Nakagawa, Y.; Takeuchi, H.; Kanamaru, N.; Tonomura, T. Proc. 10th Annual Conference of WSIA. Vol. 2, Honolulu, Hawaii, July 25-29, 1982.
5. Kurihara, M. Membrane, 1983, 8, 97.
6. Nakagawa, Y. Workshop on Membrane Separation Technology: Application to Foods and Biochemical Processes, National Taiwan Univ., March 17-18, 1983.
7. Chen, J.Y.; Kurihara, M.; Pusch W. Desalination 1983, 46, 379.
8. Ohya, H. et al. Kagaku Kogaku Ronbunshu 1981, 7, 372. 9. Ohya, H. et al. Kagaku Kogaku Ronbunshu 1982, 8, 144.
9. Ohya, H. et al. Membrane 1984, 9, 55.
10. Spencer, H.G. et al. Seminar on Membrane Separation Technology at Clemson Univ., Clemson, SC., August 2-6, 1976.
11. Hara, S. et al., Proc. of 7th International Symposium on Fresh Water from the Sea 1980, Vol. 2, 143.
12. Shimomura, T. Seminar on Membrane Development, Tokyo, June 26, 1980.
13. Ohya, H. et al. Membrane 1983, 8, 171.
14. Oyha, H. et al. International Membrane Technology Conference, Univ. of New South Wales, Sydney., Nov. 8-10, 1983.

RECEIVED February 22, 1985

23

Development of Sanitary Design Reverse-Osmosis Systems for the Pharmaceutical Industry

P. L. PARISE, B. S. PAREKH, and R. SMITH

Millipore Corporation, Bedford, MA 01730

A series of single and two-pass, high pressure operation, sanitary reverse osmosis systems have been developed to produce USP Purified Water and Water for Injection required by pharmaceutical manufacturers. The systems utilize sanitary piping, unique spiral modules and pressure vessels designed to maintain the chemical and microbiological water purity standards required by this industry. Monitors and controls have been selected to simplify process validation. Process optimization and characterization are discussed, including ion, microbe and pyrogen rejections, product flow rate, sanitization, and extractables testing. RO cartridge design (relative to standard spiral cartridge design) and its impact on maintenance of sanitary conditions are also discussed. In addition, recommended pretreatment systems are discussed and laboratory data are presented.

The pharmaceutical industry requires processed waters of different qualities for multiple uses - preparation of drug and parenteral solutions, rinsing of autoclaves and process equipment, and as formulations aids. Of the many grades of water used, two types are of primary importance, Purified Water (PW) and Water for Injection (WFI). The minimum quality requirements for these waters are defined by the United States Pharmacopeial Convention (USPC) and are listed in Table I (1). These waters are further defined by the USPC as follows (1):

"Purified Water is obtained by distillation, ion-exchange treatment, reverse osmosis or other suitable process. It is prepared from water complying with the regulations of the Federal Environmental Protection Agency with respect to drinking water. Purified water contains no added substances."

"Water for Injection is water purified by distillation or by reverse osmosis. It contains no added substances."

0097-6156/85/0281-0297$06.00/0

Table I: Water Quality Standards
(Numerical Interpretation of USPC Standards)

Parameter	Purified Water	Water for Injection
pH	5.0-7.0	5.0-7.0
Chloride	≤0.5 mg/L	≤0.5 mg/L
Sulfate	≤1.0 mg/L	≤1.0 mg/L
Ammonia	≤0.1 mg/L	≤0.1 mg/L
Calcium	≤1.0 mg/L	≤1.0 mg/L
Carbon Dioxide	≤5.0 mg/L	≤5.0 mg/L
Heavy Metals	≤0.1 mg/L as Cu[1]	≤0.1 mg/L as Cu[1]
Oxidizable Substances[2]	Passes USP Permanganate Test	Passes USP Permanganate Test
Total Solids	≤10 mg/L	≤10 mg/L
Total Bacterial Count[3]	100 CFU/ml	50 CFU/ml
Pyrogen[4]	None specified	0.25 EU/ml

Note: The USPC chemical test methods (except for pH and Total Solids) are quantitatively based on visual methods. The concentrations listed are the determined numerical equivalents for those tests.

[1]Limits for other heavy metals may be determined.

[2]Limits for specific oxidizable substances may be determined.

[3]Proposed Action Guideline for the Microbial Control of Ingredient Water as issued by the USPC, effective November 1, 1983. It should be noted the manufacturers frequently impose more stringent internal guidelines. CFU = Colony Forming Units.

[4]As determined by LAL (Limulus Amebocyte Lysate) test. EU = Endotoxin Units.

Distillation dominates all other processes in the manufacture of pharmaceutical grade waters. Past attempts to use reverse osmosis (RO) for producing pharmaceutical grade waters were not totally successful because of several reasons, including non-sanitary design of the RO cartridge and system hardware. Non-validatable system designs and a lack of understanding of the pretreatment processes to RO systems also contributed to previous problems. In this paper we describe the development of a series of single and two-pass RO systems designed specifically for the production of pharmaceutical grade waters, present operating data, and discuss pretreatment recommendations. An example of the systems developed is shown in Figure 1.

Pharmaceutical RO System Design

Sanitary Design Concept. The design of an RO system for use in pharmaceutical applications requires special considerations beyond those of a conventional RO system. It must incorporate sanitary design features in all of its components.

The "sanitary design" approach attempts to minimize the growth of microorganisms (including pyrogenic byproducts) within the system and minimize the chemical contaminants that could leach into the water by system components. In addition, the system design permits "in place" sanitization, i.e., periodic chemical sanitization of the entire system without disassembly of the system.

As an example, a conventional pressure gauge should not be used on the sanitary RO product stream. A conventional gauge contains a Bourdon tube, which fills with water to activate the gauge mechanism (Figure 2). This tube contains a stagnant volume of water in which bacteria can form colonies and is difficult to reach with the sanitizing agent. Even if the chemical sanitization fluid was somehow forced into the tube, it would then be difficult to remove, possibly leaching out into the product water during normal operation. These problems are eliminated by the use of a sanitary pressure transducer, which presents a smooth face to the product water. It measures the water pressure electronically through the deflections of its face. It traps no water and is easily flushed.

This design philosophy (no "dead legs") is applied throughout the system and component selection and design.

Process Design. Both single pass and two pass (i.e., product staged) versions of the pharmaceutical RO systems have been developed, ranging in product flow rates from 5 to 20 gpm (Table II).

Table II. Pharmaceutical RO Models

Models	Flow (gpm) 400 psid per pass 25°C water	Number of RO cartridges	Total pump hp	Typical power (KWh) at rated product flow
Single Pass		Per Pass		
MILLIRO 1100-S1	5	3	6	6.7
2200-S1	10	6	15	10.5
3400-S1	15	9	20	13.9
4500-S1	20	12	20	18.0
Two Pass		1st Pass:2nd Pass		
MILLIRO 1100-S2	5	6:3	15	15.8
2200-S2	10	9:6	30	23.0
3400-S2	15	12:9	40	29.5

The selection between single pass and two pass units depends on the RO feed water quality and the desired final product water quality. The single pass RO systems have an initial minimum rating of 93% monovalent ion rejection, 99% bacteria rejection, and 99% pyrogen rejection. The two pass systems have an initial minimum rating of 99% monovalent ion rejection, 99.99% bacterial rejection, and 99.99% pyrogen rejection.

The flow path of a single pass pharmaceutical RO system is very similar to a conventional RO (Figure 3).

Early two pass RO designs were simply two conventional single pass RO units connected in series with a buffer tank between them. The units were difficult to sanitize in place and a sanitary tank with 0.22µm vent filter was required.

The pharmaceutical RO two pass system design is shown in Figure 4. In this design, the feed water is pressurized up to 900 psi by two pumps in series and fed to the first pass. The maximum transmembrane pressure differential per pass is 450 psi. The first pass

Figure 1. MILLIRO 2200 pharmaceutical RO system.

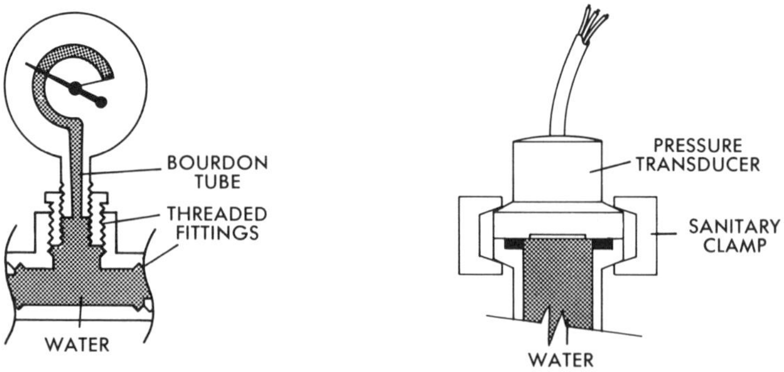

Figure 2. Conventional (left) vs. sanitary (right) pressure sensors.

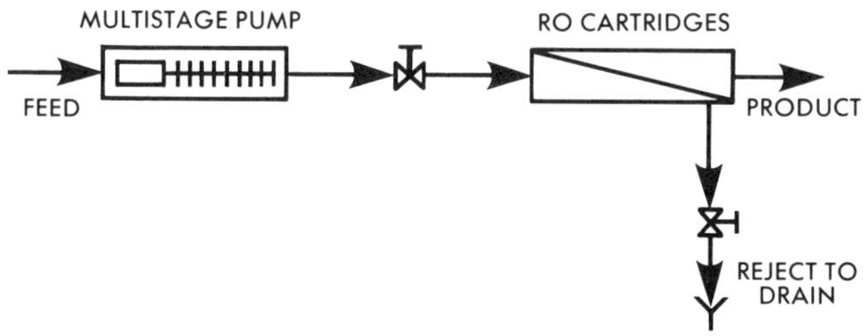

Figure 3. Single pass RO system.

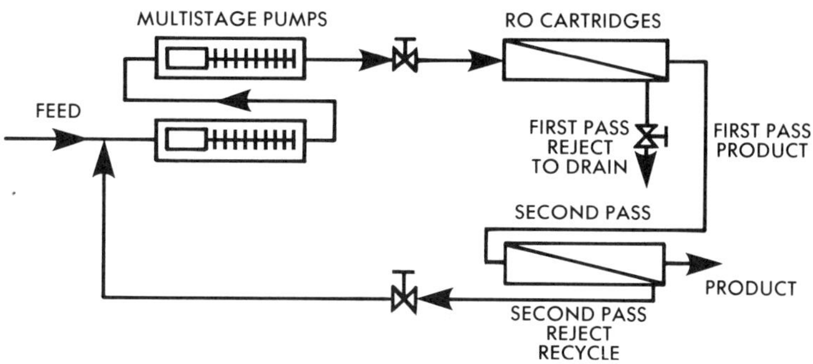

Figure 4. Two pass RO system.

product water is then fed at up to 450 psi to the second pass. This results in a very simple, clean connection between the first and second pass.

Sanitization. Sanitization is necessary to maintain the microbiological purity required for pharmaceutical grade water. Sanitization of the RO system is performed either on a periodic basis or when the bacteria/pyrogen count in the product stream exceeds predetermined limits. The system is sanitized with 10 ppm of sodium hypochlorite (NaOCl) or 100 ppm of peracetic acid (CH_3COOOH) recirculated through the system, and then flushed out. The system design permits efficient distribution and removal of the sanitization chemicals.

Component Design

Manifolding. Several steps are taken in the manifolding design to prevent excessive bacterial colonization and to allow rapid flushing of sanitization chemicals. All product piping is Type 316L stainless steel (SS) tubing polished to 180 grit inside and out, and then electropolished inside. This provides a very smooth surface down to the submicron level. The tubing is connected with high pressure sanitary fittings; instead of threaded or flanged fittings which may be easily colonized by bacteria. The water flow velocity is kept above 5 feet per second in the tubing (2). Sensor elements are located not more than 5 internal tube diameters away from the main flow stream to minimize areas of low flow velocities.

Pumps. Multistage centrifugal submersible pumps are used in series on the two pass RO systems to provide up to 900 psi feed pressure to the first pass membranes. The motors range in size from 3 HP to 20 HP, and are a liquid-cooled design. The heat transferred into the feed water by the motors increases the membrane flux slightly. The pumps rely on the feed water for lubrication and hence no oil contamination of the feed water would take place. All wetted metal parts are 316SS.

Pressure Vessels. The pressure vessels are 6" ID x 87" length, 316LSS, and hold three RO cartridges each. They are designed to operate at up to 1000 psi with a 5X safety factor. The inside of the vessels are polished to an 80 grit finish, then electropolished. The vessel endcaps are designed to minimize areas of stagnant water and have no threads on the inlet/outlet ports.

Controls. A very important factor in pharmaceutical processes is validation of the performance of the system within the allowable variation of the system. Validation confirms the capability and establishes the reproducibility of a specific process. Since validation requires collecting large amounts of operating data, a microprocessor was selected to control, monitor, and record data from the RO unit.

The microprocessor receives analog signals directly from the sensor elements: pressure, flow, temperature, resistivity, and pH. It is programmable in Basic and the program is transferred into permanent memory through the use of an EPROM. The program cannot be modified by the operators. This feature is important in pharmaceutical processes to prevent unauthorized tampering.

The microprocessor automatically scans 15 key operating parameters every two seconds and compares their values against pre-established limits. Should an alarm condition occur, the microprocessor diverts the product flow to drain and/or shuts the RO system down. It then prints out all operating parameters at the moment of the alarm condition. It also prints out all operating parameters automatically every six hours or upon request.

During normal operation, the operator can choose any six parameters to be displayed on a video screen. A remote printer or video screen can be tied into the system. The control box is NEMA 12 rated with a window to permit viewing the video screen located inside.

Sanitary Spiral RO Cartridge Design. The cellulose acetate (CA) based spiral RO cartridge has been designed specifically to minimize bacterial growth within its structure, to be easily sanitized by chemical application, and to limit organic extractables leaching into the product stream. The cartridge specifications and materials of construction are given in Table III. Materials were selected to conform with the appropriate FDA codes (3) for continuous and intermittent contact with aqueous solutions.

Table III. Cartridge Specification

Configuration	Spiral Wound
Size	6" diameter x 25" length
Transmembrane ΔP	400 psi (normal) 450 psi (max)
Monovalent/Divalent Ion Rejection	93%/96% @ 400 psid, 25°C, < 20% recovery
Particulate, Bacterial and Pyrogen Rejection	> 99%
Product Flow Rate	1.7 gpm minimum @ 400 psid, 25°C
Materials	
Central Tube, Antitelescoping Device	Polysulfone
Feed Side Spacer	Polypropylene
Product Side Spacer	Epoxy-coated Polyester
Adhesive	Epoxy
Chevron Seal	Ethylene Propylene Rubber
Outer Wrap	Fluorinated Ethylene Propylene
Membrane/Substrate	Cellulose Acetate blend/woven Polyester

A feed stream bypass has been incorporated to provide a flow of feed water within the annular space behind the chevron seal (see Figure 5). The feed stream bypass is effected by slots placed in the antitelescoping device mounted on the feed face of the cartridge. The slots allow a controlled portion of the feed flow (3-6%) to pass beneath the chevron seal to the annular space between the cartridge and housing. The outer wrap used is a heat shrinkable, fluorinated ethylene propylene (FEP) material, which, unlike tape or wrapped fiberglass coverings, is smooth, crevice-free and very low in extractables. A flow channel has been incorporated between the outer wrap and cartridge surface.

All seals between cartridges and between the cartridge and housing ports consist of a unique, taper-fit, 316SS interconnector (4) (see Figure 6). The use of a simple taper fit interconnector eliminates seal leakage associated with O-ring use.

RO Cartridge and System Evaluation

RO Cartridge. Tests were performed with the sanitary spiral cartridge to demonstrate the effectiveness of the feed stream bypass during chemical sanitization. Also, an analysis was performed to measure the total organic extractables (Dohrmann Total Organic Carbon Analyzer, Model DC-80, Detection Limit 100 ppb TC) in the product stream and to determine compliance with the USPC total oxidizable limit.

Bypass Stream Effectiveness. Testing was performed with four slots located 90° apart, placed in the feed side antitelescoping device below the chevron seal (see Figure 5). Visual tests with methylene blue dye (performed with a transparent cartridge housing) confirmed that the time required to flush the dye from the cartridge at normal feed flow rates (4-10 gpm) was less than one minute. In addition, the visual test indicated that the dye was also flushed from the channel under the outer wrap within the same period.

Tests to determine flush periods of residual sanitizing chemicals from cartridges with and without the feed flow bypass were also performed. Flush water flow rates were from 4.5 to 6.5 gpm at a feed pressure of 400 psig. The results of these tests are given in Table IV. It was also noted that for cartridges without the feed flow bypass, chemical residuals were trapped behind the seal. For these cartridges, residual sanitizing agent was determined in the product water for the first 1 to 3 minutes following system restart (system shutdown ranged from 1 to 16 hours); there were no sanitizing agents detected in the system containing cartridges that incorporate the bypass flow feature.

Organic Extractables. Testing for total organic carbon extractables in the product stream of both new and previously sanitized cartridges yielded TOC values of 0.1 to 0.4 ppm Total Organic Carbon. These samples also passed the USP Oxidizable Substances Test (permanganate oxidation).

RO System. Evaluation of the system included optimization of ionic rejection performance, evaluation of sanitization methods, determination of system ionic extractables, and long-term operation to produce WFI.

System Operation. A prototype two-pass 10 gpm (product) system (MILLIRO 2200S2) was constructed and used for laboratory testing. The following operational guidelines maximize overall system ion rejection, while maintaining the rated product flow:

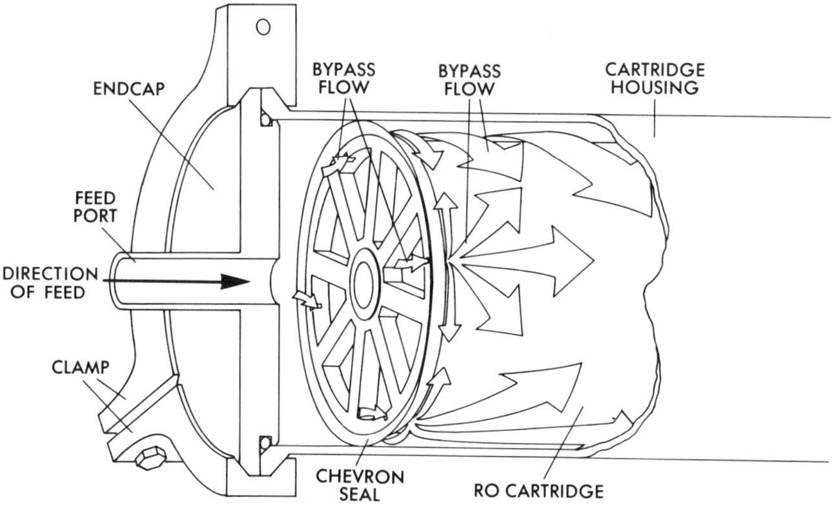

Figure 5. Feed stream bypass.

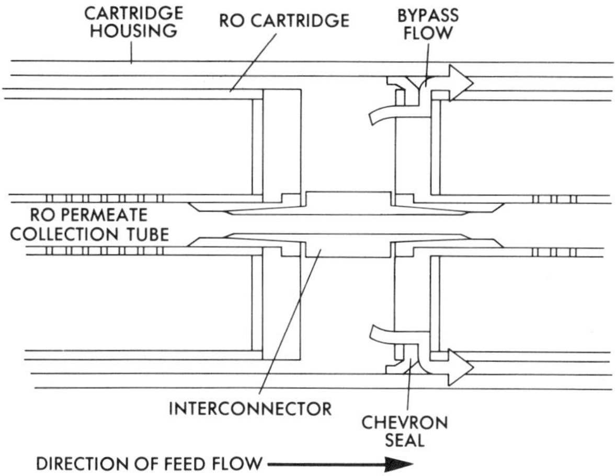

Figure 6. Taper-fit interconnector.

Table IV. Flush of Residual Sanitizing Agents

After Exposure To:	Time to Flush (to undetectable in permeate, with or without bypass slots)	Time to Flush (to undetectable behind seal) With Bypass	Time to Flush (to undetectable behind seal) Without Bypass
10 mg/L NaOCl	< 10 min	30 min	@ 1 Hr-1 mg/L present
200 mg/L CH_3COOOH	< 10 min	30 min	(no test)
2% HCHO	< 35 min	30 min	@ 1 Hr-160 mg/L present

	Detection Limits	Method
NaOCl	0.1 mg/L	APHA, "Standard Methods", Method 332, 14th ed.
CH_3COOOH	1.0 mg/L	Internally developed method based on spectrophotometric analysis of reaction with m-phenylenediamine (λ_{max} = 360 nm)
HCHO	1.0 mg/L	FORMALERT, Organon Teknika Corp., Oklahoma

- Operate the first pass at the maximum rated transmembrane pressure, to maximize the product flow rate from this pass. Maximum product output from the first pass allows the second pass to operate at the minimum transmembrane pressure and minimum recovery required to produce the rated product flow rate. (Note: Pass or system percent recovery is defined as the product flow (pass or system) divided by the feed flow (pass or system) times 100).
- Minimize the overall system recovery (within practical limits for reject water usage and disposal).
- Maintain feed water temperature as high as possible (without exceeding cartridge or system limits).

An example of optimizing system performance for the MILLIRO 2200S2 is shown in Figure 7. For a given system recovery, the maximum ionic rejection (based on 750 ppm NaCl solution challenge while maintaining a minimum 10 gpm product flow rate) is obtained by operating the first pass at 425 psid and the second pass at 275 psid.

Sanitization and Residual Flushout Test. Testing was performed to determine the distribution of sanitizing agents (sodium-hypochlorite and peracetic acid) within the system during a recirculating sanitization cycle, and to determine the time required to flush the residual sanitizing agent to a non-detectable level. The system was sanitized with either 5-10 ppm of NaOCl (pH adjusted to 6-7 with dilute HCl) or 70-100 ppm of peracetic acid (pH adjusted to 4.5 to 5.0 with dilute NaOH) for 30-45 minutes.

Sanitization and system flushout occurred at typical operating conditions (18-20 gpm feed water flow rate, 50% system recovery, first pass ΔP of 400 psid, and second pass ΔP of 300 psid). At the pH's stated above, both hypochlorous acid and peracetic acid are well distributed throughout the system as a result of the high passage (> 50%) by CA membranes for these substances. The sodium hypochlorite and peracetic acid solutions were flushed to non-detectable levels within 50 and 20 minutes, respectively. Also, no detectable levels of sanitizing agents were present upon subsequent system restarts.

System Extractables. A high resistivity feed water (2 megohm) was introduced into the system to determine if the system leaches ionic extractables. Product water resistances of 2.7 megohm (1st pass) and 3.5 megohm (2nd pass) were produced, thus indicating that feed water of this purity is not degraded by the system material during normal operation.

Operational Test to Produce WFI. A system test to produce WFI (as defined in Table I) is being conducted on Bedford MA tap water (at Millipore Corporation). Pretreated Bedford tap water is being fed to the MILLIRO 2200S2 unit (see Figure 8). The pretreatment system consists of cartridge filtration (10μm); carbon filtration for residual chlorine removal; ultrafiltration (Millipore Ultrafiltration System, Model 8817) for colloid removal; and acid addition (dilute HCl) (Liquid Metronics Chemical Feeder, Millipore Cat. No. ZRCF 11500) for pH adjustment. The system is operated continuously (except for minor maintenance shutdowns, i.e., regeneration of the UF unit) and is sanitized on a once or twice per week basis, using either 10 ppm NaOCl (pH 6-7) or 100 ppm peracetic acid (pH 4-5). Samples are collected periodically and analyzed for conformance to WFI standards. Chemical standards (including total oxidizables) have been easily met during the course of the test. Bacterial analysis (as CFU/100ml) and pyrogen determinations as (EU/ml) have been performed during the test and the results are plotted in Figures 9 and 10 respectively.

These results indicate that both bacteria and pyrogen concentrations are maintained well below the limits for WFI water after system sanitizations. This is a good indication that uncontrollable biological colonization is not occurring in the product side of the system, and the current sanitization methods are satisfactory. Based on the system operation as described above, WFI quality is produced for at least 3-5 days after sanitization. Because the frequency of sanitization will vary from site to site, the intent of this testing has not been to maximize the operating period between sanitization cycles, but to demonstrate that given the appropriate frequency of sanitization, WFI quality will be maintained.

Pretreatment

To maintain an efficient and reliable operation of the pharmaceutical RO systems, it is essential to adequately pretreat the incoming feeds. The choice and design of the pretreatment processes must address three important aspects. First, pretreatment should remove contaminants to enable the pharmaceutical RO systems to produce PW or WFI; second, it should provide adequate protection to RO system components (pump, membrane/module); and third, each pretreatment step should enhance the quality of the treated water.

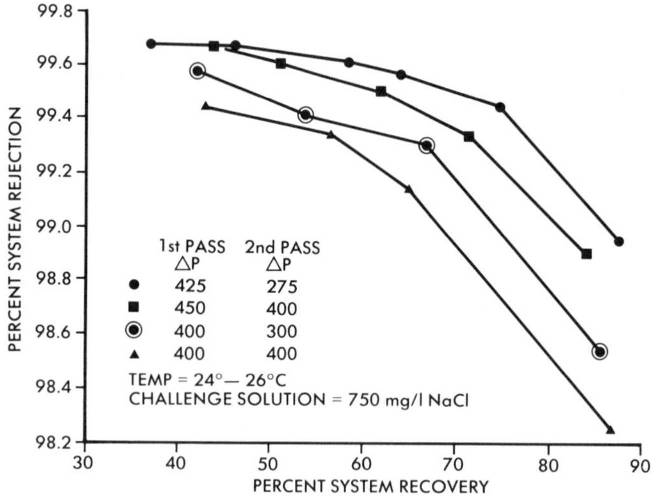

Figure 7. System rejection versus system recovery.

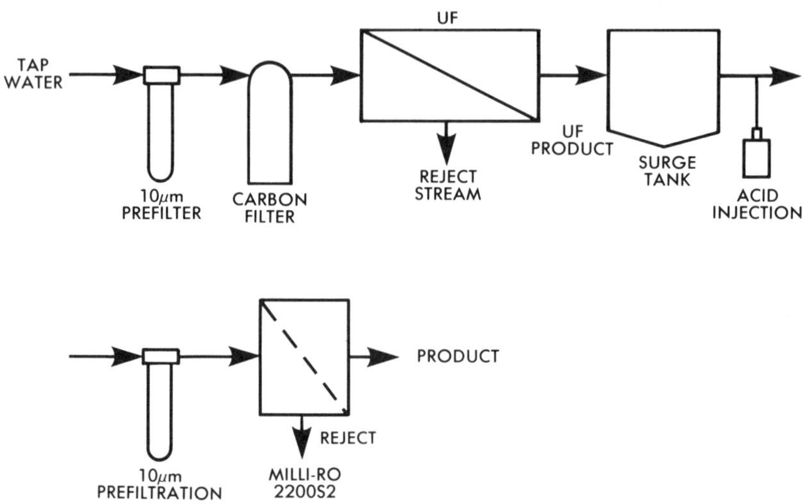

Figure 8. Pretreatment for RO during WFI test.

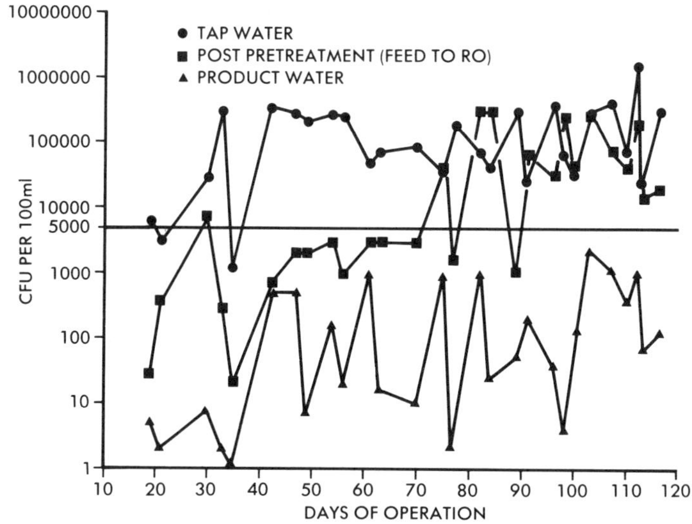

Figure 9. CFU/100 ml versus days of operation

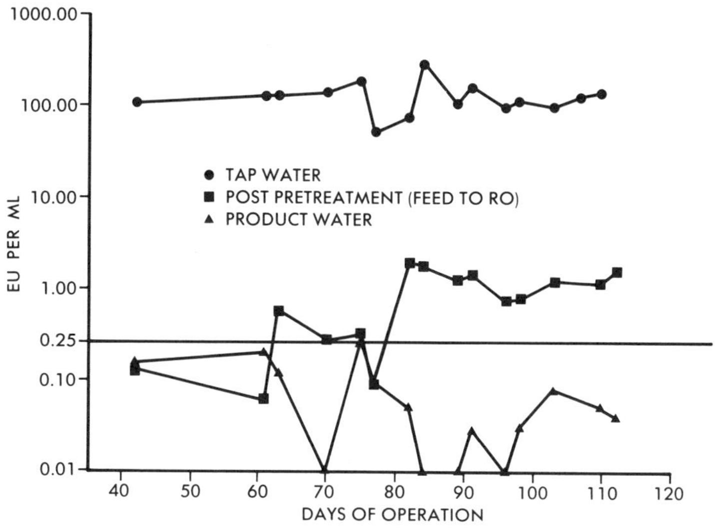

Figure 10. EU/ml versus days of operation.

Table V lists recommended feed water quality for producing WFI using pharmaceutical RO systems. The incoming limits (columns 3 & 4) are based on the expected rejection levels after one year for CA membranes. Feed waters that exceed these limits must be pretreated. For example, the WFI specification for TDS (Total Dissolved Solids) is 10 mg/L or less. To achieve this, the feed waters to one-pass systems must contain 100 mg/L or less total dissolved solids (500 mg/L or less for two-pass systems). Similarly, the WFI specification for chloride ion is 0.5 mg/L; the chloride ion content of feeds to the one-pass system must be 5 mg/L or less (25 mg/L or less for two-pass systems).

Another critical function of pretreatment is to protect RO system components from damage by feed water contaminants. Particles, colloids (< 1μm), scale-forming salts (Ca^{++}. Mg^{++}), organics, pH, Fe^{++}, Fe^{+++}, Mn^{++}, silica, and dissolved gases are key contaminants affecting the performance of the pharmaceutical RO systems. Table VI summarizes recommended pretreatment methods for the control/removal of the above contaminants.

Table V. Incoming Water Quality Required to Produce WFI Using CA RO

Contaminant	WFI Specification	Incoming Limit One Pass	Incoming Limit Two Pass
TDS	<10 mg/L	100 mg/L	500 mg/L
Chloride	<0.5 mg/L	5 mg/L	25 mg/L
pH	5-7	5-7	5-7
Sulfate	<1.0 mg/L as SO_4^{--}	12 mg/L	82 mg/L
Ammonia	<0.1 mg/L as NH_3	1 mg/L	5 mg/L
Calcium	<1 mg/L as Ca	12 mg/L	83 mg/L
Heavy Metals	<0.1 mg/L as Cu	2 mg/L	40 mg/L
Oxidizables		-	-
Pyrogens	≤0.25 EU/ml by LAL	25	2500
Bacteria	50 CFU/mL	10^6/mL	10^8/mL
Carbon dioxide	5 mg/L CO_2	5 mg/L	5 mg/L
Chlorine	No chlorine	0	0
Chloramine	No chloramines	0	0

A carefully-designed pretreatment train consisting of a suitable combination of the methods in Table VI should produce a feed of enhanced quality. One of the important considerations is to minimize the microbial growth in the pretreatment equipment by (i) efficient design and (ii) implementing sanitization protocols.

Table VI. RO Feed Contaminants and Recommended Pretreatment Methods

Contaminant	Effect	Pretreatment
Particles	Pump & module damage	10μm prefiltration
Colloids	Membrane fouling	Ultrafiltration synthetic adsorbent electromedia filtration
Scale forming salts of Ca^{++}, Mg^{++}, Sr^{++}, Ba^{++}	Membrane/ module fouling	Softening with cation exchange resin
Fe^{++}, Fe^{+++}, Mn^{++}	Membrane fouling	Green sand
High pH	Membrane attack, out-of-spec product	Cation exchange with H^+ resin or acid addition
High TDS	Out-of-spec product	Ion exchange deionization
Organic Contaminants	Low Rejection	Specific adsorption
Cl_2, chloramine	Membrane attack	Carbon adsorption
CO_2, NH_3, low mol. wt. organics	Low rejection	Degassification
Silica	Low rejection	Anion exchange resin

Discussion

The pharmaceutical RO systems have been developed to meet the specific needs of the pharmaceutical industry - production of Water for Injection and Purified Water

The organic (Table IV) and ionic extractables tests, and sanitization and flush out tests demonstrate the sanitary design of the RO cartridge and system components. The long-term test at Bedford (Figures 9 & 10) confirms the capability of the pharmaceutical RO system to produce WFI. It is important to note that proper pretreatment (Figure 8, Bedford Feed) and an adequate sanitization protocol - which are, of course, site specific - are key ingredients of the pharmaceutical RO process for producing WFI and PW.

The capital and operating costs of a distillation process and a pharmaceutical RO process (MILLIRO 1100 System) are presented in Table VII (5 gpm WFI). It is seen from the table that the initial capital cost of the MILLIRO's process is 75% of the distillation and the operating cost is less than 1/3; the payback period for the MILLIRO's process is calculated to be 1.6 years.

The unique features of these systems - sanitary design components (including the RO cartridge) and validatable process design, and lower capital and operating costs than distillation (see Table VII) now make the pharmaceutical RO systems preferable to distillation for producing WFI and PW.

Table VII. Cost Comparison: MILLIRO System vs. Distillation (5 gpm WFI)

	MILLIRO System	Distillation
Capital costs of total system*	$120,000*	$160,000*
Operating cost of total system	$3.70/1,000 liters	$12.94/1,000 liters
(260 days/yr. operation)	$14.00/1,000 gallons	$49.00/1,000 gallons
Operating cost/year	$26,208	$91,728
Capital cost/year amortized over 5 years	$24,000	$32,000
Total operating & capital cost per year	$50,208	$123,728
Savings per year	$73,520	-
Payback period	1.6 years	-
5-year savings	$367,600	

* Figures based on 1984 costs.

Total system includes all pre-treatment and ancillary equipment

Acknowledgments

The authors gratefully acknowledge the contribution of Douglas Higgins, Charles Serpa, Fran Carlson, George Witham and Barbara McLaughlin in the development work and manuscript preparation.

Literature Cited

1. "The United States Pharmacopeia" 20th Rev., July 1, 1980, United States Pharmacopeial Convention, Inc., 12601 Twinbrook Parkway, Rockville MD, 20852.
2. 3-A Accepted Practice for Permanently Installed Sanitary Product-Pipelines and Cleaning Systems, Number 605-02 effective October 25, 1967, formulated by International Association of Milk, Food and Environmental Sanitarians, United States Public Health Service, and the Dairy Industry Committee.
3. Code of Federal Regulations, Title 21 - Food and Drugs, Parts 175.300, 177.2500, 177.2600, April 1, 1982.
4. Patent Application Filed.

RECEIVED February 22, 1985

24

Ultrafiltrative Solute Rejection Behavior of Black Liquor

RAJNISH and P. K. BHATTACHARYA

Department of Chemical Engineering, Indian Institute of Technology, Kanpur-20816, India

Experimental studies were being carried out on the solute rejection behaviour of black liquor through ultrafiltration using a stirred batch cell. Influences of operating parameters viz. pressure (3-5 atm.), concentration of feed black liquor (1-5% t.d.s.) and stirrer speed (300-900 rpm) were studied on solute rejection behaviour at ambient temperature using different pore size (0.02 μm to less than 0.005 μm) cellulose acetate membranes. Concentration polarization model was used to evaluate the wall concentration by estimating the mass transfer coefficient through velocity variation method. The real rejection was always found to be more than the observed rejection and both vary similarly with pressure and concentration of black liquor.

Recovery of inorganic chemicals from spent liquor, termed black liquor, is an integral part of the Kraft Pulping Processes. Infact, the economy of the Kraft Pulping Process is tied up with the efficiency of recovery cycle. With rising energy costs, increasing capital investment and environmental regulations, it has become more and more important to find energy efficient alternative methods to concentrate and fractionate industrial effluents. Ultrafiltration (UF) is one such favourable process.

The present work was undertaken to study the separation of black liquor using UF as membrane process technique with the following objectives :

i) To study the rejection behaviour of membranes towards black liquor,

ii) To analyse both flux and rejection data,

iii) Effect of concentration polarization on transmembrane flux.

0097–6156/85/0281–0313$06.00/0

The development of high flux UF membranes (1-2) has resulted in increased importance of fluid-phase mass transfer resistances. Goldsmith (3) pointed out the importance of considering osmotic pressure effect at the membrane surface and analyzed the experimental data by solving equations from film theory :

$$J_2 = k \ln [(C'_{1w} - C''_1)/(C'_1 - C''_1)] \tag{1}$$

He calculated mass transfer coefficient k using the equation obtained from phenomenological principles, Equation 2, and then compared with theoretical values.

$$J_2 = \bar{A} [(P' - P'') - (\pi' - \pi'')] \tag{2}$$

The difference between these two values was attributed to the concentration dependency of viscosity and diffusion coefficient.

Membrane rejection, or the ability to reject or retain a given solute is defined as (4) :

$$R_{obs} = (C'_1 - C''_1)/C'_1 \tag{3}$$

The concentration at membrane surface C'_{1w} is always higher than that in bulk, C'_1, because of concentration polarization. Therefore, corrected rejection characteristics of the membrane are described by real rejection, R_{real}, as :

$$R_{real} = (C'_{1w} - C''_1)/C'_{1w} \tag{4}$$

Combining Equations 1 and 4 leads to :

$$\frac{C'_{1w}}{C'_1} = \frac{\exp (J_2/k)}{R_{obs}+(1-R_{obs})\exp (J_2/k)} \tag{5}$$

Concentration polarization modulus, (C'_{1w}/C'_1), ratio of wall to bulk concentration, signifies the extent of concentration polarization present in the system.

Velocity variation method is used to determine the mass transfer coefficient (5-6). The value of k is usually a function of Reynolds number and can be expressed as:

$$k \propto U^a = b\, U^a \tag{6}$$

In the stirred batch system, the velocity can be varied by changing stirrer speed and k can be found (5),

$$k = b\, (\omega)^{0.7} \tag{7}$$

Using relation (Equation 7), Equation 1 is rewritten as :

$$\ln\left[\frac{(1 - R_{obs})}{R_{obs}}\right] = \ln\left[\frac{(1 - R_{real})}{R_{real}}\right] + \frac{1}{b}\left(\frac{J_2}{U^a}\right) \quad (8)$$

for stirred batch cell, Equation 1 becomes

$$\ln\left[\frac{(1 - R_{obs})}{R_{obs}}\right] = \ln\left[\frac{(1 - R_{real})}{R_{real}}\right] + \frac{1}{b}\left(\frac{J_2}{\omega^{0.7}}\right) \quad (9)$$

True rejection is given by extrapolation to an ordinate axis of $\ln[(1-R_{obs})/R_{obs}]$ vs $J_2/\omega^{0.7}$ plot. Using this R_{real}, the mass transfer coefficient, k, is calculated by Equation 1 and the effects of concentration polarization can be corrected.

Experimental

Concentrated Kraft black liquor was obtained from Central Pulp Mills, Surat and was kept in air-tight bottle. It was diluted with distilled water accordingly to get required concentration. Ultrafiltration cell, with magnetic stirrer and other accessories were imported from M/s. Schleicher and Schull, West Germany. Following specifications were provided for the cell :

Model UP 110/0

Useful volume	: 500 ml;	Residual Volume	: 0.5 ml
Filter diameter	: 107.5 mm;	Effective filter area	: 70 cm^2
Maximum working Pressure	: 7 atms;	Temperature stability	: 130°C
Venting Valve	: Set for between 5-7 atms.		

Strobotac (General Electric, U.S.A.) was used to find stirrer speed. It measured the rpm by matching the frequency of its flash light with stirrer speed.

Moist type cellulose acetate membranes were obtained from M/s. Schleicher and Schull, West Germany with following specifications :

Specifications :	Type	Pore size µm	Cut-off size
Cellulose Acetate	AC-63	0.020-0.010	70,000
-do-	AC-62	0.010-0.005	20,000
-do-	AC-61	0.005-0.002	10,000

As recommended by manufacturer, unused membranes were kept in a 25-30% solution of ethanol.

Procedure. Membrane was put on a supporting porous plate, placed at the lower part of the cell. A 200 ml of black liquor of known concentration was introduced into the cell, simultaneously starting the magnetic stirrer. Black liquor feed was pressurized by pump through pressure storage vessel and was maintained, for a particular set, a constant value by adjusting the regulating valves in the pump. Permeate was withdrawn at atmospheric pressure and the volumes were collected and noted at different time intervals. After completion of the filtration, permeate concentration was determined. Cell was dismantled, washed and assembled for next run. Membranes were always thoroughly washed with distilled water to avoid clogging of pores.

Pressure (3-5 atm), concentration of black liquor (1-5% total dissolved solids) and stirrer speed (300-900 rpm) were chosen as operating parameters and were varied to study the influences on various characteristics of membrane transport along with different membranes. All the experimental data are available (7).

Results and Discussion

Selection of Black Liquor Concentration. Black liquor concentration is an important consideration to study the solute rejection behaviour through microporous membranes. It is known (8) that as the concentration of a solution increases, its activity and thus the osmotic pressure of the solution increases, resulting in lowering the solvent flux. It is also known (9) that viscosity of black liquor increases significantly beyond 5% t.d.s., which restricts the effective use of magnetic stirrer, used in the system. Black liquor, for its typical characteristics as macromolecular solutes of various sizes in alkaline solution, is a difficult solution to handle for experimental work. A little higher concentration (more than 5% t.d.s.), gave experimental difficulties like choice of membranes, clogging of pores and reuse of membranes. Hence, it was decided to use dilute black liquor of less than 5% t.d.s. (in the range of 1 to 5%) for experimental work.

Selection of Membranes. For ultrafiltrative studies, the prime consideration in selection of membranes is to obtain a balance between the characteristics of feed liquor and separation ability of membrane. Laboratory membranes (7) gave casting problems viz. non-uniformity of membrane layer, lack of reproducibility of experimental data and low solute rejection (around 32%). All the experimental work were carried out with commercially available ultrafilters for their improved solute rejection and reproducibility of results.

<u>Effect of Feed Liquor Concentration and Applied Pressure on Solute Rejection</u>. The objective of this study was to establish an analytical method of ultrafiltrative rejection data. A plot (Figure 1) was made between change in solute rejection of black liquor to pressure gradient for different feed concentrations with different membranes at a stirring speed of 300 rpm and ambient temperature. Observed solute rejection (R_{obs}) was calculated according to Equation 3. It was observed (Figure 1) that with increase in pressure gradient, solute rejection increases. However, with increase in feed concentration the solute rejection decreases significantly. It is known (<u>10</u>) that with increase in feed concentration, wall concentration increases resulting in more solute flux. So the phenomena gets varified by experimental observation. However, increase in solute rejection with pressure contradicts the usual pattern (<u>10</u>) as rejection should decrease with pressure. At higher pressure, accumulation of solute at the membrane surface increases, raising the concentration gradient across the membrane, resulting in more permeation of solute along with higher solvent flux, thus reducing the rejection. Therefore, some additional mechanism is desired to explain such experimental observation. However, this behaviour is not typical to black liquor feed only. Goldsmith (<u>3</u>) had also reported such behaviour for Cabowax (polyethylene glycol) solutions. It was explained (<u>3</u>) that the concentrated solution at membrane surface may act as a dynamic membrane. Since, black liquor constitutes organics containing lower and higher molecular weight (1,000-20,000) macromolecules of alkali legnins, a slight increase in concentration of higher molecular weight species at the membrane surface may result in increased rejection for lower molecular weight solutes. Alternatively, it can be assumed that an increase in operating pressure may lead to compaction of the membrane skin. Hence, the resultant decrease in pore size gave rise to increased rejection.

<u>Correction of Concentration Polarization</u>. The observed rejection does not take into account the concentration polarization as this only considers bulk and permeate concentrations. Therefore, determination of wall concentration becomes an important aspect for the correction of concentration polarization. It was essential to evaluate the mass transfer coefficient, k, to determine the wall concentration, $C_{1w}^{'}$, (Equation 1). It was decided to compute k by the method of variation of velocity (<u>11</u>), as no diffusion data was available for black liquor because of its complexity in nature.

For the evaluation of constant solvent flux, J_2, experiments were carried out for small amount of permeate. Around 30 ml of permeate solution was collected from a 200 ml of feed solution. It was observed that for a

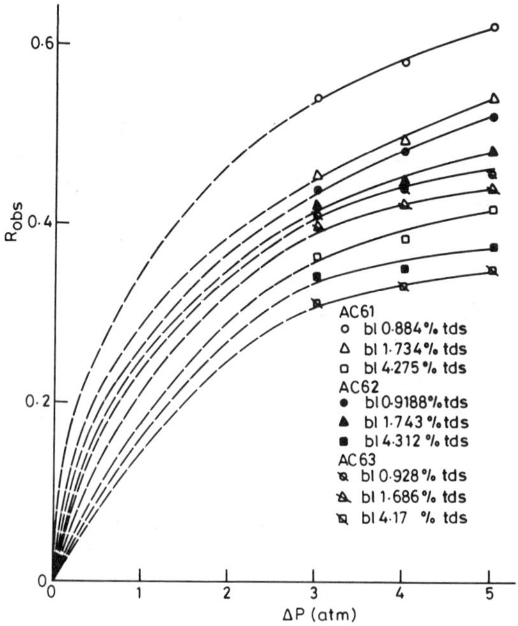

Figure 1. Effect of pressure on observed rejection at 300 rpm.

particular feed concentration and pressure, the variations in solvent flux were insignificant even with varying stirring speed. Therefore, solvent flux was assumed to be constant for a set of feed concentration, pressure and type of membrane.

A different set of experimental data at 5 atm. were obtained to compute the mass transfer coefficient, k, for velocity variation method. Equation 9 was utilized for the purpose. A plot (Figure 2) of ln $[(1-R_{obs})/R_{obs}]$ vs. $1/\omega^{0.7}$. was made. A linear relationship was observed and real rejection value, R_{real}, was evaluated from the intercept of this plot (Figure 2). This value would be independent of stirring speed and hence would represent the true rejection for a membrane. Mass transfer coefficient can now be evaluated by using Equation 1 after determining the wall concentration, C'_{1w}, from Equation 4. Once the relationship between k and ω was found for different feed concentrations and membranes, the values of R_{real} were calculated for all sets of data. The variation of R_{real} with pressure for different concentrations is shown in Figure 3. The real rejection was always found to be more than the observed which signifies wall concentration more than the bulk concentration. Hence, it concludes the presence of concentration polarization. It was also (Figure 3) clear that the real rejection increased with pressure but decreased with increase in feed concentration. This further verified the results obtained earlier (Figure 1).

Effect of Agitation on Solute Rejection. The tendency of accumulation of solute at membrane surface results in decrease in rejection. Therefore, an increase in rejection can be obtained by increasing dissipation of concentrated boundary layer into bulk solution. The effect of agitation on solute rejection is shown in Figure 4 by generating data points from Figure 2. The trend of curves indicate the increase in rejection with stirring speed. Increased agitation caused the gradual dissolution of gel layer, with subsequent increase in solute rejection. Figure 4 shows that the slopes of curves are more for high feed concentrations (4.06% t.d.s.). This specifies the pronounced effect of stirring on solute rejection for concentrated solution as compared to dilute solutions (0.869% t.d.s.). Further, observed rejection does not change appreciably with stirrer speed beyond 800 rpm (Figure 4). This specifies that at higher stirring speed R_{obs} approaches R_{real}.

Concentration Polarization Modulus. In order to study the concentration polarization further, variations in polarization modulus (C'_{1w}/C'_1) with stirrer speed were plotted (Figure 5). Higher values of polarization modulus were observed at low stirrer speeds and with increase in

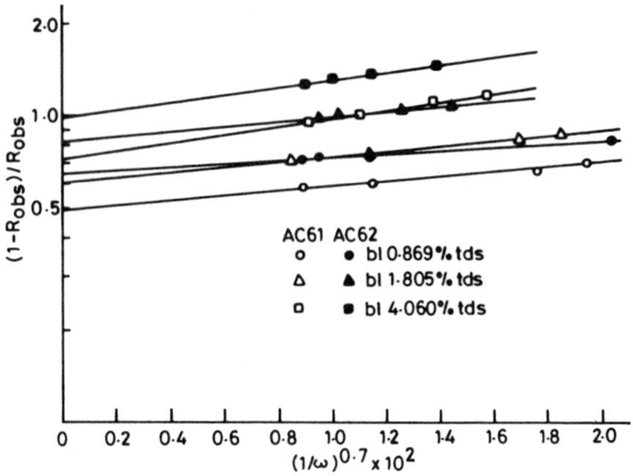

Figure 2. Relationship between observed rejection and stirrer speed.

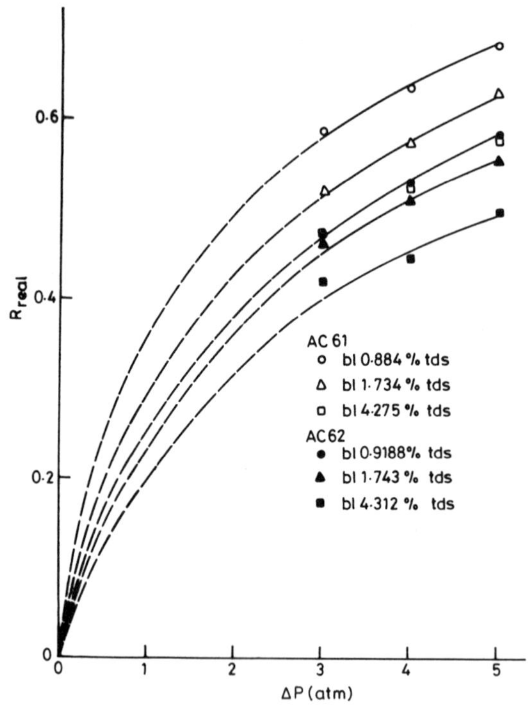

Figure 3. Effect of pressure on real rejection at 300 rpm.

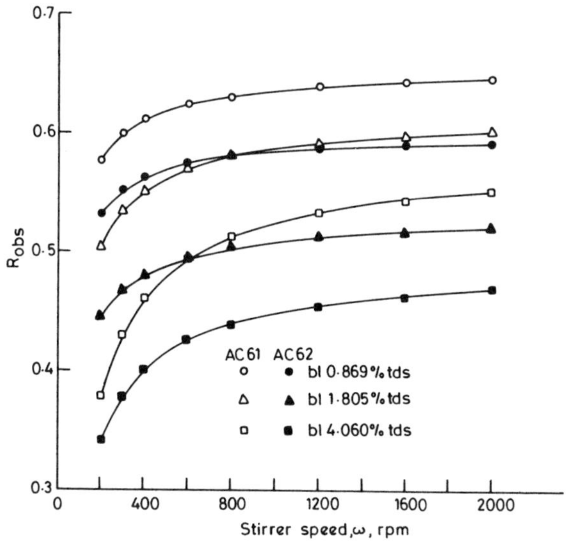

Figure 4. Effect of variation in stirrer speed on observed rejection at 5 atm.

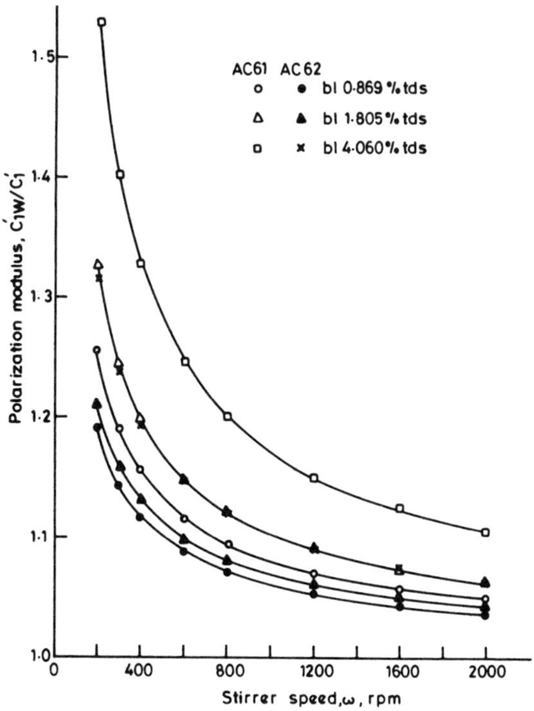

Figure 5. Effect of variation in stirrer speed on concentration polarization modulus at 5 atm.

speed it decreased gradually. On extrapolation of curves for infinite stirrer speed, the modulus would approach to unity or in otherwords the wall concentration would come closer to the value of bulk concentration. It was also observed that polarization modulus increased with feed concentration for a particular stirrer speed. This signifies the prominence of concentration polarization at higher feed concentrations. Thus verifying the inference obtained from variation of observed rejection with stirrer speed (Figure 4).

Conclusions

i) The real rejection was always found to be more than the observed rejection, signifying wall concentration to be higher than the bulk concentration, hence, specifies the presence of concentration polarization.

ii) Observed rejection does not change appreciably with stirring speed beyond 800 rpm, signifying closer approach to real rejection at higher speed.

iii) At higher feed concentration, the prominence of concentration polarization was observed from the variation of polarization modulus with feed concentration.

iv) Further work is being recommended for higher solute rejection for black liquor, using low cut-off membranes with high pressure (upto 15 atm.) ultrafiltration cell. Mass transfer properties with diffusion data estimation is to be the aim of further work on ultrafiltration studies with black liquor. Use of model compounds (globular proteins or dextrans) is suggested for the initial characterization of the membranes as a quality control measure.

Nomenclature

$\underline{a}$,b - Constants
$\bar{A}$ - Pure water permeability, cm^3/cm^2.sec.atm.
C_1' - High pressure side concentration of solute, wt%
C_1'' - Low pressure side concentration of solute, wt%
C_{1w} - Wall concentration of solute, wt%
J_2 - Solvent flux through membrane, cm^3/cm^2.sec.
k^2 - Mass transfer coefficient, cm/sec.
P - Pressure, atm.
U - Feed velocity, cm/sec.
π - Osmotic pressure, atm.
ω - Stirrer speed, rpm.

Acknowledgments

One of the authors (Rajnish) feels pleasure to acknowledge the co-operation of chemical engineering departmental staff for their assistance during laboratory work and to Mr. Raj Khanna for efficient typing work.

Literature Cited

1. VanOss, J. In "Ultrafiltration Membranes"; Perry, E.S., Ed.; PROGRESS IN SEPARATION AND PURIFICATION Vol. III; Wiley-Interscience : New York, 1970.
2. Kutowy, O.; Thayer, W.L.; Sourirajan, S. Desalination 1978, 26, 195.
3. Goldsmith, R.L. Ind. Eng. Chem. Fundam. 1971, 10, 113.
4. Kimura, S.; Sourirajan, S. AIChEJ. 1967, 13, 497.
5. Jonsson, G.; Boesen, C.E. Desalination 1977, 21, 1.
6. Kimura, S. Bull, Soc. Sea Water Sci. Jpn. 1974, 27, 295.
7. Rajnish, M.Tech. Thesis, I.I.T., Kanpur, 1983.
8. Barrow, G.M. "Physical Chemistry"; McGraw-Hill, 1966.
9. Swartz, J.N. In "Alkaline Pulping"; Macdonald, R.G., Ed.; PULP AND PAPER MANUFACTURE Vol. I; McGraw-Hill, 1969.
10. Baker, R.W.; Strathmann, H. J. Appl. Polym. Sci. 1970, 14, 1197.
11. Nakao, S.; Kimura, S. J. of Chem. Eng. of Jpn. 1981, 14, 32.

RECEIVED February 22, 1985

25

Reverse Osmosis and Ultrafiltration Applied to the Processing of Fruit Juices

DAVID J. PAULSON, RICHARD L. WILSON, and D. DEAN SPATZ

Osmonics, Inc., Minnetonka, MN 55343

Use of crossflow membrane separation technology for food and beverage manufacture has come of age. The feasibility of processing many types of fruit juices has been demonstrated and several production size units are currently operating. Applications such as replacement of troublesome diatomaceous earth (DE) filters for juice clarification have been highly successful. Employing reverse osmosis for apple juice concentration produces potential shipping cost savings of over $2000/day for a processor handling juice at a rate of 15,000 gallons per day. This paper discusses the membranes and membrane configurations available, and the steps necessary to properly develop applications for crossflow membrane technology. Apple, grapefruit, lemon, cranberry, and grape juices, as well as wine test results are presented. Clarification, pectin removal, and several concentration applications are discussed. Advances in membrane, membrane elements and hardware design and construction to allow operation under the increased temperature and pressure requirements for juice processing are also presented.

Crossflow Membrane Filtration

Crossflow membrane filtration is the separation of the components of a fluid performed by polymeric semi-permeable membranes through the application of pressure. The pressure required varies depending upon the size of the pores in the membranes. Reverse osmosis (RO) membranes have pores from 5 Angstroms (Å) to 20Å in diameter, and will reject most ions as well as most organics over 150 molecular weight. Ultrafiltration (UF) membranes have pore sizes of 10Å to 0.2 micron, rejecting larger organics such as proteins and viruses while passing most ions. Microfiltration (MF) has pore sizes ranging from 0.05 micron to 2 microns, and can replace traditional depth filtration for high volume removal of bacteria and very small suspended particulates.

0097-6156/85/0281-0325$06.00/0

Figure 1 shows a chart of separation from the ionic to the particulate range. The crossflow membrane classes overlap each other somewhat, since neither the marketplace nor the technology have yet developed sharp distinctions.

Crossflow membrane filtration is fundamentally different in design from the older technologies of depth and surface barrier filtration. In crossflow, the influent stream is separated by the membrane into <u>two</u> effluent streams: the "permeate", which passes through the membrane, and the "concentrate", which retains the solutes or suspended solids which have been rejected by the membrane. Figure 2 illustrates the process of crossflow filtration.

There are several advantages to operating in the crossflow design mode. First, the process is self-cleaning, with solutes and solids swept away by the pressurized concentrate stream which runs parallel to the membrane (hence, the term crossflow). Crossflow systems can operate economically on large fluid volumes with minimal operator attention and system downtime, since constant filter media changeouts are not required. Second, depending upon the application and process requirements, the permeate, the concentrate or both may be the product. Third, crossflow filtration membranes produce the finest separation available economically, and the most selective separation in the molecular range. Crossflow filtration systems can be designed to selectively pass some components of a feed stream while rejecting others. This "fractionation" can be performed at the ionic, molecular and macro-molecular levels.

Crossflow Membrane Polymers

The wide range of juice processing applications comprise a variety of fluid environments, some of which may be detrimental to certain polymeric materials. Operating conditions such as temperature, pressure, solution pH and chemical compatibility must therefore always be considered in relationship to the membrane. No one polymer will withstand the environments and perform adequately in all the possible juice processing applications where crossflow membrane processing can be applied, so several different membrane materials should be investigated to select the optimum for each application.

RO polymers currently available commercially include the asymmetric membranes: cellulose acetate (CA) and aromatic polyamides (PA) which are homogeneous. Also available are the newer thin-film composite membranes (TFC), usually considered polyamides but integral with a polysulfone substrate.

UF polymers currently available include: polysulfone (PS), CA-blends, a fluorinated polymer (VF), and Osmonics' patented polymer (PA type).

MF polymers commercially available are the most numerous, due to the relative ease of achieving the required pore size for MF compared to UF and RO membranes. Polypropylene, acrylonitrile, nylon and PTFE are among the more common MF polymers, and their broad chemical compatibility characteristics make the MF range of membranes the most chemically stable.

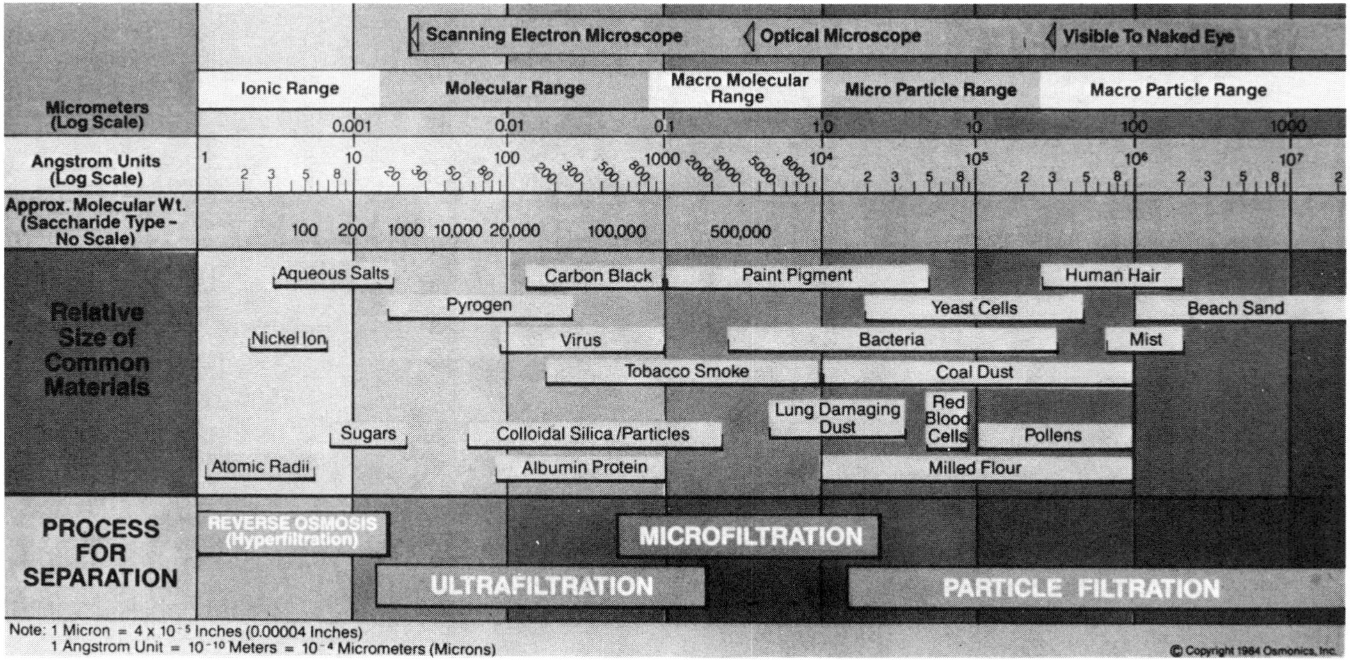

Figure 1. Filtration spectrum from ionic to particulate range. (Reproduced with permission. Copyright 1984, Osmonics, Inc.)

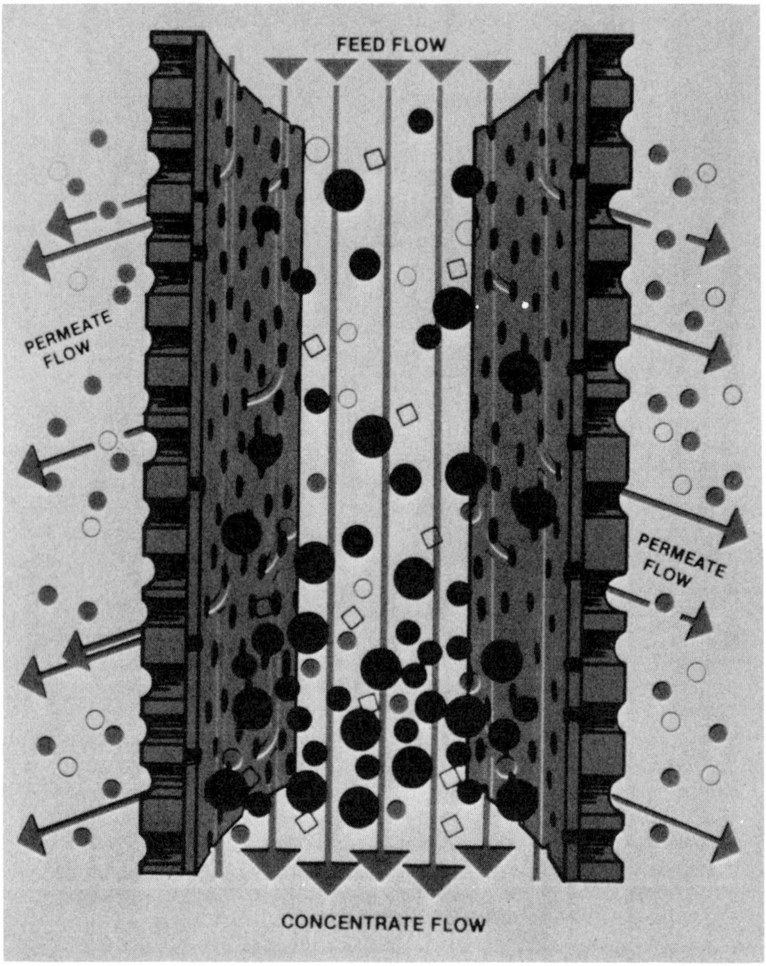

Figure 2. Artist's conception of crossflow membrane filtration.

Developmental RO polymers which show some long term promise are acrylonitrile copolymers and polyethersulfone, although neither is beyond the laboratory stage yet. TFC and grafted ultrafiltration membranes are also being investigated in the labs.

In choosing a polymer membrane, the following variables must be evaluated:

Operating Pressure. MF membranes are typically operated at transmembrane pressures (feed side pressure less permeate pressure) from 20 to 345 kPa (3 to 50 psi). UF membranes from 345-1380 kPa (50 to 200 psi) and RO membranes from 1380-6890 kPa (200 to 1000 psi). Under the higher operating pressure conditions, membranes may be subject to mechanical compression (compaction) and other mechanical deformation which can alter performance over time.

Increased temperature intensifies the compaction effect. Understanding this characteristic and selecting the most appropriate membrane for the pressure required is an important consideration.

Temperature. Since some processes in the juice industry may be performed at elevated temperatures, consideration of the polymer membrane's temperature limits is essential. Membrane separation processes are currently operating at a temperature range from 0 to 85°C (32-185°F). Membrane cleaning and steam sterilization of the system occur up to 110°C and higher. It is often preferable to operate the membrane system at elevated temperatures due to pasteurization or other processing requirements to further save process energy costs.

The best understood and least expensive RO membrane, cellulose acetate (CA), has often had an upper temperature limit assigned to it in the 35-40°C (95-104°F) range. Recent data from controlled experiments show that CA membrane retains its separation performance quite well at much higher temperatures and is effective in processes to 65°C (149°F). Polyamide (PA) type RO membranes definitely have higher temperature limits and can operate successfully at 82°C (180°F) at 400-600 psi and even higher temperatures under lower operating pressure.

UF and MF membranes, operated under much lower pressures and typically made of more temperature resistant polymers, can often be operated up to 100°C and in some cases are autoclavable. Under these conditions, it is common for the other membrane element components to be the limiting factors, not the membrane itself.

Solution pH. Although pH is often considered in the context of the overall solution chemistry, the degree of acidity or basicity itself also affects membrane life. The generally accepted pH range for CA membrane is 2 to 8, although higher and lower pH values are feasible if the economics of the application allow more frequent membrane replacement, or if ionic rejection is not critical to the application. PA type RO membranes are generally rated as compatible up to pH 11 or 12, yet many on the market are not rated below pH 3 or 4 in the acid range. For UF membranes, pH range tolerances are usually greater, with polysulfone (PS) membrane frequently rated as compatible from 0.5 to 13. MF membranes are generally even more pH tolerant.

Chemical Compatibility. Along with pH range, general chemical compatibility is also a major requirement for choosing a membrane polymer. Chlorine attack, even at very low levels, has been a major obstacle to the application of TFC-PA type membranes. Attack of PS membrane by many types of hydrocarbons has resulted in the use of CA membrane in many UF applications, despite its lower hydrolytic stability. A frequent limiting factor for MF membranes is extreme hydrophobicity, resulting in the requirement for chemical additives to lower surface tension or operation at much higher hydraulic pressures.

When chemical compatibility limitations are combined with the fact that the most desirable polymers can only be manufactured into certain types of membrane (with limited range of pore size, morphology and production rate or flux), the requirement for a wide array of membranes becomes even more apparent.

An example of how membrane and solution chemistry understanding can affect overall system performance comes from the dairy industry. Lab scale studies on the ultrafiltration of acid cheese whey showed that lowering the acid whey pH to 2, or raising the pH to 7 followed by centrifugation or 5 micron filtration, resulted in a 100% increase in membrane flux compared to whey pretreated with centrifugation at its normal pH of 4.5. The calcium and phosphorous contained in the whey are much more soluble at pH 2 and readily pass through the membrane rather than foul the membrane surface and pore walls. At pH 7, these minerals are very insoluble and therefore subject to removal by centrifugation or 5 micron filtration(1). The result is an overall system design that is less costly yet has superior membrane performance.

Membrane Element Configurations

As the membrane industry has grown and developed, several different configurations for supporting and containing the membrane were developed. With UF and MF membranes, the relatively low operating pressures allow more configurations than with RO. Commercially available configurations for UF include: tubular, larger internal-flow hollow fiber or "spaghetti bundle", plate and frame, and spiral wound.

Tubular units are generally produced with inside diameters of 12.5 or 25 mm (0.5 or 1 inch) and in lengths of 150 to 610 cm (5 to 20 ft.). The feed solution flows through the inside of these tubes, whose inside walls contain the membrane.

Hollow 'fat' fiber membranes can be thought of as self-supporting, miniature tubular units, with inside diameters of the fibers ranging from 0.5 to 1.1 mm (0.020 to 0.043 inches). Like the tubular units, the feed stream flows through the inside of the fibers and the permeate is collected on the outside.

The "plate and frame" configuration is a third style for housing membrane. Flat sheets of membrane are placed between plates with heights of approximately 0.5 to 1.0 mm (0.020 to 0.039 inches) which in turn are stacked in parallel groups. This configuration is similar to a conventional plate and frame filter press.

The "spiral wound" element also uses economical flat sheet membrane; when wound around a central permeate collection tube and separated by thin spacer materials, this design yields a high membrane packing density(2). The feed passes over the membrane through the spacer material, which also acts as a turbulence promoter to keep the membrane clean at relatively low velocities. A recent innovation for spiral elements is the use of channelized spacer materials. One new spacer material creates channels with much the same fluid dynamics as small tubes, combining the high membrane area of the spiral wound configuration with the fouling resistance of the tubular design.

Although relatively uncommon, reverse osmosis membrane is also available in the tubular configuration. As with ultrafiltration, the feed and concentrate streams flow through the inside of the tube, commonly available in 12.5 mm diameters (0.5 inch), with several tubes in a common housing.

For RO, the hollow fiber configuration has the pressurized flow on the outside of the fiber, not inside as for UF. The extremely small size of RO fibers (typically 42 microns I.D., 85 microns O.D.) results in their description as "hollow fine fibers" (HFF). This small size allows an extremely high element packing density, however, this is largely offset by the lower permeate production rate (flux values) inherent in membrane in the fiber form when compared to flat sheet membrane.

The plate and frame configuration for RO is basically the same as that applied in UF, with stronger support materials required due to higher operating pressures.

RO elements of the spiral wound configuration are of the same basic design as spiral UF. The flow channel height is generally smaller than in UF, ranging from 0.63 to 0.86 mm (0.025 to 0.034 inches). This results in increased packing density compared to UF (approximately +30%). The fluid flow dynamics are similar in the plate and frame and spiral wound configurations. For the application of pure water production, the volume of permeate produced worldwide by HFF and spiral wound elements are roughly equal(3), and both of these designs are far more common than the tubular or plate and frame. The spiral design is used in process and waste water treatment, whereas the HFF is not due to its tendency to foul or "plug" rapidly.

Crossflow MF membrane elements are the most recent entry into the crossflow membrane field and have been available commercially on a limited basis. These typically have been modified pleated cartridge filters and have been applied largely in the pharmaceutical and medical industries where high cost, low volume solutions have been processed. However, spiral-wound MF membrane elements have been in the developmental stage since early 1983. In-house application tests and field test results have shown promise for spiral wound MF use in clarification, color removal and macro-molecular fractionation applications.

The important differences between element configurations which must be considered include the pumping energy costs to maintain efficient operation (fouling resistance), pretreatment requirements, cost of replacement membrane, labor and the capital cost of the equipment. The literature cited contains more in-depth discussion of this subject.

Equipment and System Design Considerations

Quality systems engineering is as important as membrane and membrane element selection in providing the juice processor with effective long term membrane system performance. Supplying a crossflow membrane system that is the most effective design in terms of both capital and operating cost, contains the necessary controls and support equipment, meets the various FDA and USDA requirements for food contact and sanitary construction, and will meet customer separations objectives with long term system performance is the goal of both the customer and systems supplier.

Advances in product and component engineering are also enhancing the prospects for successful crossflow filtration applications in the juice industry. A new multi-stage centrifugal pump has been designed to be in conformance with sanitary standards. Such a pump should bring the efficiencies of the multi-stage centrifugal design - energy efficiency, modular design to allow custom sizing, smooth flow - to the numerous applications in the beverage industry requiring sanitary design hardware. Advances in sanitary design of other hardware components and the element itself continue. Elements, pumps, meters and plumbing hardware which can withstand higher pressures and higher temperatures are under development, driven by the needs of an increasing range of membrane applications.

Crossflow Membrane Filtration Applications in the Juice Industry

Membrane applications fall into three general categories: process, waste treatment and pure water production. In the juice processing industry, potential for all three areas exists, with the process area perhaps holding the most potential overall.

Processing. Membrane processing is often a more cost effective unit process than older technologies, totally replacing, or in some cases supplementing these existing processes. Membranes can also effect separations that are unique to crossflow filtration, or can make separations economically viable that were previously not. Replacement of evaporation as a technique for water removal is one of the larger potential areas for membrane systems. Even with the increased efficiency of multiple-effect and MVR evaporators, evaporation is an energy-intensive process which requires about 940 BTU per pound of water removed. Product dewatering with crossflow membranes requires no phase change and no added heat, not only requiring a fraction of the operating energy but eliminating heat-induced product damage. Concentration of a valuable product is performed at reduced cost while often increasing the final product quality.

Producing juice concentrates is only one of several potential "concentration" applications. Others include the concentration of water soluble dyes from grape skins, pectin and fruit flavorings for jam, jelly and flavorings manufacture, increasing concentration of dilute juices to eliminate the need for blending with concentrates (for example, taking 8-10° Brix apple juice to 12°), and removal of water to save on transportation costs.

In addition to concentration of the majority of juice constituents, flavor enhancement can be achieved by selective passage of off-tasting solutes. Tests have shown that RO membrane can remove bitterness from juices, finished wines and grape juice prior to wine-making. New RO/UF membranes can remove a portion of juice sugars to manufacture "light" beverages for marketing as low calorie alternatives. The same type membranes can increase the sugar content and reduce the acidity of juices for improved cider, wine, cognac, and other beverage production. Increasing solute concentrations can result in reduced fermentation time and also reduced waste loading by the process by-products. Figure 3 shows a production scale sanitary RO/UF system operating to fractionate a valuable beverage by-product.

Waste Treatment. The most commmon waste treatment application for crossflow membrane technology is the reduction of BOD/COD in by-product and waste rinse streams. Although the main impetus to apply membrane in these applications is pollution control, the reuse of valuable materials such as reclaimed sugars and the reuse of purified water often add an economic incentive not originally considered. Reclaimed sugars and other nutrients can be used as animal feed or as fermentation starter materials. Depending upon the molecular size of the BOD-contributing materials, both UF and RO have provided cost-effective pollution control processes.

Water Purification. RO and UF are commonly used for water purification in many industries, including medical, pharmaceutical and electronics. The benefits of salt and organic removal via crossflow filtration are well understood. For juice reconstitution, RO provides a low cost, controlled source of water which is 99.9+% bacteria-free and has virtually all undesirable taste, odor, and color causing solutes removed.

The following are several examples of applications in the juice industry which are currently in commercial production, undergoing field-site testing or appear quite viable based on application tests run on actual fluid samples provided to Osmonics by processors:

Juice Concentration. Pear and orange juice have been concentrated to $\geq$ 20° Brix with SEPA-97(CA) RO membrane. Diatomaceous earth (DE) filtered pear juice was concentrated from 11° to 20+° Brix, and the permeate stream had no detectable sugar even at the high concentration level. A slight, bitter taste was noticed in the permeate stream, indicating the passage of an off-flavor through the membrane which would be beneficial for the final product (concentrate) quality. Fresh squeezed orange juice was concentrated from 11.75° to 23.0° Brix and neither sugars nor flavors were detectable in the permeate stream, according to processors attending the test. They judged the quality of the concentrated juice as quite good. Figure 4 shows flux performance obtained during concentration of fresh orange juice.

Figure 3. Production scale sanitary RO/UF system.

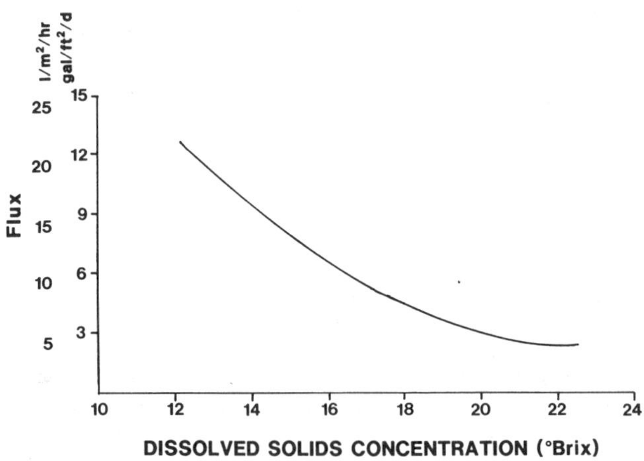

Figure 4. Concentration of fresh orange juice with SEPA-97(CA) RO membrane. Operating conditions: 600 psig operating pressure; ambient temperature; filtered to 5 micron.

'Light' Apple Juice. An apple juice processor investigating the manufacture of a "light" apple juice participated in a batch test which demonstrated good potential for the application. SEPA-50(CA) membrane (with a pore size falling between RO and UF) produced a permeate ranging from 3.3° to 4.5° Brix from an 11.2° Brix apple juice feed. The permeate contained reduced smell and flavor characteristics, but the low sugar level will allow blending with full strength juice to produce an acceptable tasting "light" product. Although not available at the time of that test, a recently developed small-pore UF membrane, SEPA-O(PA), would also be an excellent choice for performing partial removal of mono- and disaccharides.

Pectin Processing. Pectin processing is unusual in that pectin is regarded as both a contaminant and a valuable product to be separated and concentrated. Because pectin occurs naturally in apple juice but can cause post-bottling cloudiness and off-taste problems, removal is usually considered a necessary processing step. Yet pectin from citrus peels (such as lemons) is often recovered as a product after a complicated precipitation process which requires the use of several chemicals. Where essential oil recovery from citrus peels is performed, pectin again can be a contaminant. Crossflow membrane systems can be used in all pectin removal and recovery applications. Pectin removal from apple juice via membrane processing can eliminate the need to add enzymes and the required heat treatment to deactivate them.

Lemon Peel Extract. Two application tests have been run on lemon peel extract. One test objective was simple separation and concentration of the pectin as the desired end product. A "membrane scan" was run using three UF membranes of varying pore sizes and different polymers. This scan showed that UF membranes with pore sizes in the 1000-3000 molecular weight cutoff range (MWCO) performed the best for pectin separation. A qualitative analysis showed that both SEPA-O(CA) and -O(PS) membranes pass a small amount of pectin, viewed as beneficial since the permeating material is the smaller, lower quality pectin molecules and breakdown products. Despite higher pure water flow rates, the membrane scan also showed that smaller pore UF membranes maintained their flux on the process solution better than the larger pore membranes, (a result of less pore-plugging type fouling). Table I and Figure 5 show the comparative permeate flow rates of the membranes tested, both for pure water and on the process solution.

In the second test on a solution from a different extraction process, the objective was to reduce the amount of chemical addition required to precipitate the pectin, a step necessary to allow the reclamation of the essential peel oils. The three UF membranes tested all had MWCO values in the same range, so their polymer characteristics were an important factor in assessing performance. The O(PS) membrane yielded the best flow rate performance, as indicated in Table II. A concentration step increased the pectin level four-fold, according to the processor's analysis. Both tests showed the feasibility of ultrafiltration processing as a means of concentrating and purifying pectin.

Table I. Permeate Rates for UF Membrane Spiral Elements on Lemon Peel Waste - Process #1

SEPA Membrane Type:	MWCO	Permeate Rates (lph) Pure Water Q_p	Waste Solution Q_p - 5 mins.	Waste Solution Q_p - 25 mins.
20K(PS):	20,000	223	64	23
20K(CA):	20,000	209	72	40
O(CA):	1,000	102	59	49

Operating Parameters: 50 psig, 25°C
Sample Composition: 0.1% essential oils
2.7° Brix
pH 4.1

Table II. Permeate Rates for UF Membrane Spiral Elements on Lemon Peel Waste - Process #2

SEPA Membrane Type:	MWCO	Permeate Rates (lph) Pure Water Q_p	Waste Solution Q_p - 1 min	Waste Solution Q_p - 45 mins.
O(VF):	2,000-3,000	481	16	11
O(PS):	1,000-2,000	367	25	14
O(CA):	1,000	196	17	13

Operating Parameters: 100 psig, 25°C
Sample Composition: 0.5% pectin
trace lemon oil
1.1° Brix
pH 1.9

Cranberry Juice Clarification. A test was run on cranberry juice to determine performance of large pore UF membranes in the clarification step, to replace the traditional but troublesome DE filtration or the more expensive and inefficient centrifugation process. In DE filtration, filter aid addition and constant operator attention increase operating costs significantly. In addition, filter aid bleed-through during start-up can contaminate product, and the disposal of spent filter aid is an ongoing problem.

Both SEPA-50K(PS) and the larger pore HF1(PS) membrane effected good clarification, with the HF1(PS) membrane giving the highest flow rates. Crossflow MF membranes would probably perform even better, providing adequate clarification and yielding higher flow rates at lower operating pressures. This test was revealing in that despite the suspended solids level which was sufficient to blind 25 micron filter cartridges immediately, no prefiltration was performed and spiral-wound elements with the traditional mesh-spacer were used. Test results showed significant fouling and resulting reduced flux which leveled off at reasonable levels.

The new corrugated spacer spiral-wound elements should be considered in situations like this where excessive fouling from juice pulp and other suspended solids is expected.

BOD/COD Waste Reduction. Successful tests have been run on fruit juice processing waste where the objective was the reduction of BOD and COD for the purpose of waste liquid disposal. In one test, wash water from pear and peach processing lines contained 6,000 mg/l BOD and 1.5-2.0% fruit sugars (2.0° Brix). Using SEPA-97(CA) RO membrane, the solution was concentrated more than 8-fold to 16.5° Brix. Due to the high osmotic pressure resulting from the concentrated sugars, and some fouling, the expected reduction in permeate occurred. This was easily reversed with a simple cleaning cycle and 100% of the initial pure water rate was recovered. In a similar application, orange waste from an unspecified process contained 0.5% mono- and disaccharide sugars and COD in the range of 5,000 to 6,000 ppm, as well as a small quantity of peel oil. By removing most of the water (processing to high recovery), the conductive solutes were concentrated 8-fold and the refractive solutes 5-fold. Again, all permeate flow lost during processing was completely recovered with a simple cleaning cycle. BOD and COD analyses were not conducted at the time of the test, however, the very low conductivity and refractometer readings of the permeate streams indicated substantial BOD and COD reductions were made.

Flavor Enhancement. Flavor enhancement application tests on both wine and juices have shown interesting results, indicating the potential of membrane processing for improving taste. Removal of bitterness and "off flavors" in finished wine, grapefruit juice and orange juice was accomplished using RO membranes with pore sizes controlled within a small range. Flavor control is a difficult application, since the desirable and undesirable flavor and aroma bodies have close molecular size, polar and steric characteristics. However, a processor's taste-testing panel determined that a low quality finished Chenin Blanc had detectable flavor improvement after processing with a SEPA-89(CA) RO membrane.

Enhanced Second Press Apple Juice. Another concentration application is increasing the sugar and flavor levels of "second press" apple juice. Second press or "cold diffusion" juice is typically 6° Brix, but can range down to 2° Brix. Early pilot studies conducted by Agriculture Canada using SEPA-97(CA) membrane concluded RO was a viable process for achieving 12% TDS juice with minimal product quality loss and improved economics compared to evaporation(4). Commercial installations have been operating successfully in Argentina and Canada for more than two years and the 97(CA) membrane has not been replaced to date. Figure 6 shows pilot plant data from a U.S. processor comparing flux values to degree of concentration. By generating this type of test data, the processor and equipment supplier can determine the size of system appropriate to his concentration requirements.

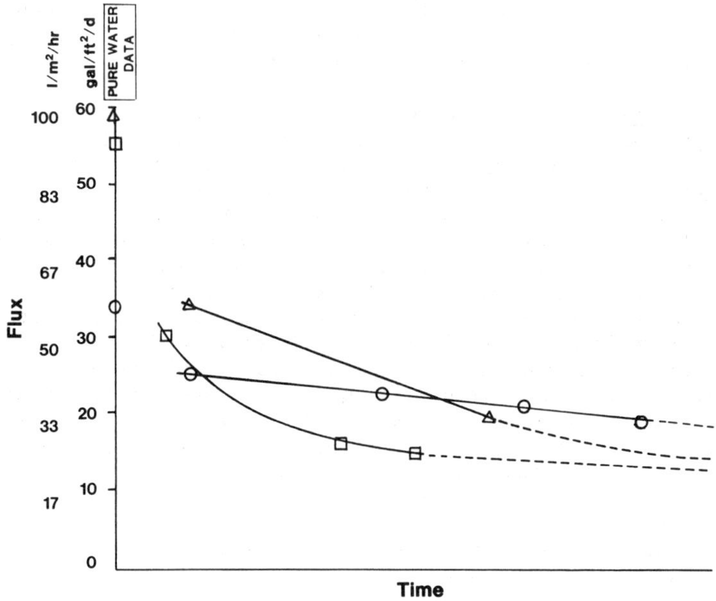

Figure 5. Flux vs. time: pectin removal from citrus waste. Key: Δ, SEPA 20K(PS) membrane; □, SEPA 20K(CA) membrane; and O, SEPA 0(CA) membrane. Operating conditions: 50 psig (345kPa) operating pressure; 2.2% total solids; 0.1% oil; and 25 °C (77 °F).

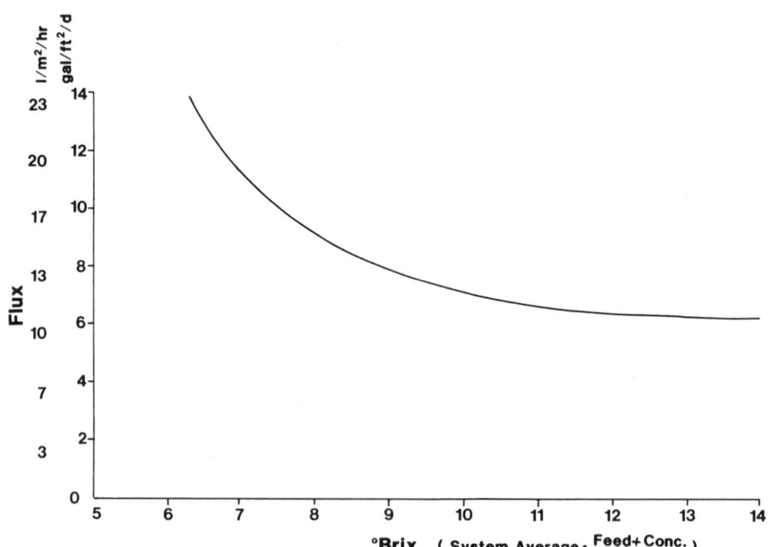

Figure 6. Concentration of second press apple juice (2-4° Brix) to full strength juice (11-12° Brix) with SEPA-97(CA) membrane. Operating conditions: 800 psig operating pressure; corrected to 25 °C; and filtered to 5 micron.

<u>Cost Savings, Payback Period Case Study</u>

RO for juice concentration is receiving increasing interest from many producers. The ability to remove water efficiently without removing the "fresh" taste greatly expands the potential fresh juice marketing area by drastically reducing shipping costs. Based on 1984 reported shipping costs, payback of an RO unit which would remove 60% of the water in juice would be less than 4 months, including both operating and capital equipment costs. Following is an economic analysis outlining costs for apple juice and the subsequent savings that would be realized.

This study is based on actual in-plant pilot testing in Washington State concentrating fresh apple juice. Shipping costs were provided by a juice processor. The DE filtered juice was concentrated from approximately 11° Brix to over 26° Brix at 800 psig operating pressure. The permeate at this high concentration had undetectable refractometer readings and a slight bitter taste. Figure 7 is a graph illustrating the relationship of the SEPA-97(CA) membrane flux to the increasing concentration of the juice, from which scale-up information was obtained.

<u>Design Case</u>. 57,000 liters per day (15,000 gallons) of clarified, fresh apple juice at ~ 10° Brix, concentrated to 22,700 l at ~ 25° Brix in 6 hours processing time.

<u>Membrane Unit Required</u>. OSMO-80B-SS97(CA) all stainless steel or food contact acceptable plastic materials. Pumps, controls, monitors, electrical logic and Clean-In-Place (CIP) system included. 800 psig operation. Machine recovery of 60% (34,000 lpd of water removed). <u>Capital Cost</u>: $110,000 Estimated Maximum

<u>Operating Costs.</u>

Electricity: $3575/year - 43.5 Hp for pump operation, 8 hours daily operation (including two for cleaning).

32.5 KWH x 8 hours/day = 260 KWH

260 KWH x $0.055 KWH =

$14.30/day x 5 days/week x 50 weeks/year = $3575

Cleaning Chemicals: $15,000/year estimated maximum (based on daily cleaning with acid and enzyme cleaners, and chemical sanitization).

Membrane Replacement: $40,000/year estimated maximum (based on complete membrane replacement every year, the guaranteed maximum frequency - actual Osmonics units in the field have achieved membrane life of over 2 years)

Labor: 3 hours/day at $20/hour x 5 days/week x 50 weeks/year = $15,000

<u>Total Yearly Operating Cost</u>: $73,575

<u>Reconstitution Cost</u> - $188,243

Assume $13.21/1000 liters cost ($0.05/gallon) x 57,000 liters/day x 5 days/week x 50 weeks/year = $188,243.

<u>Total First Year Cost</u> - $371,818

Shipping Costs From Washington State(5)

To Dallas, TX - \$0.19/liter (\$0.72/gal.)
To Midwest (Average) - \$0.131 liter (\$0.50/gal.)

34,000 liters/day x 5 days/week x 50 weeks/year = 8,500,000 liters/year water removed.

Assume 20% sales to Dallas, 80% to Midwest

1,700,000 liters x \$0.19/liter = \$ 323,000
6,800,000 liters x \$0.13/liter = \$ 884,000

Total Shipping Costs Saved: \$1,207,000

First Year Savings - \$945,182

\$1,207,000	Shipping Cost Savings
- 261,818	Operating & Reconstitution Costs
\$ 945,182	

Payback Period: less than 4 months:

$$\frac{\$371{,}818 \text{ Total First Year Cost}}{\$100{,}583 \text{ Monthly Shipping Cost Savings}} = 3.7 \text{ Months}$$

General Application Development Procedures

For applications in the juice industry other than simple water purification, there are a series of steps which are very useful in developing the proper membrane system for a particular application. Short-circuiting these steps may result in an underdesigned system, so understanding the steps is important for all potential membrane users. Due to the similarity in some juices from one processor to the next, the juice industry has numerous opportunities for eliminating some of these steps when there is already a body of successful test data and processing experience. Care must be taken to assure juices or waste solutions and process conditions are similar before eliminating these steps.

If juice processors have their own laboratory, they may wish to undertake initial investigation of membrane performance in-house. Samples of commercial and some developmental flat sheet membranes can be purchased from element manufacturers. Traditional laboratory equipment to run these tests includes the "stirred cell", which can yield a rough idea of the separation possible, but is not a crossflow apparatus and does not provide accurate data on flux rates or the degree of concentration feasible. Far superior for lab testing are the crossflow test cells available on the market. These devices test small pieces of different membranes quickly and inexpensively, and with the important crossflow fluid dynamics. A feed, permeate and concentrate stream are all generated in small quantities. Figure 8 shows a

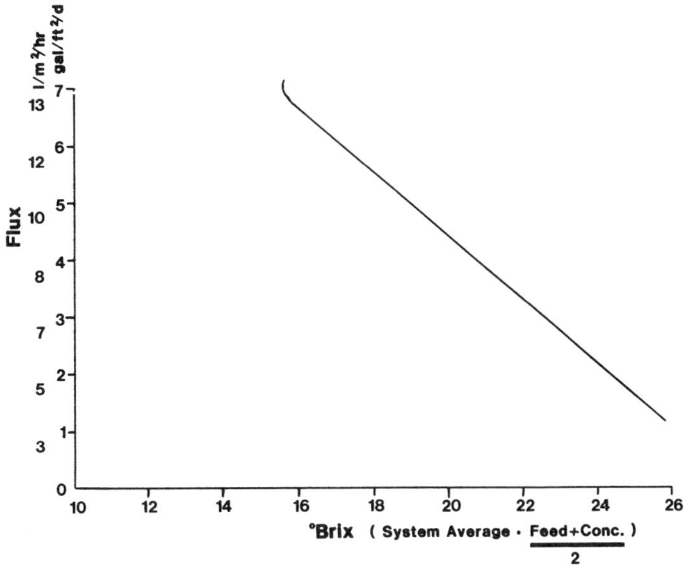

Figure 7. Concentration of fresh, DE filtered apple juice with SEPA-97(CA) RO membrane. Operating conditions: 800 psig operating pressure; corrected to 25 °C.

Figure 8. Stainless steel membrane test cells.

stainless steel membrane test cell capable of 20 to 1000 psig (140-6900 kPa) operation. Lower pressure, less expensive plastic cells are also available.

A laboratory scale sanitary membrane element housing, such as that shown in Figure 9, when coupled with a pump, is a superior test apparatus for creating bench scale-up data. Since the crossflow fluid dynamics are the same as production scale equipment, more realistic concentration, membrane fouling and compatibility data can be generated.

For assessment of overall membrane system performance, the most accurate testing procedure for the juice industry is batch processing conducted with Process Evaluation Systems (PES) equipped with full instrumentation and all the basic components of production scale equipment. Systems such as that in Figure 10 are available for rent or sale for use at the processor's facilities. They are also used for applications testing at the membrane suppliers' facilities. Such systems can quickly scan numerous membrane elements to determine optimum performance. This approach yields the best short term data for selection of membrane type, element configuration and general equipment hardware.

The next step is long-term testing with a prototype pilot system at the processor's facility. Pilot plant test equipment, complete with pretreatment, is commercially available to run on batch process solutions, small process streams or slip streams from larger processes. Testing of two to three months, with daily performance monitoring, will provide accurate performance data and the necessary information on membrane/hardware interaction with the solution to allow a membrane systems manufacturer to guarantee the production scale system.

Experienced and qualified membrane system vendors should work with juice processors to determine the application testing approach best suited to their fluid characteristics and separations objectives. The result should be a full scale crossflow filtration system with guaranteed performance and assurance of long term successful operation.

In conclusion, we would like to state our belief that when adequately evaluated, juice processors will find that crossflow membrane technology holds a great deal of promise for reducing process energy costs, creating new or superior products, dewatering juices, providing a superior source of pure make-up water, and solving waste treatment and disposal problems in ways which often provide extra economic returns as well. Technical processing personnel should work closely with membrane equipment suppliers in testing, evaluation and scale-up of any crossflow membrane filtration system on untested juices or waste streams applications. An experienced understanding of the technology, access to a full range of membranes and hardware and sound engineering practices will all combine to yield superior juice processing methods with this exciting new technology.

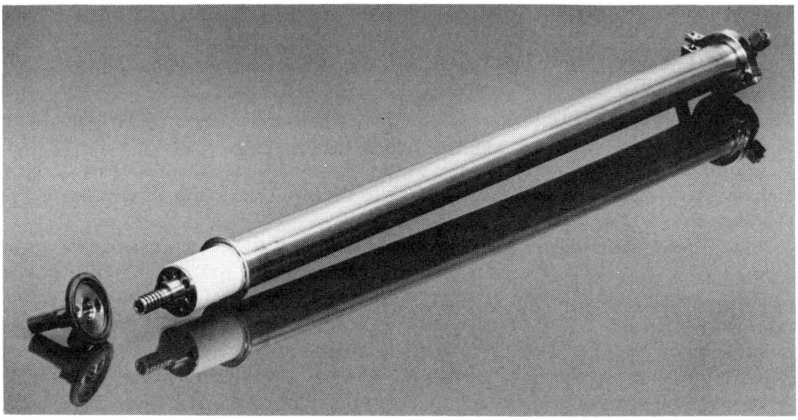

Figure 9. Lab scale membrane element and element housing.

Figure 10. Process Evaluation System (PES).

Literature Cited

1. Kuo, K.; Cheryan, M. J. Food Sci. 1983, 48,113
2. Paulson, D.J.; Wilson R.L.; Spatz, D.D. Food Technology. 1984, 38, 77-87.
3. Kremens, S.S. Proc. 12th Annual Conf. WSIA, 1984
4. Timers, G.E. "Concentration and Recovery of Second Press Apple Juice," Engineering Research Service #7235, Agriculture Canada, 1973
5. Huck, R., personal communication.

RECEIVED March 26, 1985

26

A Mechanistic Study of Halogen Interaction with Polyamide Reverse-Osmosis Membranes

JULIUS GLATER and MICHAEL R. ZACHARIAH

School of Engineering and Applied Science, University of California, Los Angeles, CA 90024

Degradation and subsequent failure of polyamide reverse osmosis membranes in response to aqueous chlorine and bromine species has been reported in the literature and observed in the field. The process is clearly pH dependent showing a different pattern of sensitivity related to membrane type. Work described in this study has been directed toward better understanding the nature of halogen attack on membrane polymer. Experiments were conducted with du Pont Aramid B-9 membranes. Halogen exposed samples were examined by infra-red and NMR spectroscopy. Studies with benzanilide, a model compound, are also reported. Experimental evidence shows aromatic ring halogenation to be the dominant process. Membrane failure appears to result from alteration in hydrogen bonding modes within the polymer.

Polyamide reverse osmosis membranes are known to be chlorine sensitive as reported from both field and laboratory data (1,2,3). Sensitivity varies not only with membrane type but is also related to feed water pH. Considerable test data on membrane halogen interaction has been published by the UCLA Group (4,5,6).

During the course of this work, we became interested in studying the nature of chemical interaction at the molecular level. These studies were undertaken for academic interest and, in addition, with the intent of providing guidelines for polymer synthesis of halogen resistant membranes.

Initial experiments were carried out by attempting to obtain infra-red spectra of various membrane types. Most of this work with thin film composite membranes was abandoned due to insolubility of these polymers in organic solvents.

Our investigation focused on du Pont B-9 Aramid membrane. This material is sensitive to chlorine at low pH and to bromine over a wide pH range (6). Both unexposed and halogen exposed membrane are readily soluble in dimethyl sulfoxide. In addition, this patented membrane has a well-defined polymeric structure (7). Certain experiments were also conducted with benzanilide, a model compound, which simulates the monomeric B-9 unit.

0097-6156/85/0281-0345$06.00/0

Results reported in this paper suggest mechanisms for both halogen attack and subsequent membrane failure. These chemical models were developed from analysis of infra-red (IR) and nuclear magnetic resonance (NMR) spectra of halogen exposed membrane samples under conditions of accelerated testing.

Experimental Procedures

All membrane exposures were conducted at fixed halogen concentrations and pH levels for varying time periods. Following exposure, membrane samples were dissolved in dimethyl solfuxide (DMSO) and re-cast into films of optical quality, on the order of 25 microns thick. IR spectra were taken by direct transmission using a Perkin-Elmer 521 double beam grating spectrophotometer. NMR spectra were taken with membrane solutions in deuterated DMSO solvent using a Bruker WP-200 operating at 200 MHz. Details of experimental procedures and equipment have been described in previous publications (6,8).

Results and Discussion

Several structures for aromatic polyamide membranes are described in the previously mentioned du Pont Patent (7). One listed polymer, the commercial B-9 Aramid membrane consists of a chain of six membered aromatic rings connected by peptide linkages in the meta [1,3] position. The polymer is linear with some hydrogen bonding between chains and with alternate rings bonded on each side to either carbonyl or NH groups. The hydrophilicity of the polymer has been improved by addition of sulfonic acid groups to approximately 10% of the aromatic rings. A general structure for du Pont B-9 membrane is shown in Figure 1.

Figure 2 shows an IR spectrum of unexposed polymer. This trace clearly verifies aromatic character, amide linkages and 1,3 aromatic ring substitution. Evidence for hydrogen bonding is also established from N-H and C-N stretch absorptions. Assigned frequencies used in this analysis are given in Table I (9,10,11,12,13). IR spectra of chlorine exposed membrane were obtained at various time intervals. Figure 3 is a typical spectrum of B-9 membrane following exposure to 30 ppm chlorine at pH 3.0 for 112 hours. Note the pronounced changes, with appearance of new bands along with shifts in previously existing spectral details.

Since spectral interpretation of polymeric materials is difficult, it was decided to select a simple pure compound which would model B-9 polymer chemistry and thus simplify spectral details. The compound chosen was benzanilide ($\phi CONH \phi$), which contains essential structural features of the polymer without the added complexity associated with macromolecules.

Halogenation of benzanilide was conducted with bromine only. This halogen was chosen since it is easier to maintain steady concentration levels in a small system. In addition, both chlorine and bromine have been shown to react similarly with B-9 polymer (3). IR spectra of benzanilide and its reaction products were taken as KBr pellets, which makes sample preparation considerably easier than work-up on polymer films.

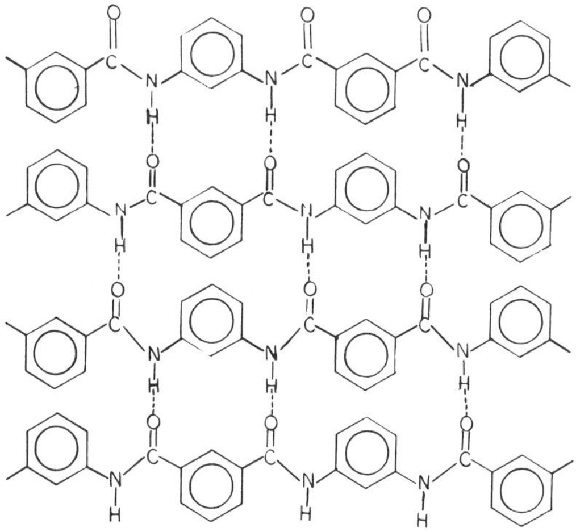

Figure 1. Representative Structure of du Pont Aramid B-9 Polymer Showing Intermolecular Hydrogen Bonding.

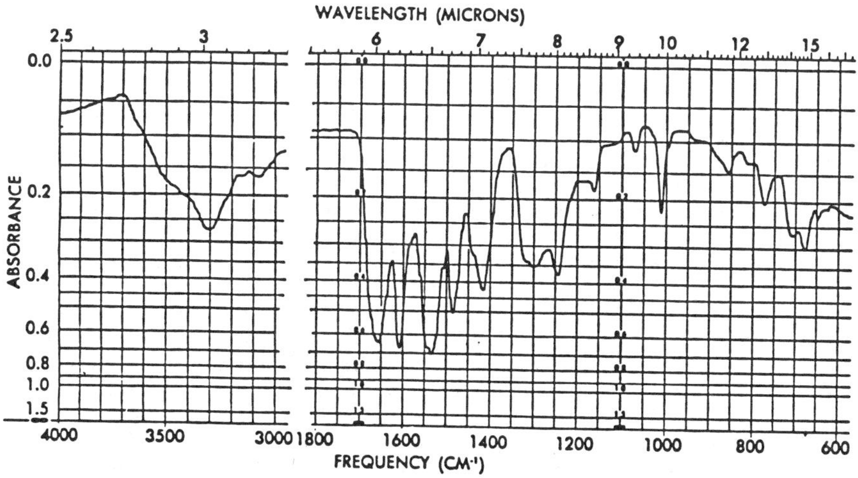

Figure 2. IR Spectrum of Unexposed B-9 Polymer (Thin Film).

Table I. Infrared Frequency Assignments for B-9 Polymer Frequency (Cm^{-1})

N-H Stretch	3310	(H bonded trans)
N-H Stretch	3500	(is/trans from N-H Groups)
Amide I (C=0 Stretch)	1655	
Amide II	1535	
Amide III	1245	or 1300
C-N Stretch	1420	
=C-H Stretch	3070	
Aromatic ring breathing	1610	and 1485
1,3 Aromatic substitution	775	(main), 680 and 1030
1,3,5 " "	860	
Amide I out of plane deformation mode	710	

Bromine exposed benzanilide samples showed an increase in weight due to halogen uptake. Weight increases of 25.4% after 40 hours and 37.3% after 260 hours are in reasonable agreement with mono and di-substitution products (theoretical percentage weight increases of 28.9% and 44.8%). Halogen uptake data will be discussed later in this section.

IR spectra of unexposed and bromine reacted benzanilide show similar trends to B-9 membrane. The spectrum of unexposed benzanilide is shown in Figure 4. Changes in the aromatic regions and in hydrogen bonding character are clearly evident. These two areas of interest (3500 Cm^{-1}) and (800 Cm^{-1}) are examined in time composite spectra shown in Figures 5 and 6. Spectral regions selected for these figures were most amenable to interpretation based on the literature (9,10,11,12,13,14,15).

Note the peak at 3340 Cm^{-1} which steadily decreases coupled with the appearance and development of a second band at 3280 Cm^{-1} upon continued bromine exposure. These absorptions are identified with N-H stretch and the shift can be ascribed to a hydrogen bonding effect. This phenomenon will be discussed further as it applies to our model for membrane failure.

The low frequency end of the spectrum is examined in Figure 6 which provides information on aromatic substitution patterns. A sharp mono-substitution peak of benzanilide for both rings of unexposed compound occurs at 740 Cm^{-1}. As bromination progresses, this peak disappears and is replaced by a well-defined para substitution band at 810 Cm^{-1}. The new absorption is attributed to a 1,2,4 aromatic substitution pattern.

NMR spectra have also been run on brominated benzailide samples using deuterated DMSO solvent. These spectra show extensive changes in the aromatic region of exposed sample but the multiplets are far too complicated for detailed analysis. Certain interpretations, however, may be derived from examination of the low field end of the spectrum shown as a time composite in Figure 7. The prominent peak in unexposed benzanilide is evidently due to the amide proton. With progressive bromination, the peak intensity diminishes and a new peak appears and develops at higher field strength. Movement of this N-H proton peak is indicative of increased proton shielding corresponding to diminished hydrogen bonding. The NMR spectrum shows no evidence

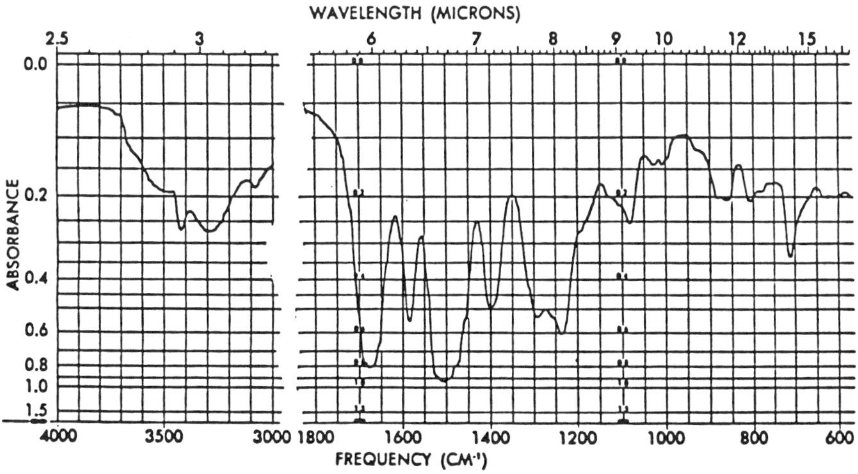

Figure 3. IR Spectrum of B-9 Polymer Exposed to 30 ppm Chlorine at pH 3.0 for 112 Hours (Thin Film).

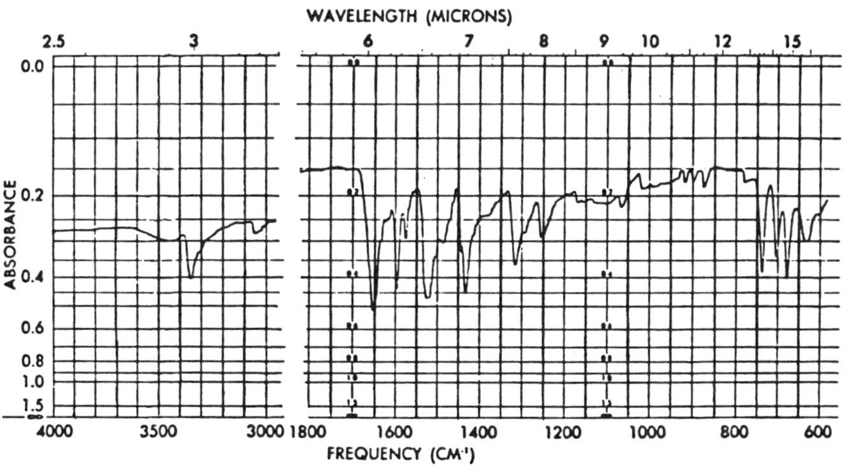

Figure 4. IR Spectrum of Unexposed Benzanilide (KBr Pellet).

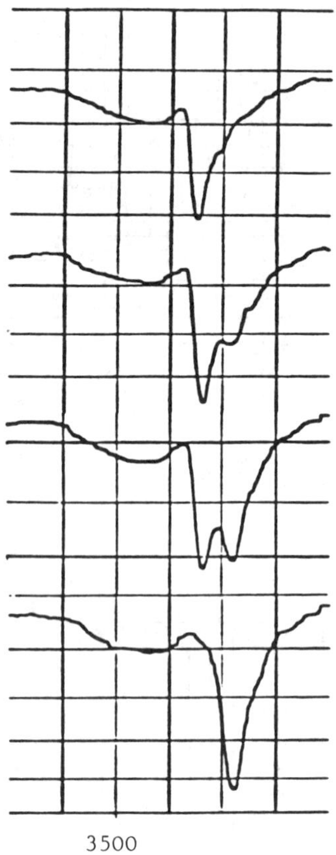

Figure 5. IR Spectra of N-H Stretch Region. All spectra are Time Composites Taken on Samples of Benzanilide Exposed to 6000 ppm Bromine at pH 3.0.

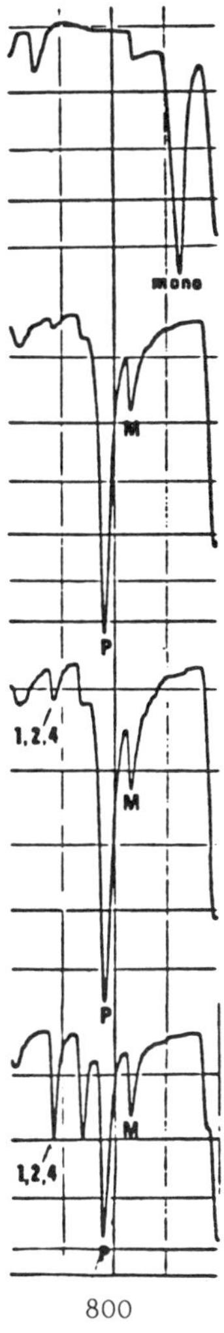

Figure 6. IR Spectra of Aromatic Substitution Pattern. All spectra are Time Composites Taken on Samples of Benzanilide Exposed to 6000 ppm Bromine at pH 3.0.

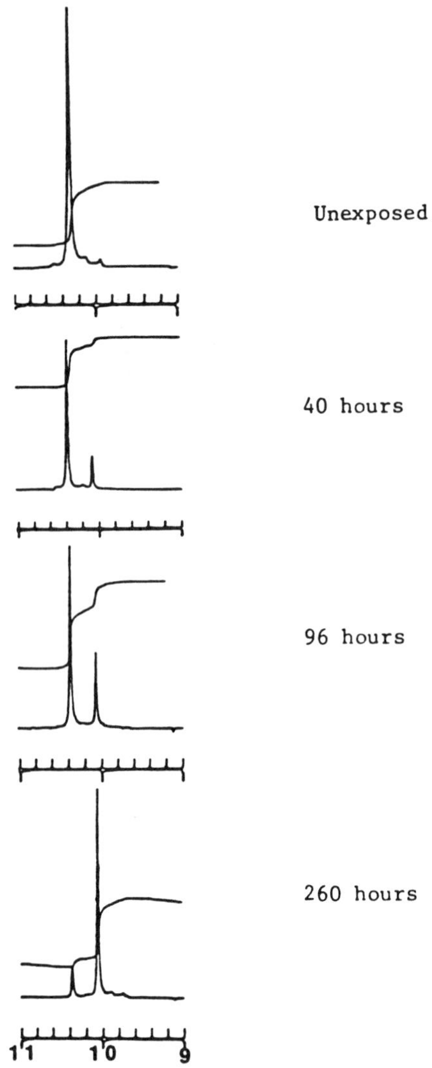

Figure 7. NMR Spectra of Amide Proton (d_6DMSO). All spectra are Time Composites Taken on Samples of Benzanilide Exposed to 6000 ppm Bromine at pH 3.0.

of direct bromine attack on peptide linkages or polymer chain cleavage.

We conclude from model compound studies that halogen substitution occurs on aromatic rings. Evidence for halogen ring substitution of aromatic amides is discussed in the literature by Zabicky (16) and Orton (17). According to Orton, halogen atoms first attack amide nitrogen followed by rearrangement involving ring substitution. We are not certain whether aromatic halogenation takes place directly on the ring or via the Orton rearrangement. We would tend to favor direct halogenation since N-H stretch frequencies in both IR and NMR spectra are shifted but not weakened following halogenation.

Regardless of the mechanism, our results indicate a mixture of meta, para, and 1,2,4 substitution products. Alternate benzene rings of benzanilide may be considered attached to a different type of aromatic substituent. One ring is activated by the -NHCOϕ group (ortho-para director) while the adjacent ring is deactivated by the -CONHϕ (meta director).

The theoretical substitution pattern for benzanilide with complete halogen (X) saturation of each ring is shown in Figure 8. Because of these directive influences, each type of ring should theoretically have a different substitution pattern. In addition, substitution rates should differ since substitution kinetics are significantly faster in aromatic rings containing activating groups (18).

From the IR spectra, (Figure 6) para substitution is observed and evidently occurs on the activated ring. Further exposure causes diminution of the para band with appearance of a 1,2,4 substitution peak. This evidently corresponds with ortho and para substitution on the activated ring. The meta substitution peaks evolve at much slower rates since the deactivated ring is undergoing halogenation. Figure 9 suggests a general sequence for ring substitution of benzanilide at intervals within the experimental time frame. Note that formation of essentially mono and disubstitution products is consistent with the halogen uptake data discussed previously.

Evidence from IR and NMR spectra also shows an increase in hydrogen bonding of the amide proton with increased ring halogenation. This may be interpreted as follows. The majority of benzanilide molecules, as observed from NMR spectra, are in the trans configuration such that hydrogen bonding may only take place intermolecularly as illustrated in Figure 10a. Following halogenation, however, para or meta position halogen atoms protude from the molecule at close proximity to N-H groups. The resulting hydrogen-halogen bonds are thus favored over hydrogen-oxygen bonds (Figure 10b). These bonds should also be more stable because of a reduction in steric hindrance imparted by large ring structures on either side of the amide group.

IR spectra of exposed B-9 polymer provides some insights into aromatic substitution patterns. Analysis of spectral data shows evidence of 1,2,3 (1580 and 1080 Cm^{-1}) and small amounts of 1,3,5 aromatic substitution (810 Cm^{-1}) replacing the original 1,3 substitution (775 Cm^{-1}) found in unexposed polymer. This information compares favorably with spectral changes observed in model compound studies. Probable substitution patterns on B-9 polymer are proposed by the formulas shown in Figure 11. These structures were developed by piecing together spectral evidence with theoretical prediction based on ring activation or deactivation.

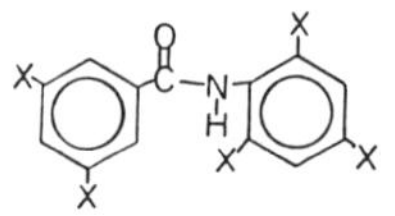

Figure 8. Maximum Theoretical Halogen Substitution Pattern of Benzanilide.

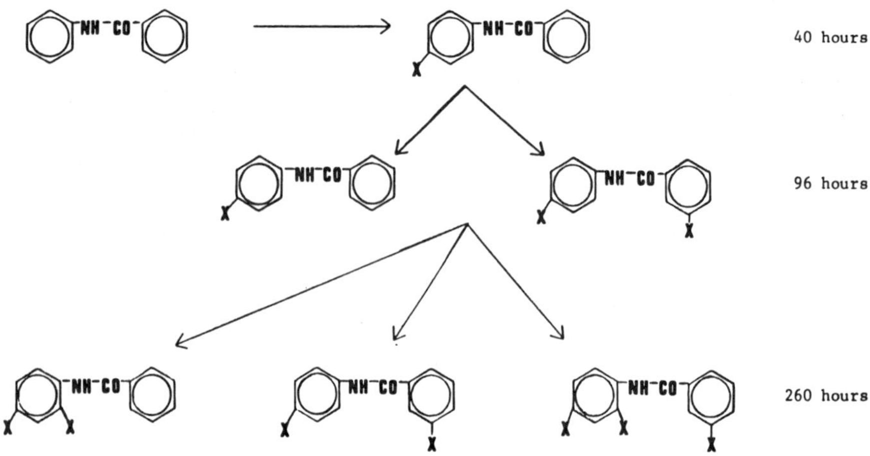

Figure 9. Proposed Halogen Substitution Patterns of Benzanilide Exposed to 6000 ppm Bromine at pH 3.0 for Varying Time Periods.

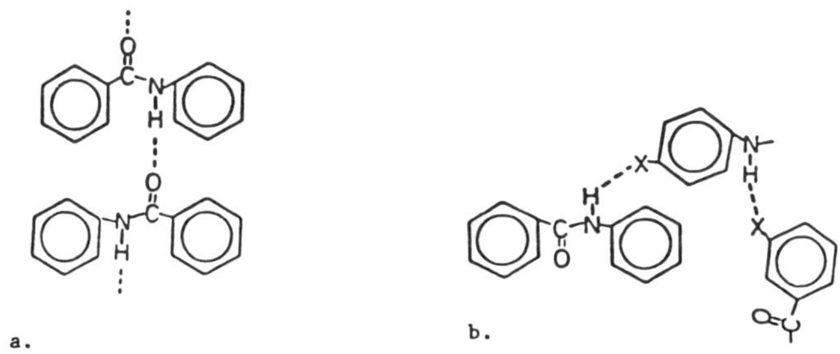

Figure 10a, b. Hydrogen Bonding Modes in Unexposed (a) and Halogenated Benzanilide (b).

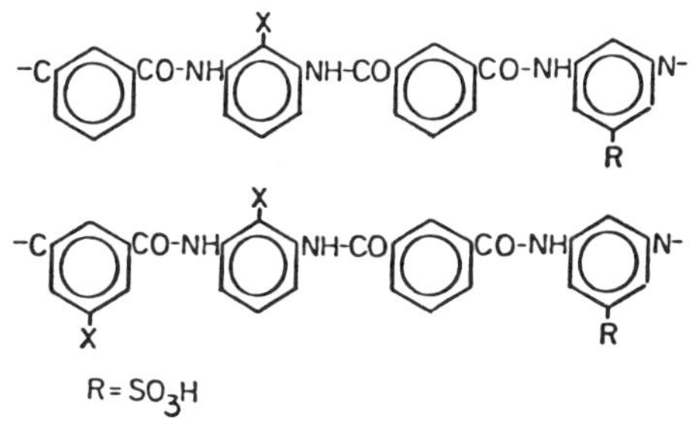

Figure 11. Proposed Halogen Substitution Patterns for B-9 Polymer.

The R substituents in Figure 11 symbolize sulfonic acid groups which are found on approximately 10% of the aromatic rings. These groups occur only on the N-H substituted aromatic rings. In B-9 membrane they are usually in the form of the calcium sulfonate salt (19) but become protonated at low pH.

Halogenation of the N-H substituted rings containing sulfonic acid groups is unlikely because of ring deactivation by this m-direction group. The three available sites for substitution would thus be blocked by antagonistic action of N-H and sulfonic acid groups.

Interaction between $-SO_3H$ groups and chlorine have been considered and are suggested from literature. It is proposed that $-SO_2C\ell$ groups may be formed on the ring or that the whole sulfonic acid group is replaced by chlorine. The first possibility seems more plausible since aromatic substitution of one electronegative group by another is not very likely.

In any event, the 2,4,6 positions on these rings would remain inactive toward further halogen substitution. It is our feeling that sulfonic acid groups have little or no bearing on the model for membrane failure proposed in this paper.

IR spectra also show the presence of intermolecular hydrogen bonds in unexposed polymer as indicated by the strong absorption band at 3280 Cm^{-1}. This is in good agreement with studies by Bellamy (9) on simple proteins with an absorption band at 3300 Cm^{-1}. Bellamy attributes this band to intermolecular hydrogen bonding in trans configuration. Note that B-9 peptide linkages resemble bonding in simple Proteins.

Comparison of Figures 2 and 3 also show a decrease in hydrogen bonding as indicated by weakening of the 3280 Cm^{-1} band and appearance of a new band at 3420 Cm^{-1}. This shift to higher frequency suggests a transition to weaker intramolecular hydrogen bonds involving ring halogens. A proposed structure for portions of the halogenated polymer chain is given in Figure 12.

The favorable (ortho) halogen position enables the amide proton to hydrogen bond intramolecularly. This phenomenon probably causes aromatic rings to be drawn out of plane. The polymer may then pucker or coil back on itself by hydrogen bonding at different positions along the chain. The resulting deformation provides a physical basis for membrane failure accompanied by increases in water and salt transport.

The observed increase in hydrogen bond strength with halogenation of benzanilide seems to contradict observed weakening of hydrogen bonds in halogenated B-9 polymer. This apparent contradiction may be explained by considering differences in polymers compared with molecules containing the same functional groups.

The model compound being a small highly mobile molecule is able to move into positions which enhance hydrogen bonding. The strong intermolecular halogen-hydrogen bonds shown in Figure 10b form readily with no stereochemical restrictions.

By contrast, B-9 polymer chains are strongly intermolecularly hydrogen bonded as shown in Figure 1. It is proposed that ring halogenation causes chain separation with accompanying steric hindrance between polymer chains. Because of the large size and rigidity of these chains, hydrogen bonding is restricted to relatively weak intramolecular interactions with para position halogens as shown in Figure 12.

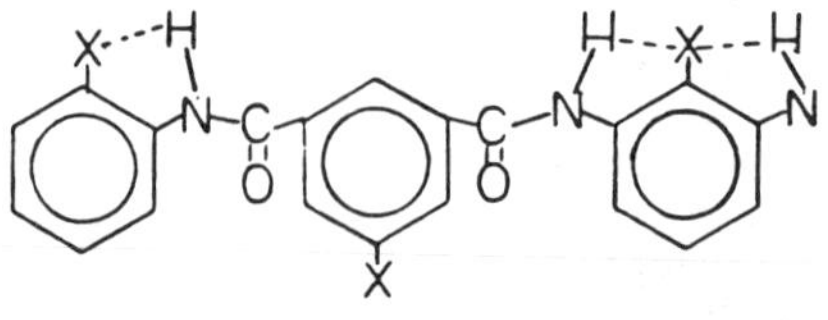

Figure 12. Intramolecular Hydrogen Bonding in Segment of B-9 Polymer Chain Following Halogenation.

From this study, it appears that B-9 membrane failure takes place in three concerted stages as follows:

1. Ring halogenation according to allowable substitution patterns.
2. Disruption of intermolecular hydrogen bonding with consequent formation at intramolecular hydrogen bonds to halogen substituents.
3. Poly-peptide chain deformation resulting in puckering and/or coiling of chain segments.

Additional experimental evidence on polymer chain deformation would be desirable and would help better establish this proposed mechanism.

Conclusions

Exposure of du Pont B-9 aromatic polyamide membrane to chlorine or bromine under accelerated testing conditions results in membrane failure. The halogen reacts irreversibly with the membrane polymer causing structural changes which bring about increasing passage of water and salt.

Evidence generated in this study suggests that halogen attack occurs directly on aromatic rings by electrophilic substitution. This evidence was developed from analysis of IR and NMR spectra of halogen exposed membrane samples. The substitution pattern is dependent on time of exposure and on whether the ring is activated or de-activated by attached peptide linkages. Polymer chain cleavage was not indicated from examination of the spectra.

Both IR and NMR data also suggest that aromatic halogenation causes an alteration on hydrogen bonding patterns within the polymer. We propose that hydrogen bonding shifts from inter to intramolecular within a chain. This tends to cause chain deformation followed by alteration in gross polymer properties. Loss of salt rejection and enhanced water transport result from these structural changes.

The reaction mechanism suggested for du Pont B-9 Aramid membrane is not necessarily applicable to other types of aromatic polyamides. It seems reasonable, however, to assume that a dominant process in failure of aromatic polyamide membranes involves ring halogenation.

Based on these findings, we feel that halogen resistance of membranes can be improved by protecting chemically reactive sites on aromatic rings. Results reported here may be of interest to groups engaged in membrane polymer synthesis.

Editorial Comment

It appears from du Pont Patents that several types of membrane materials may have been used at various times in Permasep reverse osmosis modules. The reader should be aware that mechanisms of halogenation and membrane failure proposed in this paper may have relevance to only one of these variants.

Acknowledgments

The authors wish to acknowledge the support of this research provided by the Office of Water Research and Technology, U.S. Department of the Interior, Washington, D.C., under Grant No. 14-34-0001-0470. Partial support was also provided by the State of California Saline Water Research Funds administered by the Water Resources Center at the University of California, Davis, California. We also express our thanks to E.I. du Pont de Nemours & Co. for providing membranes used in this research. Thanks are also due to Drs. Lynn Applegate and Herman Pohland of the du Pont Permasep division for their constructive suggestions.

Finally, we would like to dedicate this paper to the memory of the late Professor J.W. McCutchan - a pioneer and leader in the development of reverse osmosis technology.

Literature Cited

1. Progress Report by Fluid Systems Division of U.O.P. on Contract No. 14-30-0001-3303, to the Office of Water Research and Technology, U.S. Department of the Interior, 1975.
2. Progress Report by Fluid Systems Division of U.O.P. on Contract No.14-34-0001-6516, to the Office of Water Research and Technology, U.S. Department of the Interior, 1976.
3. Glater, J., McCutchan, J.W., McCray, S.B., Zachariah, M.R. In "Synthetic Membranes" Turbak, A.F., Ed. ACS SYMPOSIUM SERIES No. 153, American Chemical Society, Washington, D.C. 1981; p. 171.
4. Glater, J., McCutchan, J.W., McCray, S.B., Zachariah, M.R. Proc. Water Reuse Symposium II, 1981, p. 1399.
5. Glater, J., Zachariah, M.R., McCray, S.B., McCutchan, J.W. Desalination, 1983, 48, 1.
6. Final Report by UCLA-Engineering on Contract No. 14-34-0001-0470 to the Office of Water Research and Technology, U.S. Dept. of the Interior, 1983.
7. Richter, J.W., Square, K., Hoehn, H.H. U.S. Patent 3 567 632, 1971.
8. Zachariah, M.R. M.S. Thesis in Engineering. University of California at Los Angeles, Los Angeles, 1982.
9. Bellamy, L. "The Infrared Spectra of Complex Molecules," Wiley, London, 1975.
10. Badger, R.M., and Rubalcava, H. Proc. N.A.S., 1954, 40, 12.
11. Flett, M. Spectrochimica Acta, 1962, 18 (2), 1546.
12. Letaw, H., Gropp, H. J. Chem. Phys., 1953, 21, 1621.
13. Richards, R.E., Thompson, H.W., J. Chem. Soc., London, 1947.
14. deKlein, W.J., Plesman, A.R. Spectrochimica Acta, 1972, 28A, 673.
15. McLachlan, R.D., Nyquist, R.A. Spectrochimica Acta, 1964,20,1397.
16. Zabicky, J. "The Chemistry of Amides," Wiley Interscience, New York, 1970.
17. Orton, K.J.P., Frederick, G.S., Williams, G. J. Chem. Soc., London, 1982, p. 998.
18. March, J. "Advanced Organic Chemistry," McGraw-Hill, New York, 1977.
19. Sundet, S.A. U.S. Patent 4 385 148, 1983.

RECEIVED February 22, 1985

27

Characteristics of Reverse-Osmosis Membrane Fouling at the Yuma Desalting Test Facility

JOHN W. KAAKINEN and CHARLES D. MOODY

Division of Research and Laboratory Services, Engineering and Research Center, Bureau of Reclamation, Code D–1523C, Denver, CO 80225

From extensive investigations it was concluded that membrane fouling and mechanical deformation of spiral-wound cellulose-acetate membrane-element components were responsible for excessive decline rates of permeate water and solute transport during RO (reverse osmosis) proof tests of two contractors' equipment at the Yuma Desalting Test Facility. Effects of fouling were significant even though stringent criteria for the RO feed water quality including turbidity and plugging factor were met or exceeded by the proof test pretreatment system and the mass of material found fouling the membrane surfaces seemed slight. Analyses of membrane sediments and the results of RO experiments have delineated the causes for the changes in RO transport, which differed for each of the two proof units even though the units were generally similar.

In the application of RO (reverse osmosis) for desalting it is necessary to include sufficient initial equipment capacity in the plant design to compensate for the gradual losses with time in water production rates that result from membrane fouling and mechanical deformation of membrane element components. If there are greater losses in RO performance than predicted by the design and if any loss caused by fouling cannot be reversed sufficiently by cleaning, the membrane replacement rate required to maintain the design capacity will be greater than expected which, in turn, will increase the cost of producing desalted water. Membrane fouling as considered in this paper is the adsorption onto the membrane surface or absorption into the membrane of materials from the RO feed water, which is responsible for a change in the RO transport properties. The amounts and composition of materials that foul RO membranes as well as the RO performance response to the foulants are a function of the raw water composition, the pretreatment's effectiveness to remove fouling materials from the raw water and to control biological growth, the design of the RO equipment including RO element materials and configurations, and RO operational parameters including flow rates and desalting recoveries.

The high applied pressure (approximately 1.7 to 6.9 MPa) required to drive RO desalting can cause mechanical deformation of the plastic RO element components, which includes membrane compaction and in the case of spiral-wound elements the compression of the product-water channel material or deformation of the membrane into the tricot product-water channels (1). Such deformations cause a greater hydraulic resistance to product-water flow, and thus, a drop in RO product flow capacity. The rate of compaction of cellulose acetate membranes is increased by RO operation on feed water containing large concentrations of certain organic solutes that plasticize the membrane material causing the membrane to lose some of its asymmetry (2). Another effect is the preferential absorption of some organic solutes in the membrane causing a decrease in water and ionic transport, but this process may be reversible as initial RO performance may again return when the organic solutes are absent from the RO feed water (3). It is generally difficult to impossible to differentiate among these potential causes of RO performance decline from only the water and solute transport data of an RO plant because the performance responses to various types of membrane fouling and mechanical deformation are usually similar.

This paper describes the nature of membrane fouling and RO performance changes as the result of membrane fouling and mechanical deformation observed at the YDTF (Yuma Desalting Test Facility). Included are descriptions of RO feed water quality and the compositions of sediments that fouled RO membrane surfaces at the YDTF. Also included is a summary of RO experiments conducted with different pretreatment processes to identify the causes of excessively high decline rates of product water capacity of RO proof test units at the YDTF, which were operated for the final acceptance of desalting equipment for the YDP (Yuma Desalting Plant). These RO experiments were able to differentiate between membrane fouling and mechanical deformation of RO elements as general causes of water and solute transport rate changes and to further identify RO performance responses to specific classes of membrane foulants.

The Yuma Desalting Plant and RO Proof Tests

The YDP, under construction by the Bureau of Reclamation near Yuma, Arizona, presently scheduled for completion and initial operation in 1989, will desalt the agricultural return flow from 30 000 hectares of agricultural land. The YDP product will blend with the Colorado River to lower the salinity of the river flowing into Mexico as required by an international treaty between the United States and Mexico. The YDP will be the worlds largest RO plant with a design product-flow capacity of 3.17 m^3/s (72.4 million gal/d) operating at about 70 percent recovery on a feed water containing 3.0 kg/m^3 of TDS (total dissolved solids). The U.S. Government contracted with FSD (Fluid Systems Division of UOP, Inc.) to supply equipment to produce 2.2 m^3/s of desalted water and HWS (Hydranautics Water Systems) to supply equipment to produce 1.0 m^3/s of desalted water. Annual cost of the YDP was estimated in 1977 to be $20 million (July 1982 dollars), including amortization of capital, chemical, energy, labor, and membrane replacement costs (4). For expected membrane lives of 3.0 years for FSD equipment and 3.75 years for HWS equipment, $3 million or 15 percent of the total YDP cost equal to $3 million is the estimated annual cost for membrane replacement.

Contracts between the Government and the two RO equipment suppliers required the successful completion of operational proof testing of a representative segment of the contracted RO equipment of FSD and HWS. Both FSD and HWS had previously provided smaller RO units that were tested for several years at the YDTF prior to the prepararation of their YDP bids, and both used their prebid YDTF RO performances to develop their YDP proposed bids. Each PTU (proof test unit) included a first stage of four parallel vessels and a second stage of two parallel vessels fed from the reject of the first stage. The equipment of both suppliers contained spiral-wound cellulose-acetate RO membrane elements. Each FSD element was 0.30 m in diameter and 1.5 m long and contained cellulose-diacetate membranes. The HWS supplied elements, manufactured by DSI (Desalination Systems, Incorporated), were 0.20 m in diameter, 0.63 m long, and contained cellulose-di/triacetate-blend membranes.

Results of the proof testing were mixed. Both suppliers' equipment demonstrated adequate mechanical strength. The salt rejections of 96.5 to 97.5 percent were higher than contracted for and, contrary to expectations, increased with time, especially initially. However, the temperature-corrected product flows of both PTU's decreased at faster rates than projected by the contractors such that both units failed to meet their warranted desalting capacities. The water-transport coefficient A(t), which corrects product flux to conditions of constant temperature, feed pressure, feed and product salinity (osmotic pressure), and recovery, was expected to decline with time according to the power relationship:

$$A(t) = A(1)\ t^{m} \tag{1}$$

where t is the operating time in h, A(1) is the water-transport coefficient at an operating time of 1 h, and m is the log-log slope or flux-decline rate. Both YDP contractors projected m = -0.02 for the expected YDP feed water temperature range of 18.3 °C to 29.4 °C with an applied pressure of 2.76 MPa. Each contractor's proposal stated that the power model with m = -0.02 should take into account the expected declines due to both mechanical deformation of their elements plus membrane fouling, although FSD also provided for increasing the RO feed pressure to 3.10 MPa to be able to compensate further for decreases in membrane productivity, as required. Regression analyses of water-transport coefficient versus operating time of both manufacturers' equipment did give m = -0.020 for the operating period of 1 to 500 h, but with greater operating time the best curve-fit values of m progressively increased, for example, up to m = -0.040 for the operating period of 1 to 2000 h. The increasing magnitude of m with time indicates that Equation 1 would underestimate the actual drop in product flow capacity at operating times that are longer than that of the data used in the curve fit. Membrane cleanings by solutions specified in the contracts failed to adequately reverse the water-flux declines.

To satisfy the Government that their equipment will meet their guaranteed product flow capacities in the YDP and to continue with their contracts it was agreed that both FSD and HWS will provide additional membrane equipment than originally specified in their YDP Proposals, 25 percent for FSD and 12 percent for HWS, at no additional cost to the government. Before and after these agreements numerous studies into the causes of the poor proof test performance were conducted by the Government (5) and the two contractors. Because the contractors' test results regarding mechanical deformation of their membrane elements are proprietary, their results are not included here. However, the present paper describes some of the Government's key RO experiments conducted at the YDTF to study membrane fouling and its significance compared to overall mechanical deformation of the contracted membrane elements.

RO Feed Water Quality and Pretreatment

The raw water consists of agricultural return flow from 30 000 hectares of land in the Wellton-Mohawk Irrigation and Drainage District pumped from 106 drainage wells and conveyed up to 140 km to the desalting site. Because the canal system is uncovered, windblown dust and soil particles enter and become suspended in the canal water. The open canals also support high biological activity including planktonic algae (diatoms) and aquatic weeds, which enter the pretreatment system. Decaying plant and animal materials form humic and fulvic acids in the canal water. The amounts of suspended sediments and biological activity in the canal vary with the weather and season. The dissolved species in the canal water remain relatively constant in composition.

The raw water required extensive pretreatment to meet RO feed water criteria in the YDP contracts. Pretreatment steps operated at the YDTF simulate those to be used in the YDP, which include:

1. Primary chlorination to prevent biological growth in the pretreatment and desalting systems, also serving to oxidize dissolved iron and manganese for subsequent removal during lime softening.
2. Sedimentation of large suspended particles in a grit basin.
3. Partial lime softening in a solids-contact reactor-clarifier to remove about half of the calcium ion to avoid gypsum scaling in the RO systems; to remove the majority of suspended particulates including algae, clays, and iron and manganese hydroxides, which would quickly foul the membrane surfaces; and to remove alkalinity, thereby reducing the subsequent acid dosage needed to lower the pH of the RO feed water. (The cost savings of buying sulfuric acid alone justify the total cost of partial lime softening in the YDP, aside from the other benefits.)
4. Acidification of the clarifier effluent to prevent calcium-carbonate cementation of the filters.
5. Anthracite-sand gravity filtration to remove additional particles.
6. Secondary chlorination to maintain 1.0 g/m^3 of free chlorine in the clearwell to prevent microbiological growth.
7. Sulfur dioxide addition to the clearwell effluent to control the free chlorine residual at 0.5 g/m^3 in the RO feed water.

8. Sodium hexametaphosphate addition of 1.5 g/m^3 to the RO feed water to provide a further safety factor against scaling of the membranes by gypsum and other sparingly soluble salts.
9. Addition of sulfuric acid to lower the pH of the RO feed water to 5.5, which is near the pH of the minimum rate of cellulose-acetate hydrolysis.

During RO proof testing at the YDTF, the above pretreatment system performed reliably in terms of meeting all RO feed water criteria agreed to by FSD and HWS in the contracts (5). Calcium, strontium, barium, silica, iron, and manganese were kept at sufficiently low levels in the RO feed water to prevent precipitation of sparingly soluble salts concentrated in the RO reject-brine stream. Plugging factors, measured by an automatic monitor (6), averaged less than 15 percent (corresponding to a 15-min silt-density index of less than 1.0), well below the 65 percent criterion in the YDP contracts for acceptable RO feed water and also less than the plugging factors usually measured at other RO desalting sites. Turbidities were below 0.8 JTU during 99 percent of the proof-test operating time. The contractors stated that feed water turbidities up to 1.0 JTU have been found acceptable at other RO desalting sites. The pretreatment chlorination was very effective in controlling biological activity in the pretreatment and desalting systems as there was never any biologically related operational problem apparent during normal pretreatment operation. Furthermore, weekly measurements of bacterial plate counts of the RO feed water gave insignificantly low values, and there were also only insignificant accumulations of polyanionic slimes of perifitic bacteria on glass slides that were suspended for different exposure times in flowing RO feed water (measured by Dr. Harvey Winters, biology professor at Fairleigh Dickinson University).

Although not included as part of the RO feed water quality criteria of the contracts, characterization of aquatic humic substances at the YDTF by the U.S. Geological Survey (7) indicated that there were 5 g/m^3 of DOC (dissolved organic carbon) in the raw canal water and 3 to 4 g/m^3 of DOC in the RO feed water, of which approximately 25 percent were humic and fulvic acids. Because these YDTF DOC levels were less than an average value of 7 g/m^3 for surface water in the U.S. (7), they cannot be considered unusual. The chlorination of the organic materials in the raw water resulted in the formation of halogenated organic compounds, primarily organic bromides rather than chlorides, due to the presence of about 0.7 g/m^3 of bromide in the raw saline water.

Membrane Sediment Characteristics

Initial efforts to try to determine the causes of the high product waterflow declines were to inspect membrane surfaces for the presence of material fouling the surfaces and to try cleaning procedures to determine if RO performance could be restored by removal of any foulants present. The cleaning solutions used contained sodium tripolyphosphate and sodium EDTA, and citric acid. None were effective at permanently restoring the RO performance.

Just prior to the cleaning attempts the lead membrane element from vessel number 4 of the first stage of each proof test unit was removed, dissected (the outer shell cut open), and unwrapped to expose the spiral-wound components for inspection. In both the FSD and DSI elements there were seemingly small amounts of extremely fine, brownish colored material on the membrane surfaces, distributed unevenly in a pattern apparently affected by the ribbed brine-channel material (Vexar) adjacent to the membrane surface. There was more material visible on the DSI membranes than the FSD membranes. Every membrane leaf in each element was scraped with a rubber squeegee and rinsed with deionized water to transfer as much of the membrane sediment as practical into plastic sample containers. The sediment samples were analyzed chemically and mineralogically by methods described elsewhere (7).

Results of the sediment analyses are given in Table I. Also included at the top of Table I are some RO performance data for the corresponding vessels of the proof test units for the operating periods prior to element removal. Note that although the two PTU's operated on essentially the same pretreated water for approximately the same operating time before element removal, there was about twice as much sediment per membrane area on the DSI membranes as on the FSD membrane. Note also that the DSI membrane element had a 46-percent greater average product water flux, a 53 percent greater feed-end brine-channel velocity, and thus, nearly the same ratio of water flux to brine-channel velocity in comparison to the corresponding data for the FSD membrane element. The analyses results summarized in Table I show that the predominant size fraction of the sediment samples is less than 2 μm and consists mostly of smectite clays (montmorillonite). Although neither the surface area nor the charge of the sediment samples was measured, it has been reported (8) that montmorillonite has a very high external specific surface area of approximately 250 m^2/g and a high negative charge according to its high specific cation-exchange capacity of about 0.9 meq/g. Some of this negative charge is probably neutralized by the coatings of iron and manganese oxides and organic material, which would tend to promote coagulation of the otherwise stable clay particles.

A very approximate estimate can be made from the data in Table I of the potential ability of the sediments found on the proof-test membranes to coat the membranes in layers and thereby possibly affect membrane transport rates. Such an estimate provides perspective on the importance of the small mass of sediment found on the membrane surfaces. It is assumed that the sediment has a specific surface area of 200 m^2/g and a plate-like structure as does montmorillonite (8). It is further assumed that one side of the plate or half of the external surface is in contact with the membrane surface as if the montmorillonite plates were lying parallel along the membrane surface. The results of the calculation given at the bottom of Table I show that the estimated sediment masses on the membranes after less than 3000 h of operation had sufficient total surface area to potentially cover the FSD membranes 4.5 times and the DSI membranes 10.5 times. Thus, although the sediment had a relatively small mass per membrane area and was not uniformly distributed on the membrane surfaces, there was reason to suspect that the sediment played a role in the observed poor RO performance because of the sediment's large estimated surface area. However, specially designed RO experiments were required to identify the relative importance of membrane-surface fouling as compared to mechanical deformation in causing the changes in RO water and solute transport observed in the YDTF proof tests.

Table I. RO Performances of Vessel No. 4 of Two Proof Test Units Prior to Removal of Their Lead Elements, and Analyses of Membrane Sediments Removed from These Elements

	Membrane manufacturer	
RO vessel performance up to element removal	FSD	DSI
Operating time, h	2674	2965
Number of elements per vessel	4	10
Active membrane area per element, m^2	102	20.9
Brine-channel cross-sectional area, m^2	0.022	0.010
Average feed flow rate entering vessel, 10^{-3} m^3/s	4.2	3.0
Total volume of feed water entering vessel, 10^3 m^3	39	32
Average product flow rate per element, 10^{-3} m^3/s	0.50	0.15
Total volume of product per element, 10^3 m^3	4.8	1.6
Average water flux through membranes, 10^{-6} m/s	4.9	7.2
Average feed-end brine-channel velocity, m/s	190	291
(Flux)/(feed-end brine-channel velocity) X 10^9	26	25
Decrease in water flux during operation, percent	17	15
Decrease in salt transport during operation, percent	29	44
Membrane sediment analyses results		
Mass of freeze-dried sediment recovered from membrane surfaces, g	1.53	1.08
Estimated efficiency of recovering sediment from membrane surfaces, percent	33	50
Estimated total mass of sediment fouling membrane surfaces, g	4.6	2.2
Estimated sediment mass per membrane area, g/m2	0.045	0.105
Composition of material coating inorganic sediment		
Organic carbon, percent of sediment	14.2	20.0
Iron oxide coating, percent as Fe	1.3	4.0
Manganese oxide coating, percent as Mn	0.64	0.96
Size fractionation of inorganic sediment, percent by mass		
2 mm - 60 μm (sand-size fraction)	0	0
60 μm - 2 μm (silt-size fraction)	8	15
less than 2 μm (clay-size fraction)	92	85
Composition of clay-size fraction, percent by mass		
Smectite (montmorillonite plus mixed-layer swelling clays)	65	70
Amorphous clay minerals	10-15	5-10
Kaolinite	5-10	5-10
Illite (clay-size mica)	5-10	5-10
Quartz and feldspars	10	10
Est. total surface area of sediment (200 m^2/g), m^2	920	440
Sediment surface area/(2 X membrane surface area)	4.5	10.5

Note: Analyses were conducted by the U.S. Geological Survey (7).

Clean Water Test

An experiment called the CWT (clean water test) was designed and operated at the YDTF to differentiate between the effects on RO performance of mechanical deformation and of surface membrane fouling for the FSD and DSI equipment. In the CWT new proof-test-specification RO elements of both manufacturers were operated on a feed of presumably foulant-free, "clean water," which was the RO permeate from the FSD PTU. The original concept for the CWT came from the compaction test loop used by Gulf General Atomic (10). Water should be essentially free of colloidal particles and larger organic molecules immediately after passing through an effective RO membrane. Thus, the changes in RO performance when operating on clean water should be due to mechanical deformation of the elements rather than membrane surface fouling. However, some organic solutes in the RO feed water, including halogenated organic byproducts of chlorination such as bromoform, readily pass through CA RO membranes and were present in the PTU permeate feeding the CWT unit. Such dissolved organic compounds may affect RO performance by absorbing in the membrane making it more hydrophobic and/or by causing an increase in the rate of mechanical deformation of the plastic RO element components.

Experimental Equipment and Procedures. To provide experimental controls for the CWT both PTU's were operated on the standard pretreated water. PTU instrumentation included interstage temperature sensors, feed and differential pressure transducers and gauges, ultrasonic flowmeters on the feedline to each vessel, paddle-wheel flowmeters on the product from each vessel and the combined reject, and conductivity sensors on the product of each vessel, PTU feed, interstage, and reject. The CWT equipment included one PTU-specification pressure vessel for each manufacturer, a high-pressure pump and piping to supply repressurized RO permeate to the two vessels, and instrumentation similar to that in the PTU's.

New PTU elements from each manufacturer were divided into two matched sets according to the manufacturer's initial test data for the elements. One of each manufacturer's matched sets was installed in vessel number 4 of that manufacturer's PTU and the other set was installed in a CWT vessel to eliminate bias due to differences in element characteristics between the CWT and the PTU control. During the CWT both the PTU and CWT equipment were operated at an applied feed pressure of 2.76 MPa. The reject flows were maintained to obtain 70 and 72.5 percent overall recovery for the two-stage FSD and HWS PTU's, respectively. The reject flow of each CWT vessel was set equal to the reject flow of its corresponding PTU vessel 4.

PTU and CWT sensor transmitters were connected to a data acquisition system to obtain readings every 5 min and to calculate hourly average values for each sensor. These averaged data were transmitted daily to a main frame computer, which calculated hourly values for each PTU and CWT vessel of water-transport coefficient; TDS concentrations of the feed, product, and reject from their specific conductivities; TDS-transport coefficient; and salt rejection. Weekly samples of all feed, reject, and product streams were collected and analyzed for major anions and cations. These ionic concentration data were combined with operational data to calculate individual ion transport coefficients for each vessel. The equations used to calculate the water and solute transport coefficients per vessel (rather than per unit membrane area) account for net osmotic pressure, log-mean concentration difference between the bulk feed-reject and product streams, and temperature corrections to 23.9 °C.

Results. The changes in RO performance of the new elements in vessel 4 of each PTU, which provided the experimental controls for the CWT, duplicated the trends of the earlier proof test data. Immediately after startup of the CWT the water transport coefficients of each PTU dropped as shown on Figure 1a. The least squares fit of slopes in Equation 1 were m = -0.029 for FSD and m = -0.026 for HWS for the operating time of 10 to 1260 h. The sharp increases in transport coefficients in Figures 1a and 1b at an operating time of 1263 hours occurred as a result of a 9-day equipment shutdown caused by an electrical failure. The water-transport coefficients of the FSD CWT vessel operating on RO permeate also decreased with m = -0.017 (Figure 1a), approximately 60 percent of the log-log slope of its control. Thus, mechanical deformation of FSD element components, perhaps aided by organic solutes penetrating the CA RO membranes, contributed somewhat more than did the membrane surface fouling to the FSD water-transport decline. However, the HWS CWT vessel containing DSI membrane elements showed no decrease in the water transport coefficient at all (m = 0.0001 in Figure 1a) and also no decrease in the Na or Ca transport coefficients (Figure 1b). For the HWS PTU vessel 4 the Na and Ca transport coefficients (Figure 1b) decreased at a faster rate than did the water-transport coefficients (Figure 1a), which is why the salt rejection increased with time as shown in Figure 1c. Thus, the HWS performance shows that the entire water transport declines of the DSI membranes during the proof tests were caused by materials that could be removed by an RO membrane, presumably materials which foul membrane surfaces such as those described in Table I, and the DSI membranes had no measurable mechanical deformation of element components including no membrane compaction when operated with an applied pressure of 2.76 MPa. However, by comparing differences between the changes in water transport coefficients of the CWT's and PTU's of the two manufacturers, these data also show that the HWS equipment was more than twice as sensitive to surface membrane fouling by the same feed water as the FSD membranes. Explanations of the CWT are given in the Discussion following the presentation of data from an additional RO experiment conducted at the YDTF.

Foulant Component Study

To elucidate the information on the effects of membrane fouling on RO performance gained from the CWT a further experiment was conducted called the FCS (Foulant Component Study). The FCS was designed to give data on the RO performance responses to different types of membrane fouling materials contained in the YDTF RO feed water and to suggest methods for improving the YDP pretreatment. In the FCS, RO elements were operated on feed water consisting of RO reject from a PTU plus different additional pretreatment steps selected to remove specific classes of fouling materials. A flow schematic of the FCS is given on Figure 2a. RO reject was used instead of RO feed water

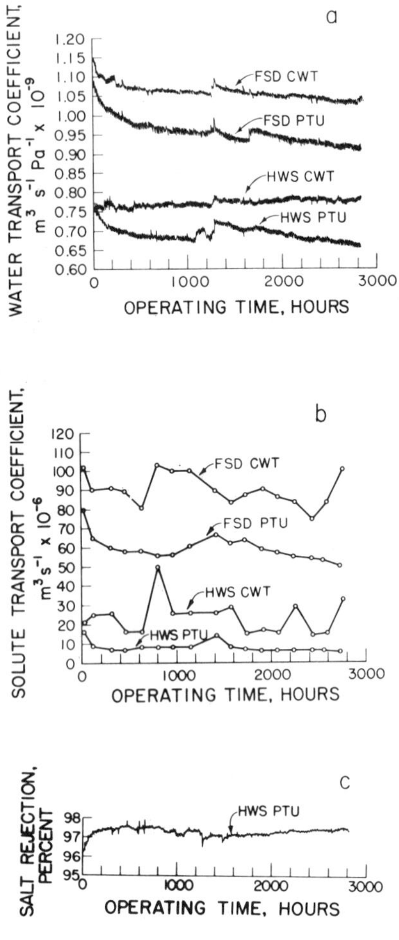

Figure 1. Clean water test results including changes with operating time of (a) water transport coefficients, (b) total ionic solute-transport coefficients estimated from specific conductivities, and (c) salt rejections of vessels number 4 in the FSD and HWS PTU's and of the CWT vessels.

to increase the concentrations of and response rate to membrane surface-fouling materials, which are rejected by RO, and to simulate RO operation at recoveries similar to the interstage of the PTU's. The additional pretreatment steps of the FCS included

1. UF (ultrafiltration) pretreatment with a pore size small enough to remove most colloidal material but coarse enough to allow DOC including dissolved humic and fulvic acids to pass without rejection to the following RO,
2. AS (aeration stripping) to remove air-purgeable organics,
3. AC (activated carbon) in packed beds followed by ultrafiltration to remove most organic and suspended material,
4. RO (reverse osmosis) to repeat the CWT by removing all materials rejected by a CA RO membrane.

The FCS was performed in three phases. Phases I and II included operation with new FSD RO elements. Phase III included RO operation with new DSI RO elements. Phase I was run using a feed water of 70 percent recovery reject from two stages of a PTU, and phases II and III used 50 percent recovery reject from one stage of a PTU. Only the third phase included all four additional pretreatment steps and five experimental RO sections. During Phase I only the control, UF, and AS segments were successfully operated. An RO pretreatment leg was not included until Phase II, and there were numerous difficulties in operating the AC leg during Phases I and II.

Experimental Equipment and Procedures. Each of the five RO sections shown on Figure 2a included two pressure vessels in parallel (in series following the AC plus UF). Each pressure vessel contained a single 64-mm-diameter spiral-wound RO element made of the same component materials as the proof test elements to simulate the PTU performances. A pair rather than a single element was used in each section to replicate RO performance. The UF consisted of two pressure vessels in parallel, each containing a 157-mm-diameter spiral-wound porous (50-nm nominal pore size) polysulfone element obtained from FSD. These UF elements should perhaps be more accurately called microfilters because of their relatively large pore sizes, which were much larger than the measured average size of the humic and fulvic acids [0.64-nm radius of gyration by low angle X-ray scattering (7)] in the RO feed water at the YDTF. The AS system was designed to remove at least 90 percent of bromoform. There was much difficulty operating the AC columns because the chlorine residual was removed by the carbon, which allowed biological growth in the carbon. Only in the third phase was the operation of the AC successful in controlling biological activity when the water following the AC was rechlorinated and treated further by ultraviolet light prior to the UF as an additional safety factor to guard against possible biological contamination from the AC.

RO data and water samples were collected manually every 8 hours. A volumetric container and a stopwatch were used to accurately measure flow rates. Water samples during Phases II and III were analyzed in the YDTF chemistry laboratory for specific conductivity, sodium, chloride, and magnesium. Values of water and ionic transport coefficients for the RO membrane elements were calculated for each observation from the raw data entered manually into a computer.

Data Analysis. The objective of analyzing the performance data from the different CWT and FCS RO units was to quantify the decreases in flux decline caused by the additional pretreatment processes, and thereby to indicate the effects on RO performance of the different classes of fouling materials and of element mechanical deformation. A mathematical method was developed to relate the relative improvement in RO performance caused by an additional pretreatment step in terms of the log-log flux decline slopes of the RO following that pretreatment step and of the control RO. The relative decrease, e_i, in the water-flux-decline rate of the RO system following pretreatment i is defined by

$$e_i = \frac{\dfrac{A_c(t) - A_c(1)}{A_c(1)} - \dfrac{A_i(t) - A_i(1)}{A_i(1)}}{\dfrac{A_c(t) - A_c(1)}{A_c(1)}} \qquad (2)$$

where $A_i(t)$ is the water-transport coefficient of the RO system following pretreatment i at operating time t and $A_c(1)$ is the water-transport coefficient of the control RO system at 1 operating hour. Thus, e_i provides a measure of the efficiency of the additional pretreatment system i to lower the water-flux-decline rate of the control RO. Because e_i is also affected by mechanical deformation of the RO elements, the maximum e_i possible if pretreatment removal of membrane foulants is perfect is equal to one minus the relative decrease in the water-flux rate caused by mechanical deformation.

The power model (Equation 1) can be written as the series:

$$A(t) = A(1)\ (1 + m \ln t + \frac{(m \ln t)^2}{2} + \ldots) \qquad (3)$$

The second order and higher terms of Equation 3 are neglected. The truncated series (Equation 3) is substituted for the A's in Equation 2. Following algebraic manipulation the final simple expression is

$$e_i = 1 - \frac{m_i}{m_c} \qquad (4)$$

Values of m_i were obtained by nonlinear regression of $A_i(t)$ versus t for each FCS RO system. Each m_i provided a single parameter to characterize RO performance behind pretreatment step i for the given RO element design and feed water composition. Although m became increasingly more negative with greater operating times during the proof tests, during the FCS when operating times were less than 2000 h, the fits of the power model were adequate. Furthermore, because each m_i was ratioed to m_c of its control RO for the same operating period, any systematic errors introduced by an imperfect fit of the power model and the truncation of Equation 3 would tend to cancel out. This cancelling out was observed in additional performance comparisons between pretreatment and control RO sections using regression estimates of semilog and linear slopes in place of the log-log m's in Equation 4. For all three empirical models the calculated values of e_i were essentially the same for the data of each FCS phase. Thus, despite the fact that the power model did not provide confident projections of long term RO performance at the YDTF, it was valuable for fitting and comparing $A_i(t)$ of RO units operated in parallel in the FCS and CWT for less than 1300 h.

Pretreatment Performance Results. RO feed water quality parameters for the control (50-percent recovery PTU reject) and following each additional FCS pretreatment process during Phase III are summarized in Table II. Results were very similar during Phase II. During Phase I the materials rejected by the PTU were in approximately 50 percent higher concentrations than those given in Table I because the PTU recovery during Phase I was 70 percent.

Table II. FCS Pretreatment Systems Performances during Phase III

FCS pretreatment	Foulants removed	Turbid. JTU	Plugging factor,%	TTHM. mg/m^3	TOX mg/m^3	TOC g/m^3	HFA g/m^3
Control	None	0.18	13.6	130	500	6.8	0.98
UF	Colloids	0.15	7.4	120	500	6.8	0.95
AS	Air-purgeable organics	0.19	16.2	6.5	420	6.0	0.90
AC + UF	Organics & colloids	0.10	-0.7	0.7	77	0.8	0.04
RO	Colloids & large organics	0.18	4.3	140	100	0.5	0.04

The data in Table II indicate that each additional pretreatment step performed in accordance with the FCS experimental design, although turbidity readings were not substantially changed by additional pretreatment except by the AC plus UF. The UF removed colloids as indicated by the lower UF-effluent plugging factor compared to that of the control. (The standard deviations of plugging factor ranged from 1.8 to 2.6 for the mean values given in Table II, each representing approximately 400 observations.) In accordance with its selected pore size the UF allowed dissolved organic compounds to pass without rejection including HFA (humic and fulvic acids), TOX (total organic halogen), and TOC (total organic carbon, which is equal to DOC when there is negligible suspended organic material present). The AS removed 95 percent of the TTHM (total trihalomethanes) and 16 percent of the TOX corresponding to the portion of TOX that was air purgeable including the TTHM's. The AS removed insignificant amounts of TOC, HFA, and colloids as expected. The AC lowered organic solute concentrations to near or below their detection limits. Moreover, the AC followed by UF apparently made the water so free of colloids and large organic compounds that the average plugging factor was slightly negative (statistically significant at the 95 percent confidence level). The RO pretreatment performed as in the CWT by removing colloids and large organic compounds but slightly concentrating TTHM's.

RO Performance Results. Sample plots of normalized corrected product flow (equivalent to a normalized water-transport coefficient) versus operating time for DSI RO elements with and without additional UF pretreatment preceding are given on Figure 2b. The reason that the data plotted in Figure 2b had been normalized to (divided by) their respective 100 h values was to clearly demonstrate relative performances, thereby to prevent the appearance of different performances caused only by differences in the initial performances of individual elements. During each FCS phase the individual elements in each pair of RO elements in a section had virtually identical RO performances in every instance (except for unexplained different performances only of the two FSD elements following RO pretreatment during Phase II). Thus, the RO performance of an individual element was used as being representative of that pair. The good fit of the power model to the data as shown on Figure 2b is typical of the power model fits of the other FCS and CWT RO data. For the example data (Figure 2b) the control RO had a water-transport decline rate (m in Equation 1) of m_c = -0.0305 while the RO following the UF had m_i = -0.0071, which means that according to Equation 4 the UF removed fouling materials which contributed to 77 percent of the water-transport decline of the control RO. Table III summarizes the results of the CWT and the three phases of the FCS regarding water-transport decline rates of the FSD and DSI membrane elements.

Table III. Percentage Decrease in the Water-Transport Decline Rate Caused by Added Pretreatment Steps for the FCS and CWT, Plus Related Data

Added pretreatment	e_i X 100, % decrease in -m due to pretreatment: FSD RO elements, CWT	FSD RO elements, FCS-I	FSD RO elements, FCS-II	DSI RO elements, CWT	DSI RO elements, FCS-III
UF		17±5	21±9		77±6
AS		15±4	19±7		-13±8
RO	39±1		62±7	110±2	79±5
Concentration factor	1.41	3.46	1.93	1.61	1.93
m_c in Equation 1	-0.0284	-0.061	-0.051	-0.0266	-0.030
Uncertainty in mc	±0.0004	±0.003	±0.003	±0.0004	±0.001
Operating hours regr.	1241	1829	519	980	484
Number of observations used in regression	1229	220	64	966	61

Note: Values of e_i are calculated from Equation 4.

The following comments apply to the data in Table III. The uncertainty values are 95 percent confidence limits calculated from the errors of the fits of Equation 1. The concentration factors are for totally rejected solutes and were calculated for a reject stream from the fractional product water recovery R (the ratio of product flow to feed flow) using the expression 1/(1-R) and for an average value within a single vessel using -ln(1-R)/R. The percentage decreases in -m for FCS phase I only were estimated from linear decline slopes for the operating period 390 to 1830 h because operational difficulties in starting the experiment caused greater errors in the data before 390 h. The FSD CWT results (m_i = -0.017) are more in agreement with the results of the RO section of FCS-II (m_i = -0.019) than the respective percentages (39 and 62) in Table III seem to show because of the different m_c's of their respective control RO sections. Moreover, as demonstrated by the data in Table III, m_c increased in magnitude in accordance with the average brine-concentration factor.

Data for the RO performances following AC plus UF during Phase III are not included in Table III because the performances were not consistent with the other RO data of the FCS and, therefore, could not be directly compared with their control RO performances. Water, sodium, and chloride transport coefficients increased with operating time (m = 0.08 for water transport in Equation 1) while magnesium transport dropped for the first 200 h and then increased. These same behaviors repeated with a new pair of RO elements operated following the AC plus UF following the replacement of the first pair after 380 h of operation with a second pair operated for 320 h until Phase III was completed. Frequent measurements of plate counts found no significant biological activity in the rechlorinated water following the AC. Apparently, some unknown, uncontrolled variable was responsible for the puzzling RO performance of this FCS segment, but so far, no satisfactory explanation for these data has been found. These findings indicate that operation of RO following AC can be difficult to accomplish successfully.

Figure 2c shows how the RO permeate concentrations of total hardness (primarily divalent calcium and magnesium) varied versus time for three legs of the FCS. For the control RO and the RO following AS the permeate total hardness concentrations dropped dramatically - a factor of 10 during the first 100 operating hours. However, for the RO operating behind UF the permeate total hardness concentration did not change substantially. A likely explanation of these data are given in the following Discussion.

Discussion

The power model (Equation 1) did not provide an adequate fit of water-transport coefficients versus operating time for long term RO proof-test data collected at the YDTF. Initial declines in water transport rates follow the power model reasonably well. However, the power model predicts a continuous leveling off of the decline with increased operating time, but the YDTF data for operating times greater than 1000 h indicate a nearly constant decline rate. The power model (a straight line on a log-log plot) has been successfully used to fit water-transport declines that result from membrane compaction (10), but its use to predict changes of RO performance due to effects of membrane fouling, as included in the supplies contracts, is not supported by the YDTF proof test data. Based on their proof test performances and as required by their contracts the contractors will provide additional RO equipment in the YDP at no additional cost

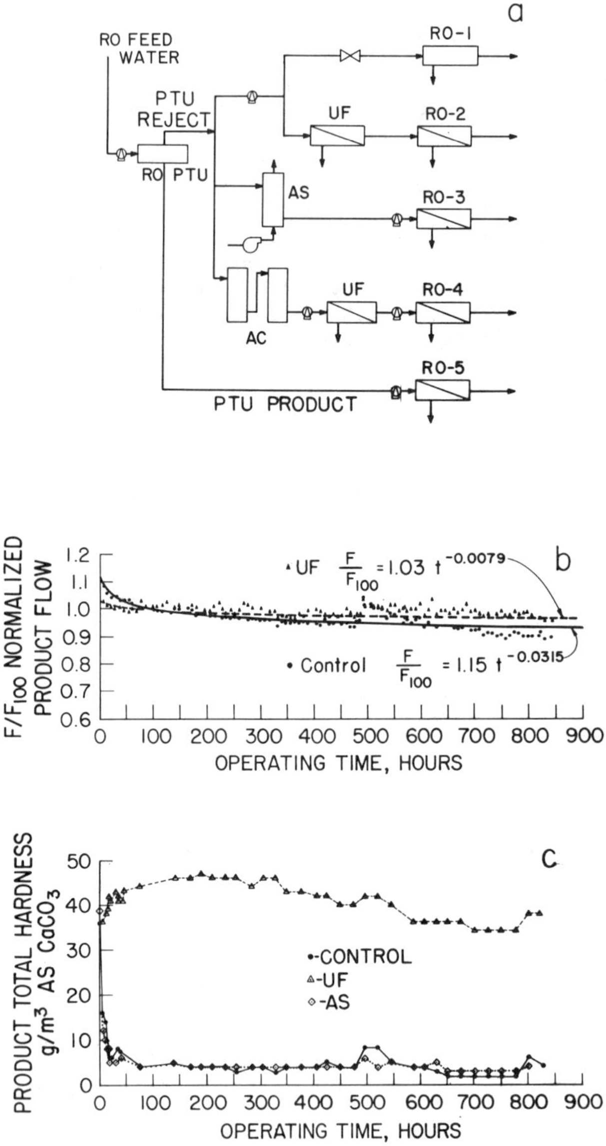

Figure 2. FCS during Phase III including (a) flow diagram of experimental equipment, (b) corrected product flow normalized to 100-h value versus operating time for RO's following UF and for control, and (c) concentration of total hardness in RO product versus operating time for control RO and RO following UF and AS.

to the Government to compensate for RO desalting capacity which decreased faster than they predicted based on the power model with m = -0.02. However, for more effective operation of the YDP, an improved model to predict RO performance long term is needed.

Using the percentages in Table III it is possible to attribute portions of the water-transport decline rate to different types of fouling materials and mechanical deformation of the membrane elements. The percentages of the declines attributed to colloidal material correspond to the percentages of declines removed by UF, 19 percent on average for the FSD elements and and 79 percent for the DSI elements. The percentages of the declines attributable to air-purgeable organics correspond to the percentages of the declines removed by AS, 17 percent on average for the FSD elements and none for the DSI elements. The percentages of the declines from dissolved organic compounds with molecular weights over about 1000 are assumed to be the differences between the percentages of the declines removed by RO and by UF, which was 41 percent for the FSD elements and 23 percent for the DSI elements. The different performance responses to membrane fouling of the two cellulose-acetate membranes during the FCS and CWT were at first surprising because the proof test performances of the two contractors were so similar. However, the causes for the different CWT and FCS performance responses can be explained by the different properties of the DSI and FSD membranes.

Membrane Surface Fouling by Colloids. The mechanisms of membrane surface fouling by colloidal particles, perhaps in combination with large molecular weight organic compounds, which occurred at the YDTF can be understood qualitatively using principles of colloid stability (11). Stable colloids are those whose charges of the same sign (negative for clays) on individual particles cause a repulsive electrostatic force among those particles preventing their coagulation, thereby maintaining indefinitely their suspension in solution in random Brownian motion when not acted upon by external forces. Conversely, unstable colloids are those which tend to coagulate with one another to form larger particles. The smaller diffusion coefficients of larger particles resulting from coagulation cause greater concentration polarization (the larger particles diffuse more slowly away from the membrane surface in opposition to convective flow toward the membrane) and, therefore, make the larger particles more likely to foul the membrane.

Based on the principles of colloidal chemistry the following are the expected behaviors of colloids passing through the YDTF pretreatment system. The colloids suspended in water leaving a lime treatment system are predominantly stable because the unstable colloids would have been coagulated and collected in the calcium carbonate sludge during the 90-min typical retention time of partial lime softening-clarification. Lowering the pH of the water following lime treatment causes greater colloid instability because it lessens the negative and increases the positive charges of the metal oxides and organic compounds coating the clay surfaces, which tend to neutralize the clays' negative charges and thereby the mutually repulsive forces that keep colloids stable. Thus, the final lowering of the pH of the RO feed water to 5.5, done to minimize cellulose-acetate hydrolysis rates, may also cause greater rates of colloidal fouling within the RO--to what degree is presently unknown.

Within an RO system there are three factors which contribute to the instability of colloids (11) and large dissolved organic molecules in the RO feed water leading to RO membrane surface fouling.

1. Colloids are concentrated in the bulk flow as they pass from the feed to reject of the RO system because the colloids are rejected by the membrane as product water permeates through the membrane.
2. Ions are concentrated in the reject stream of an RO element. An RO membrane generally rejects multivalent ions to a greater degree than monovalent ions. The increased concentrations of ions, especially multivalent, cause compression of the double layers of charge surrounding individual colloids, which contribute to colloid instability and coagulation (11), sometimes called the "salting out" effect.
3. Concentration polarization (12) causes a greater concentration of rejected solutes and colloids near the membrane surface than in the bulk feed-reject stream. The greater the flux, the greater will be the amount of concentration polarization resulting in greater colloid concentrations near the membrane surface. Because the amount of concentration polarization is greater with material that has a lower diffusion coefficient, colloids because they are much larger than ions have a higher tendency than do ionic solutes to concentrate near the membrane surface.

The higher flux-decline rate (m_c) with higher concentration factor of rejected material (higher desalting recovery) demonstrated by the data in Table III is consistent with factors 1 and 2 above. However, because factors 1 and 2 were similar for the FSD and DSI membrane elements at the YDTF, number 3 is the most likely cause for the different tendencies of the two brands of CA membrane elements to be fouled by colloids. The approximately 50-percent higher flux of the DSI membranes in the HWS PTU would produce greater particle concentrations at the membrane surfaces relative to that of the FSD PTU.

Apparently, the 53-percent higher brine-channel velocity of the DSI elements did not decrease the mass transfer boundary layer sufficiently to offset the effects of the higher flux. As listed in Table I the ratios of permeate flux to brine-channel velocity were nearly identical for both types of RO element tested and, therefore, were not a factor contributing to the different rates of fouling. Although the comparison of the fouling rates of the FSD and DSI membranes at the YDTF does not constitute a controlled experiment of RO fouling rates as a function of permeate flux and brine-channel velocity rates, interpretations of the YDTF findings disagree with results of other experiments where tubular membranes were fouled by sewage effluents under a variety of flow conditions, and it was concluded that the log-log flux-decline slope was proportional to the one-third power of the ratio of membrane flux to brine-channel velocity. (13) Of course, the colloidal foulants at the YDTF have much different flocculation and fouling properties than do sewage effluents. Furthermore, the brine-channel velocities of the FSD and DSI spiral-wound elements tested at the YDTF were up to 100 times greater than those of the tubular membranes just cited.

The charge of the membrane may also be important as to whether colloids approaching the clean membrane will be drawn to and adhere to the membrane surface. The greater the negative charge of the membrane surface the greater would be the forces tending to repel negatively charged colloids away from the membrane surface thereby causing a lower rate of colloidal fouling, other factors being equal. However, the colloidal fouling at the YDTF was sufficient to coat the membrane multiple times. Therefore, after start of operation when the initial layer of foulant coated the new membranes, the charge on the surface should have been determined more by the highly charged, adsorbed foulant than by the weakly charged cellulose-acetate. Thus, it is unlikely that any possible differences between the weak charges of the FSD and DSI cellulose-acetate membranes were related to their different propensities for colloidal fouling.

The rapid initial decrease in ionic solute transport observed for new membranes at the YDTF is attributed to plugging of membrane imperfections by some of the colloids. This mechanism is consistent with the divalent ion product concentration data of the FCS (Figure 2c), which shows an immediate large initial drop for the control RO and RO following aeration but no drop for the RO operating on colloid-free water following UF. A similar but less dramatic decrease in the ionic transport coefficient based on conductivity (predominantly monovalent ions) occurred during the CWT (Figure 2b). A relatively greater change in the permeate concentration of divalent ions is expected for the mechanism of the plugging of membrane imperfections because divalent ion diffusion through CA is at a much lower rate than it is for monovalent ion diffusion (10), whereas convective transport of solutes through imperfections would be independent of ionic charge number but proportional to solute concentrations near the feed side of the membrane. Thus, plugging of convective flow through membrane imperfections would cause a greater net decrease of divalent transport than monovalent ion transport which was observed.

The continual decreases in ionic solute-transport rates of the PTU's (Figure 1b), especially the data of the HWS PTU because its membranes were not subject to compaction, indicate that the colloidal material fouling the RO membranes acts like a dynamic RO membrane providing a series resistance to ionic transport in addition to that of the celluose-acetate membrane. In fact, some of the materials making up the sediments on the surfaces of the PTU membranes (iron oxides, clays, and humic acids listed in Table 1) are the same as those used to form salt-rejecting dynamic membranes on porous solid supports (14, 15). The increases in the rates of water and solute transport following a temporary shutdown of the RO equipment (for example, the several day shutdowns at 1263 operating hours shown on Figures 1a and 1b and at 488 operating hours on Figures 2b and 2c) imply a temporary disturbance of the fouling layer on the membrane, which is apparently restored some hours after further operation on the same feed water. In the case of a membrane subject to compaction such as the FSD membrane some restoration of transport rates following a shutdown could also be due to a partial reversal of membrane compaction occurring when the applied pressure to the RO is temporarily relaxed. The decreasing ionic solute-transport rates and increasing salt rejections caused by the membrane fouling material at the YDTF is opposite to the description of membrane fouling by

Brunelle (16) who claims that salt passage commonly increases and salt rejection decreases when colloids foul RO membranes.

Effects on RO Membranes by Organic Compounds. FCS results indicate that the log-log flux decline slopes for FSD RO elements operated following aeration stripping were significantly less than those of the controls. The decrease in -m due to aeration was 17 percent during Phase I and 19 percent during Phase II (Table III). The only known effect of aeration on the quality of the RO feed water (other than the possible introduction of a small amount dust from the air despite the fact the air was filtered, which would presumably increase rather than decrease the decline rate if it fouled the membrane) was the removal of halogenated biproducts of chlorination at concentrations of approximately 100 mg/m^3 entering the control RO (Table II). A possible explanation for these data is that the halogenated organic compounds, which are preferentially absorbed in the CA, plasticized the control CA membranes, which contributed to an increase in membrane compaction. It is also possible, but not demonstrated, that the organic compounds could plasticize the product-water-channel tricot, which would cause a loss of element efficiency. Membrane plasticization was demonstrated in dramatic fashion in previous work involving the exposure of CA membranes to high concentrations of organic solvents (for example, 25 kg/m^3 of phenol) (2), which greatly increased rates of membrane compaction, in extreme cases densifying the membranes and completely destroying the asymmetric nature of the CA membranes.

In contrast to the FSD membranes, the DSI membranes showed no decrease in water-transport coefficient at all during the CWT (Figure 1a) while operating on RO product water containing organic compounds, some air purgeable such as bromoform, that passed through the PTU membranes. There also was no improvement in the RO performance of DSI membranes whether or not they were preceded by aeration during the FCS (Table III). The reason that the FSD membranes but not the DSI membranes experienced membrane compaction and were affected by purgeable organics may lie in the relatively different properties of their respective CA membrane polymers. The FSD membranes were of cellulose diacetate but the DSI membranes were a blend, presumably of cellulose diacetate and cellulose triacetate. Cellulose triacetate is much more resistant to dissolution in most organic solvents and has a higher glass-transition temperature than does cellulose diacetate (16), which means that the triacetate may be more resistant to plasticization and membrane compaction in the presence of such organic solvents.

Decreased RO water-transport rates can also occur because an organic solute in the RO feed water is preferentially absorbed by the membrane (3), which, in effect, makes the membrane more hydrophobic. Such experiments were conducted with bromoform in laboratories of FSD supervised by Bob Riley and repeated at the Bureau of Reclamation Chemical Engineering Laboratory. In these experiments the presence of some 100 g/m^3 of bromoform in the RO feed water caused a decrease in the rate of RO water transport, but original RO performance returned when the bromoform was removed. Such a reversible drop in RO water transport cannot explain the decrease in the log-log flux decline slope following AS.

Measurements of RO Feed Water Quality. Because RO feed water plugging factors and turbidities at the YDTF were at low levels relative to those reported at other sites but yet water-transport decline rates due to membrane surface fouling were unacceptably high, it can be concluded that neither of these feed water quality measures is sufficient to be used to accurately predict water transport decline rates in lieu of RO performance tests. During the FCS the turbidity was not even changed by passage of water through an RO membrane (no significant difference between the turbidities of the RO feed and RO product in Table II), which means that turbidity was particularly useless to monitor the colloid concentrations present in the YDTF feed water. It is not surprising that turbidity was ineffective because its measurement depends upon measuring the scattering of visible light but the colloidal particles in the RO feed water at the YDTF were so small that they could not even be observed with a light microscope with side illumination by a laser, which precluded the measurement of the colloids' zeta potential. However, plugging factors of the different FCS legs (Table II) correlated highly with RO performance decline attributed to colloids (Table III). Thus, while plugging factor may not be an adequate absolute measure in particular, automated plugging factor measurements (6) can be an effective relative predictor of colloidal fouling of RO for a particular feed water where the characteristics of the fouling material is relatively constant, for example, as a preliminary measure to compare the efficiencies of alternative pretreatment processes to remove colloids at a particular site.

Pretreatment Improvements. The YDTF results indicate the extreme importance of an effective pretreatment system to prevent RO membrane fouling. Membrane replacement at the YDP is expected to be about $3 million, an estimate made before the proof tests, which may be low based on present information (5). Higher than expected water transport declines in the YDP would require greater membrane replacement rates, which would increase plant operating costs. Based on the CWT and FCS data the maximum potential for improvements in pretreatment to remove all fouling materials at the YDP would prevent up to 85 percent of the water-transport decline observed at the YDTF and potentially save over $2 million annually in membrane replacement costs assuming that membrane life were controlled only by losses in capacity due to fouling. However, other factors such as membrane hydrolysis also determine membrane life. The costs of any additional pretreatment need to be balanced against savings in membrane replacement costs, and some pretreatment processes would be more expensive than the potential savings. For example, ultrafiltration proved to be effective at controlling colloidal fouling but it is relatively expensive, and an ultrafiltration system with sufficient capacity to pretreat the entire YDP feed water flow has never been built. Research at the Bureau of Reclamation is continuing on methods to economically improve the pretreatment of the existing YDP design.

Conclusions

The experiments at the YDTF have yielded the following conclusions:

1. The overall RO performances of the FSD and HWS proof test units, both containing spiral-wound cellulose-acetate membranes, were

similar. The RO product-water fluxes continued to drop more rapidly than predicted by the power model with m = -0.02, which according to both contractors' proposals should have been sufficient to account for flux declines resulting from mechanical deformation of the elements plus membrane fouling. The salt-transport rates also decreased, which resulted in increasing salt rejections with time. Prior to additional RO experiments that proved otherwise, membrane fouling as a major cause of the flux decline seemed unlikely because the turbidity and plugging factor readings of the RO feed water exceeded criteria agreed to by the contractors, cleanings prescribed in the contracts were ineffective at restoring RO capacity, and the mass of colloidal clays found fouling the RO membrane surfaces appeared to be minor. Only from the results of additional pretreatment-RO experiments was it possible to conclusively delineate the causes for the poor proof test RO performances, which turned out to be relatively different for the two contractors' membrane elements.

2. Mechanical deformation of the RO elements (excluding that possibly caused by purgeable organic compounds described in 3 below) contributed to about 20 percent of the flux decline of the FSD proof test elements. This estimate from Phase II FCS results is based on the flux decline remaining after accounting for the decline prevented by AS plus RO pretreatments. The YDTF experiments were not designed to nor were they able to clearly identify the specific nature of the mechanical deformation, but it probably consisted primarily of membrane compaction with perhaps some contributions from compaction of the tricot product-water channel and intrusion of the membrane into the channel. The DSI membranes in the HWS proof unit demonstrated no mechanical deformation during nearly 3000 h of operation, which clearly shows that membrane compaction is not inevitable for cellulose-acetate membranes operated with an applied pressure of 2.8 MPa. Different cellulose acetate polymers in the FSD and DSI membranes may explain the sucseptibility of the FSD but not the DSI membrane to compact.
3. Improvement in the FSD flux-decline rate following aeration implies that about 20-percent of the total decline was contributed by organic compounds that are air purgeable. Experimental evidence of other researchers (_2_) suggest that the organic compounds plasticized and accelerated mechanical deformation of one or more plastic components of the RO elements, most likely the cellulose-diacetate membranes.
4. The flux-decline rate due to surface fouling increased with concentration factor of the reject brine, which is attributed to the higher concentration of colloids and organic compounds present near the RO membranes and the increased ionic concentration causing a greater tendency of the colloids and dissolved organic materials to coagulate.
5. The DSI membranes accumulated more than twice as much surface foulants as the FSD membranes during similar operating times. Material scraped from the membrane surfaces consisted primarily of smectite clays coated by iron and manganese oxides and organic compounds. The flux decline due to surface fouling of the DSI membranes was also more than twice that of the FSD membranes and, in fact, accounted for all of the performance decline of the DSI

membranes. The greater sensitivity to surface fouling of the DSI membranes is attributed to their nearly 50-percent greater flux causing greater concentration polarization (higher foulant concentrations) near the membrane surface. The relative hydrodynamics as characterized by the ratio of membrane flux to brine-channel velocity were nearly identical for the FSD and DSI membrane elements tested and, thus, do not explain the membranes' different sensitivities to surface fouling. The surface fouling acted like a dynamic RO membrane to increase salt rejections and appeared to plug imperfections in new membranes during their initial 50 hours of operation on YDTF feed water.

6. The methods of turbidity and plugging factor were unable to predict the seriousness of the RO membrane fouling by colloids of submicron sizes. Differing concentrations of submicron colloidal particles at the YDTF gave essentially the same turbidity readings. Plugging factors, however, were lowered by different pretreatment steps, which also resulted in correspondingly lower RO flux-decline rates. A better method than now exists is needed to monitor the concentrations, sizes, and charges of colloids that cause RO fouling. Until such a method is developed and experimentally validated, there will continue to be no confident substitute for field testing of representative equipment at a site to determine pretreatment requirements and long term RO performance for sizing requirements in the design of an RO desalting plant.
7. Pretreatment consisting of lime-softening clarification followed by conventional filtration was unable to remove submicron-sized colloidal clays that fouled the RO membranes at the YDTF. Cleaning was ineffective at permanently reversing the effects of surface fouling. Thus, more effective methods need to be developed to remove extremely fine colloids from RO feed waters. Such new methods must be less expensive than the corresponding savings in RO membrane replacements.

Acknowledgments

Former employees of Planning Research Corporation, which operated the YDTF under a Government contract, played key roles in conducting the RO experiments described herein and include Paul Laverty (who also reviewed a draft of this paper), James Lozier, Cindy Hoeft, and Shirley McAdams. The Bureau's YDTF resident engineer was Chuck Goodner. Paul McAleese was the Bureau's YDTF facility engineer. Dr. Ron Malcolm and others of the U.S. Geological Survey provided characterizations of humic and fulvic acids and RO membrane sediments.

Literature Cited

1. Bray, D. T. In "Desalination by Reverse Osmosis"; Merten, U., Ed.; M.l.T. Press: Cambridge, Mass., 1966; Chap. 6.
2. Anderson, J. E.; Heyde, M. E.; Plummer, H. K., Jr. Desalination 1981, 37, 307-311.
3. Sourirajan, S.; Matsuura, T. In "Reverse Osmosis and Synthetic Membranes"; Sourirajan, S., Ed.; National Research Council Canada: Ottawa, 1977; Chap. 2.
4. "Status Report, Plans and Estimates Data, April 1977, Colorado River Basin Salinity Control Project, Title 1 Division, Desalting Complex Unit, Arizona," Bureau of Reclamation, 1977.
5. "Final Technical Report, Yuma Desalting Test Facility," Planning Research Corporation, Yuma, Arizona, 1982.
6. Eisenhauer, R. J.; Goodner, C. G. "Development of an Automated Plugging Factor Monitor," Bureau of Reclamation, Denver, 1981; REC-ERC-81-12.
7. Malcolm, R. L.; Wershaw, R. L.; Thurman, E. M.; Aiken, G. R.; Pinckney, D. J.; Kaakinen, J. "Reconnaissance Samplings and Characterization of Aquatic Humic Substances at the Yuma Desalting Test Facility, Arizona, U.S. Geological Survey Water Resources Investigations 81-42; Denver, 1981.
8. Olphen, H. van. "Clay Colloid Chemistry," Interscience: New York, 1963.
9. Forman, G. E.; Kremen, S. S.; Wight, W. W.; Wolfe, T. D. Office of Saline Water Research and Development Progress Report No. 880; 1973.
10. Merten, U.; Lonsdale, H. K.; Riley, R. L.; Vos, K. D. Office of Saline Water Research and Development Progress Report No. 208; 1966.
11. Hiemenz, P. C. "Principles of Colloid and Surface Chemistry"; Marcel Dekker: New York, 1977; Chap. 9.
12. Brian, P.L.T. In "Desalination by Reverse Osmosis"; Merten, U., Ed.; M.I.T. Press: Cambridge, Mass., 1966; Chap. 5.
13. Thomas, D. G.; Mixon, W. R. Ind. Eng. Chem. Process Des. Develop. 1972, 11, 339-43.
14. Johnson, J. S.; Kraus, K. A. Office of Saline Water Research and Development Progress Report No. 508; 1970.
15. Minturn, R. E. Office of Saline Water Research and Development Progress Report No. 903; 1973.
16. Brunelle, M. T. Desalination 1980, 32, 127-35.
17. "Cellulose Ester Selector Chart"; Eastman Chemical Products, Inc.: Kingsport, Tenn., 1978.

RECEIVED February 22, 1985

28

Particulate Membrane Fouling and Recent Developments in Fluid Mechanics of Dilute Suspensions

GEORGES BELFORT, ROGER J. WEIGAND, and JEFFREY T. MAHAR

Department of Chemical Engineering and Environmental Engineering, Rensselaer Polytechnic Institute, Troy, NY 12180

Fluid dynamic analysis is used to better understand the mechanism of particulate membrane fouling from dilute suspensions. The fluid mechanics of a neutrally buoyant particle moving in laminar flow in a slit with two porous walls is compared with that in a porous tube of similar dimensions. Trajectory calculations are used for several commercial membrane modules to estimate the propensity of one and ten micron radius particles to foul. Under the assumptions of the theoretical models, and all things being equal, tubular systems capture more particles from dilute suspensions than slits with two porous walls. Also, higher flux membranes lead to particle capture and fouling at shorter axial lengths for both tube and slit configurations.

Although pressure-driven membrane processes are gaining wide acceptance in many new laboratory and industrial applications, and are even replacing traditional separation techniques such as distillation, ion exchange and centrifugation, they all suffer from a potentially serious and limiting phenomenon called concentration polarization (cp). Complications such as membrane fouling (mf), thought to be a direct result of cp, have proved to be extremely difficult to model theoretically or even predict experimentally. Well-defined and simple systems have thus been actively studied in order to gain a better understanding of mf. By arbitrarily defining mf as a two-step process: **transport** and **solute association**, Belfort and Altena (1) were able to separate the fluid mechanics of foulant materials in a membrane duct with the complicated physicochemical and electrokinetic phenomena that occur when these materials enter the solution- membrane interfacial (or near-field) region. They and others (2) have argued that a detailed understanding of each step will eventually lead to a better understanding of mf.

Recently, several researchers have reported on solute association studies in which the effects of a model solute such as

0097-6156/85/0281-0383$06.00/0

bovine serum albumin (bsa) on membrane performance has been evaluated (3-7). Rapid irreversible adsorption of the solute onto and probably into the membrane surface has been reported. Also, studies on how suspended particles such as monodispersed polystyrene colloids behave in the presence and absence of dissolved salt or in the absence of a moving fluid have appeared in the literature (8,9).

Transport processes or cp **near** the membrane-solution interface have been modeled using mass balances. Michaels (10) and later Blatt, et al. (11) were the first to present and analyze the gel-polarization (gp) model. Belfort and Altena (1) give a detailed analyses of the limitations and refinements of the gp model. Modifications of the theory in the near-field region include osmotic effects (5,6,9,12) concentration-dependent viscosity and diffusivity effects (13,14), and lateral migration effects (15,16).

Only recently has some work been reported in the far-field region in which hydrodynamic forces dominate (17-19). Their results were used to explain the inability of the gp model to predict the correct permeation flux and power relationship between flux and axial velocity for suspensions. Analysis in a slit with one porous wall and in a porous tube have been reported.

The work we report in this paper compares the fluid mechanics of a particle moving in laminar flow in a slit with **two** porous walls with that in a porous tube of similar dimension. Several commercial membrane modules are available with both these configurations and it is of practical as well as theoretical interest to compare these systems.

After reviewing in the next section the Altena and Belfort (17) and the Weigand, et al. (18) theories for particle motion in a slit and tube, respectively, the results are presented. The significance of these results are then discussed especially in terms of currently available commercial modules. The paper closes with the conclusion that under the assumptions of the theoretical models, and all things being equal, tubular systems will capture more particles from dilute suspensions than slits with two porous walls.

Theory

Flow in a Porous Duct. Lateral migration of spherical rigid neutrally buoyant particles moving in a laminar flow field in a porous channel is induced by an inertial lift force (tubular pinch effect) and by a permeation drag force due to convection into the porous walls. Altena and Belfort (17) extended the analysis of Cox and Brenner (20) for particle motion in a non-porous duct to include the effect of wall porosity. They considered non-interacting particles in dilute solutions at some distance from the duct walls. Criteria were established under which the inertial and permeation drag force in the lateral direction can be vectorially added. Thus, when

(i) $\lambda \sim Re_p\kappa^2$, superposition is feasible since these terms are of the same order.

(ii) $\lambda << Re_p\kappa^2$, permeation is negligible and lift drag dominates in essentially a non-porous duct.

(iii) $\lambda >> Re_p\kappa^2$, inertial lift is negligible and permeation drag dominates in a very porous duct.

Solution of the Navier-Stokes equations impose additional restrictions on the system,

$$Re_p << \kappa << 1 \tag{1}$$

and

$$\lambda << 1 \tag{2}$$

where Equation 1. implies

$$Re_p << 1 \tag{3}$$

For the case of a spherical particle of radius, a, moving in a laminar fluid flow field in a slit of half-height, ℓ, or a porous tube of radius, R, where the wall flux is constant and independent of the axial coordinate z, (Figure 1) the convective flow term of order Re_w is, in dimensional form, given by (21,22)

$$v_p = v_w\ [g(\beta) + Re_w h(\beta)] \tag{4}$$

where

$$g^s(\beta) = \frac{\beta}{2}(3-\beta^2) \tag{5a}$$

Slit

$$h^s(\beta) = -(2\beta - 3\beta^3 + \beta^7)/280 \tag{6a}$$

and

$$g^t(\beta) = 2\beta - \beta^3 \tag{5b}$$

Tube

$$h^t(\beta) = -(-4\beta + 9\beta^3 - 6\beta^5 + \beta^7)/36 \tag{6b}$$

Cox and Brenner (20) derived a general form for the lateral velocity due to inertial effects. Vasseur and Cox (23) and Ishii and Hasimoto (24) obtained the detailed coefficients of the general form given by Cox and Brenner for the case of a plane and tube Poiseuille flow using intergal and Fourier-Bessel transform techniques, respectively. Since the boundary conditions for the disturbance flow at the porous wall is effectively zero at low relative wall permeabilities [$Re_w << 1$], the Green's function for the porous wall problem was taken idential to the Green's function for the non-porous case (17). Consequently, for the inertial particle velocity the expression derived by Vasseur and Cox (23) and Ishii and Hasimoto (24) were used. For a neutrally buoyant particle (which is allowed to rotate) the lift velocity in dimensional form is given by:

$$v_\ell = \frac{1}{4} Re_p \kappa^2 U_m f(\beta) \qquad (7)$$

where $f(\beta)$ is the result of a numerical integration involving the undisturbed velocity profile and the Green's function. The function $f(\beta)$ has been calculated using the expressions of Vasseur and Cox for a slit $[f^s(\beta)]$ and by Ishii and Hasimoto for a tube $[f^t(\beta)]$, and presented graphically in Altena and Belfort (17) and in Weigand, et al. (18) and in tabular form in Otis (25). The tabular data was used by Otis in a regression analysis to yield the following sixth order polynomial fit for a slit:

$$f^s(\beta) = 1.532139 - 12.182786\ \alpha + 21.652283\ \alpha^2 + 4.495068\ \alpha^3 - 28.176666\ \alpha^4 + 10.950694\ \alpha^5 + 0.198042\ \alpha^6 \qquad (8a)$$

where $\alpha = (1+\beta)/2$, and a 5th order polynomial fit for a tube (26):

$$f^t(\beta) = 1.601512\ \beta - 0.860212\ \beta^2 - 0.70634\ \beta^3 - 2.734119\ \beta^4 + 1.382202\ \beta^5 \qquad (8b)$$

The functions $g(\beta)$ and $f(\beta)$ from Equations 5.-8. are shown in Figure 2.

The maximum shown in $g^t(\beta)$ arises from neglecting terms of order Re_w and higher order terms in Re_w in Equation 4. Only the leading order term is used as a good approximation for the total permeation velocity. Two major differences between the two-dimensional slit and the three dimensional tube are: (1) the tube shows higher wall suction for all values of β, i.e., $g^t(\beta) > g^s(\beta)$, and (2) the tube shows higher inertial lift away from the center-line but **lower** inertial lift away from the porous wall, w. Because of this the tube and slit have different equilibrium positions of $\beta^* = 0.71$ and $\beta^* = 0.62$, respectively. Thus, for $\beta < \beta^*$ lift and permeation suction drag forces will enhance each other, while for $\beta > \beta^*$ they will oppose each other.

Particle Trajectories. Under the condition of superposition, when $\lambda \sim Re_p\kappa^2$, the permeation and inertial drag forces are added vectorially to give the net motion in the β-direction. Terms of order Re_w and higher (small wall velocities) and the pressure drop in the z-direction are ignored. Then, from Berman (21) for a slit and Yuan and Finkelstein (22) for a tube, the particle moves in the z-direction with a velocity:

$$\frac{dz}{dt} = U_m^* (1-\beta^2) \quad \text{(slit or tube)} \qquad (9)$$

In the β-direction, adding Equations 4. and 7.

$$\ell\frac{d\beta}{dt} = \frac{1}{4} Re_p^* \kappa^2 U_m^* f(\beta) + v_w g(\beta) \qquad (10)$$

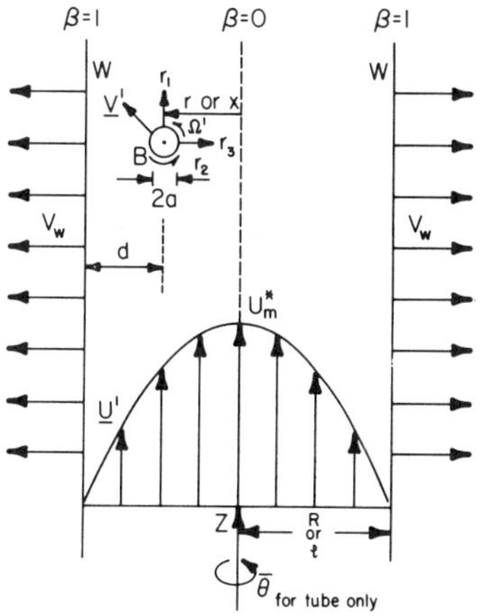

Figure 1. Spherical particle, B, suspended in Poiseuille flow in a porous tube or a slit with two porous walls.

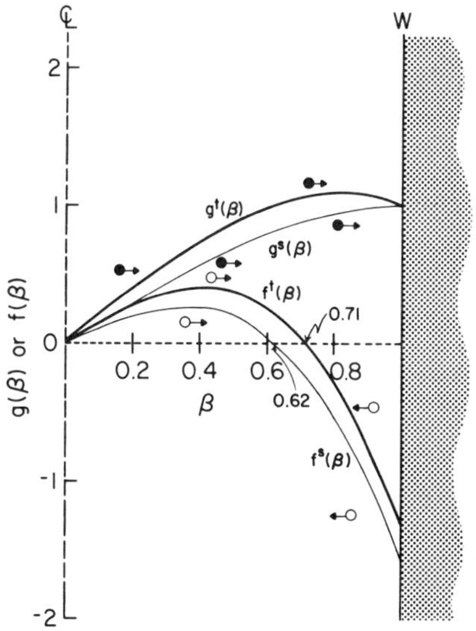

Figure 2. Permeation functions g(β) and lift functions f(β) for a porous tube (t) and a slit with two porous walls (s).

where Re_p^* is the local Reynolds number $Re_p^* = aU_m^*/\nu$, ℓ is replaced by R for a tube, and the appropriate functions for $g(\beta)$ and $f(\beta)$ are used. Dividing Equation 9. by Equation 10. we obtain:

$$\frac{dz}{d\beta} = \frac{U_m^*(1-\beta^2)\ell}{\frac{1}{4}\,Re_p^*\kappa^2 U_m^* f(\beta) + v_w g(\beta)} \tag{11}$$

Substituting for U_m^* from a material balance:

$$U_m^* = U_m - \eta v_w z/\ell \tag{12}$$

where $\eta = 4$ (tube) and $\eta = 3/2$ (slit) and ℓ is replaced by R for a tube, we obtain after rearrangement:

$$\frac{d\beta}{dz/\Theta} = \frac{1}{4}\left(1 - \frac{\eta\lambda z}{\ell}\right)\frac{f(\beta)}{(1-\beta^2)} + \frac{\gamma}{\left(1 - \frac{\eta\lambda z}{\ell}\right)}\;\frac{g(\beta)}{(1-\beta^2)} \tag{13}$$

which obtains the desired particle trajectory in a tube ($\ell \equiv R$, $\eta = 4$) or in a slit ($\eta = 3/2$). Equation (13) gives the derived particle trajectory from the coordinates of the initial position of the particle entering the duct (z_o, β_o) to some later position (z,β). The well-behaved trajectories were solved numerically using an IBM 3033 computer with Euler's method of integration. In the above equations γ is a measure of the ratio of permeation drag to initial lift drag, and is given by

$$\gamma = \frac{v_w}{Re_p\;\kappa^2\;U_m} = \frac{\lambda}{Re_p\;\kappa^2} \tag{14}$$

and Θ is a characteristic dimensional distance:

$$\Theta = \frac{\ell}{Re_p\kappa^2} \tag{15}$$

Results and Discussion

Particle Trajectories. In Figure 3 trajectories are presented for laminar flow in a slit (a) and in a tube (b) for initial positions $\beta_o = 0.3$ and $\beta_o = 0.9$ (with $z_o = 0$ in all cases) with γ as a parameter. It takes a particle about 6.0 and 8.0 dimensionless distance units (z/Θ) to reach the equilibrium positions of $\beta^* = 0.71$ and 0.62 in a non-porous tube and slit, respectively. It is important to note that at the low values of γ shown in Figure 3 the particles are captured at the permeable wall at the same axial position, $(z/\Theta)_{\beta_o}^{Cap}$

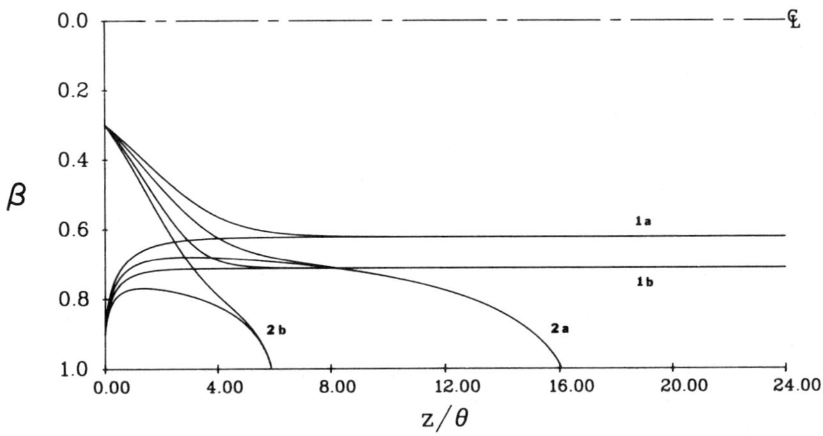

Figure 3. Particle trajectories in a slit (a) and tube (b) for 1. γ = 0.0 and 2. γ = 0.03.

independent of entrance radial position for $\beta_0^c < \beta_0 < \beta_0$-a where β_0^c is some critical position near the centerline. For this to occur the trajectories from each entrance position must converge to a unified pseudo-equilibrium trajectory prior to capture.

For high values of γ however, the particles move toward the wall with separate trajectories and are captured at different axial positions dependent on entrance radial position. This is shown in Figure 4 for $\beta_0 = 0.3$ and 0.9.

For $0.1 \leq \beta_0 \leq 0.9$ a critical flux value, γ_c, is defined where for any value of $\gamma > \gamma_c$ the particles follow separate trajectories for different values of β_0. Figure 5 shows the critical capture distance $(z/\Theta)_{\beta_0}^{Cap}$ as a function of γ for a porous tube and a slit with two porous walls with $0.1 \leq \beta_0 \leq 0.9$. Figure 5 shows that $\gamma_c = 0.02$ for a tube and $\gamma_c = 0.05$ for a slit. It is also apparent that for any γ for a given β_0 the particle will reach the wall in a shorter axial distance $(z/\Theta)_{\beta_0}^{Cap}$, in a tube than in a slit. This arises from the fact that the permeation function $g(\beta)$ is higher for a tube than a slit in all cases. Also the parameter η from the material balance in Equation 12. is higher for a tube than a slit indicating a greater area for permeation. At starting positions near the wall ($\beta_0 \geq 0.7$) the curves for a slit and tube seem to approach one another as γ increases. This is due to the fact that at high values of γ the particle reaches the wall in very short axial distances for both the tube and slit.

In summary Figure 5 shows that, under the constraints of the theory, a membrane in the porous tube configuration will foul at shorter axial lengths than a membrane in the slit configuration. Also higher fluxes lead to fouling at shorter axial lengths for both the tube and slit.

Commercial Relevance. The above analysis can be used to test the design of commercial membrane modules for the treatment of dilute suspensions of neutrally buoyant particles in laminar flow. Design data were chosen for the membrane filtration modules of seven manufacturers including Abcor (USA), Berghof (Germany), DDS (Denmark), Dorr-Oliver (USA), Enka/Membrana (Germany), Romicon/Amicon (USA), and Schleicher and Schüll (Germany). Table I gives the product name, design configuration and membrane type for ultra-, hyper- and microfiltration units studied (27-33). Also in Table I are the physical parameters needed for a trajectory analysis of each unit. In many cases the necessary data were incomplete in the commercial literature, therefore estimates were made for some of the values (34,35).

Figure 6 is a two dimensional velocity plot of maximum axial entrance cross-flow velocity, U_m, versus membrane permeation velocity, v_w. For the above trajectory theory to hold the stipulations $Re_p < 1$ and $Re_w < 1$ must both be met. These are both shown graphically for particles of 1 µm and 10 µm radii in Figure 6. Also shown are the lines of 100% recovery for the tube and slit configurations. Along these lines all of the fluid entering the duct permeates the membrane. Practical operating conditions with sufficient flow exiting the unit requires an operating region to the

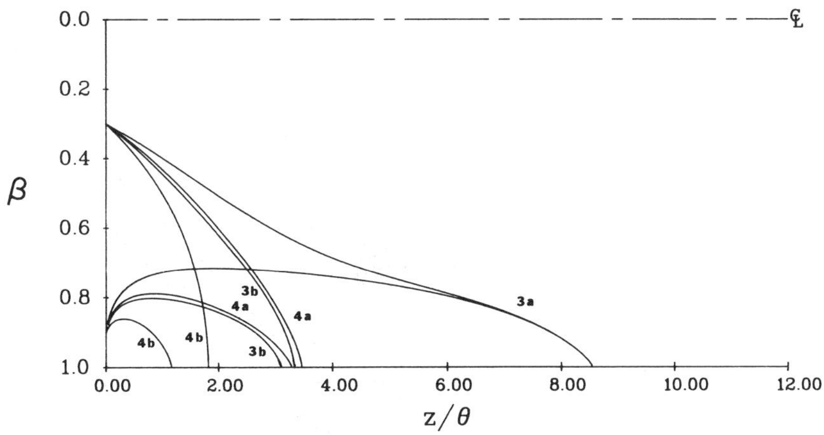

Figure 4. Particle trajectories in a slit (a) and tube (b) for 3. γ = 0.05 and 4. γ = 0.10.

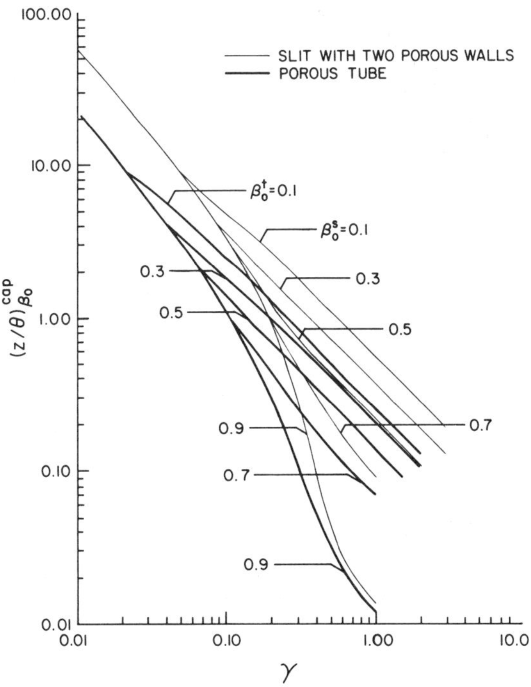

Figure 5. Critical capture distance, $(z/\theta)^{Cap}_{\beta_0}$, as a function of γ for a tube and a slit.

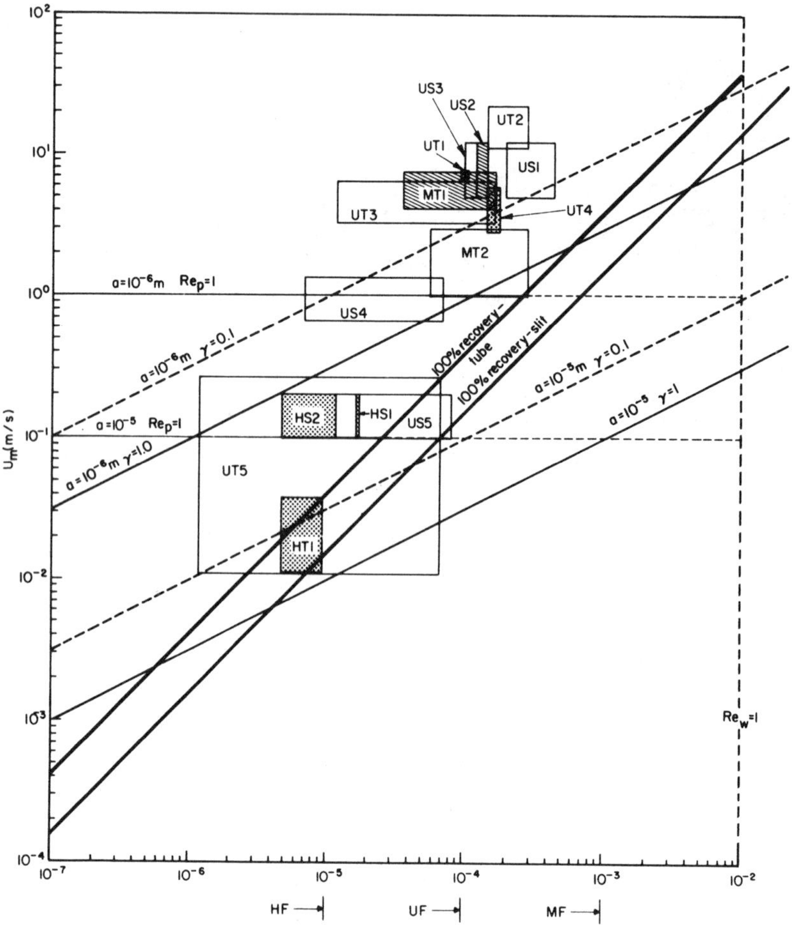

Figure 6. Two dimensional velocity plot of maximum entrance crossflow velocity, U_m, versus permeation velocity, v_w, showning the operating ranges of commercial membrane filtration units by code (see Table I.).

left of this line. Also shown are lines of constant γ which apply for the tube and slit. The operating ranges of the various commercial units studied are shown (see Table I for product code) and for the most part lie within or slightly outside the area in which the theory applies for model particles of 1 and 10 μm radii.

Tables II and III list parameters κ, γ, Θ and L/Θ calculated for each unit based on a 1 μm and 10 μm particle size, respectively. Also given are the critical capture distances, $(z/\Theta)^{Cap}_{\beta_o}$, for starting positions β_0 = 0.1 and 0.9. By comparing the capture distances with the membrane unit length, L/Θ, it may be determined whether the system will capture particles and foul and to what degree fouling will take place. If $(z/\Theta)^{Cap}_{\beta_o} < L/\Theta$ for both the case of β_0 = 0.9 and 0.1 the unit will be fouled by most of the particles entering the duct. If, however $(z/\Theta)^{Cap}_{\beta_o} < L/\Theta$ for β_0 = 0.9 but not for β_0 = 0.1 only some of the particles will be captured and cause fouling. If $(z/\Theta)^{Cap}_{\beta_o} > L/\Theta$ for all β_0 then the unit will not be fouled by particles of the particular size considered.

In Figure 6, the different regions of operation of hyper-, ultra- and micro-filtration are shown. They show that, in general, as the membrane permeability increases, axial velocity is increased. Some question remains regarding the applicability of the theories presented above to these results. Measurements of single particle trajectories (25) have shown that the theory applies for laminar flow above channel Reynolds numbers of one. How far above a value of one is not clear at present. What is clear is that the flow must be laminar. For porous tubes and slits, the onset of transition flow from laminar to turbulent occurs at a Reynolds number of about 4000 as compared to 2100 for flow in a smooth non-porous tube (36). Thus, the theory is probably not applicable to several of the entries shown in Table I (i.e., US1, US3, US5, UT1, UT2, and MT1). The entrance Reynolds numbers listed in Table I are probably on the high side since the entrance velocity used is a maximum and decreases along the flow path in the module. Also, the viscosity value used (water) is lower than that usually encountered and increases along the flow path.

For the case with the 1 micron particle in laminar flow (Table II) capture and fouling occurs for the modules HT1, UT3-5, and MT2, while for the modules HS1 and US4 all the particles escape capture and exit the module downstream. When the γ-value was very large (> 10^7), the critical capture distance was extremely short and could not be estimated with the current computer program. For this case the symbol NC - not calculable, is used in Table II.

For the case with the 10 micron particle in laminar flow (Table III) capture and fouling occur for only the HT1 and UT5 modules while for all the other modules with laminar flow the particles escape capture and exit the module downstream.

In summary, the data shown in Tables II and III suggest that for the standard operating conditions most commercial modules will not capture 10 micron diameter particles. Several modules will

Table I. Design Data for Commercial Membranes Modules

Configuration	Company, Location	Product	Code[a]	Membrane Type	U_m (ms^{-1})	Height or ID (mm)	L (m)	Membrane Surface Area (m^2)	V_w ($m\ s^{-1}$)	Re $x10^{-3}$[c]
				Hyperfiltration						
Flat Plate	DDS, Denmark	Lab Unit 20	HS1	Composite	0.10-0.20[b]	5.19[b]	0.076	0.018	$1.67x10^{-5}$	0.78
Flat Plate	Schliecher and Schull, W. Germany	CA	HS2	Cellulose acetate	0.10-0.20[b]	6.75[b]	0.4	0.0415	$4.63x10^{-6}$ $1.63x10^{-5}$	1.01
Tube	Abcor, USA	HFA	HT1	Celluose acetate	0.0189[b]-0.0379	12.7	3.0	0.127	$4.72x10^{-6}$ $9.44x10^{-6}$	0.36
				Ultrafiltration						
Flat Plate	Dorr-Oliver, USA	Ioplate	US1	Poly-sulphone	5.0-12.2	2.54	0.37	0.067	$1.89x10^{-4}$ $4.25x10^{-4}$	21.84
			US2	Poly-vinylidene fluoride	5.0-12.2	2.54	0.37	0.067	$1.18x10^{-4}$ $1.42x10^{-4}$	21.84
			US3	VC-AN copolymer	5.0-12.2	2.54	0.37	0.067	$0.95x10^{-4}$ $1.18x10^{-4}$	21.84
Flat plate	DDS, Denmark	Lab Unit 35	US4	Poly-sulphone	0.68-1.36[b]	5.90[b]	0.22	0.150	$0.69x10^{-5}$ $6.94x10^{-5}$	6.02
		Lab Unit 20	US5	Poly-sulphone	0.10-0.20[b]	5.19[b]	0.076	0.018	$1.16x10^{-5}$ $8.10x10^{-5}$	0.78

Tube	Abcor, USA	HFA	UT1	Celluose acetate	7.60[b]	2.54	3.05	0.204	9.43×10^{-5}	19.30
		HFM	UT2	Poly-vinylidene fluoride	11.36[b]-22.7	2.54	3.05	0.204	1.41×10^{-4} 2.82×10^{-4}	43.26
Tube	Romicon-Amicon, USA	PM	UT3	Poly-sulphone	3.32-6.48	1.1	0.635	0.022	1.16×10^{-5} 1.62×10^{-4}	5.39
		XM	UT4	Modacryl polymer	2.84-6.02	1.1	0.635	0.0021	1.39×10^{-4} 1.74×10^{-4}	4.87
Tube	Berghof, Germany	BM	UT5	Polyamide-imide	0.011-0.266	0.6	0.3	5.65×10^{-4}	1.16×10^{-6} 6.94×10^{-5}	0.083
				Microfiltration						
Tube	Enka/Membrana, Germany	Dyna-Sep Sampler	MT1	Poly-propylene	4.21-7.97	5.5	1.83	0.032	3.65×10^{-5} 1.62×10^{-4}	33.44
		Micro-dyn	MT2	Poly-propylene	1.0-3.0	1.8	0.56	1.2	5.56×10^{-5} 2.78×10^{-4}	3.60

[a] HT = hyperfiltration tube, HS = hyperfiltration slit, UT = ultrafiltration tube, US = ultrafiltration slit, MT = microfiltration tube.

[b] Estimate

[c] $\nu = 10^{-6}$ m^2/sec for water at 25°C

Table II. Comparison of Module Length with Critical Capture Distance for Particles of Radius 1 Micron

System Code[a]	$\kappa \times 10^3$	γ	Θ (m)	L/Θ	$(z/\Theta)^{Cap}_{\beta_o}$ $\beta_o = 0.1$	$(z/\Theta)^{Cap}_{\beta_o}$ 0.9
HS1	0.385	5007	1.17×10^5	6.51×10^{-7}	1.14×10^{-4}	2.0×10^{-6}
HS2	0.296	2.717×10^7	2.17×10^9	1.84×10^{-10}	NC[b]	NC
HT1	0.157	356,000	9.07×10^6	3.31×10^{-7}	1.61×10^{-6}	2.7×10^{-8}
US1	0.787	6.70	238.43	0.00155	0.085	0.0016
US2	0.787	2.83	238.43	0.00155	0.201	0.0039
US3	0.787	2.34	238.43	0.00155	0.250	0.0049
US4	0.339	319.50	2.517×10^4	8.74×10^{-6}	1.78×10^{-3}	3.1×10^{-5}
US5	0.385	13,880	1.17×10^5	6.51×10^{-7}	4.15×10^{-5}	7.2×10^{-7}
UT1	0.787	2.64	269.80	0.0113	0.093	0.0040
UT2	0.787	1.18	120.62	0.0253	0.22	0.010
UT3	1.82	1.09	33.89	0.0187	0.24	0.011
UT4	1.82	2.42	37.48	0.0169	0.11	0.0042
UT5	3.33	165.80	194.63	0.00154	0.0015	5.75×10^{-5}
MT1	0.364	20.20	3.41×10^3	5.37×10^{-4}	0.0121	0.0005
MT2	1.111	37.78	364.5	0.00154	0.0073	0.0003

[a] See Table I for code.

[b] NC = Not calculable.

Table III. Comparison of Module Length with Critical Capture Distance for Particles of Radius 10 Micron

System Code[a]	$\kappa \times 10^3$	γ	Θ (m)	L/Θ	$(z/\Theta)^{Cap}_{\beta_0}$ $\beta_0 = 0.1$	$(z/\Theta)^{Cap}_{\beta_0}$ 0.9
HS1	3.85	5.01	116.7	0.0007	0.12	0.0021
HS2	2.96	27,200	2.17×10^6	1.84×10^{-7}	2.09×10^{-5}	3.67×10^{-7}
HT1	1.57	356	9,070	0.0003	1.60×10^{-3}	2.72×10^{-5}
US1	7.87	0.0067	0.238	1.55	82.43	82.43
US2	7.87	0.0028	0.238	1.55	202.64	202.64
US3	7.87	0.0023	0.238	1.55	309.32	309.32
US4	3.39	0.3195	25.17	0.0087	1.17	0.24
US5	3.85	13.88	116.7	0.0006	0.041	0.0008
UT1	7.87	0.0026	0.270	11.3	115.49	115.49
UT2	7.87	0.0012	0.121	25.3	236.33	236.33
UT3	18.2	0.0011	0.034	18.7	236.33	236.33
UT4	18.2	0.0024	0.037	16.9	115.33	115.33
UT5	33.3	0.166	0.195	1.54	1.55	0.43
MT1	3.64	0.0202	3.41	0.534	9.50	9.48
MT2	11.11	0.0338	0.365	1.534	6.65	5.86

[a] See Table I for code.

however capture 1 micron diameter particles and most of these are of the tubular configuration.

Conclusions

In this paper, we consider only the fouling of membranes resulting from the capture of particles from dilute suspensions. The fluid mechanics of a neutrally buoyant particle moving in a slit with two porous walls in laminar flow is compared with that in a porous tube of similar dimension. Trajectory calculations are used for several commercial membrane modules to estimate whether one and ten micron radius particles are captured. Under the assumptions of the theoretical models, and all things being equal, tubular systems will capture more particles from dilute suspensions than slits with two porous walls. This is expected since the fractional recovery for a tube is 2.67 times that for a slit, when v_w/U_m, and L/R and L/ℓ are kept the same.

Nomenclature

a	particle radius, m
$f(\beta)$	dimensionless lift velocity function
$g(\beta)$	leading order permeation velocity function
$h(\beta)$	higher order permeation velocity function
ℓ	slit half height, m
L	duct length, m
R	tube radius, m
Re	Reynolds number based on channel entrance flow
Re_p	Reynolds number based on particle radius at channel entrance
Re_w	Reynolds number based on permeation velocity
t	time, s
U_m	maximum fluid velocity at entrance of tube or slit
v_ℓ	inertial or lift velocity, m/s
v_p	permeation velocity as a function of β, m/s
v_w	permeation velocity at wall, m/s
z	axial distance from channel entrance, m

z_o initial axial position, m

$(z/\Theta)^{Cap}_{\beta_o}$ dimensionless axial distance for particle capture from initial position β_o, z_o

Greek Symbols

α dimensionless distance function in Equation (8a)

β dimensionless distance from the centerline of slit or tube

β_o initial β

β^c_o critical initial β

β^* dimensionless equilibrium position

γ dimensionless ratio of permeation to lift

γ_c critical γ

κ ratio of particle radius to tube radius or slit half height

λ ratio of permeation velocity to crossflow velocity (v_w/U_m)

η coefficient in Equation (12) characterizing permeation area

Θ characteristic axial distance, m

Superscripts

s slit with two porous walls

t porous tube

* local value at axial distance z

Acknowledgments

The authors thank Frank W. Altena, Donald A. Drew, George Green, Jane R. Otis, and Jeffrey A. Schonberg for technical discussions and contributions to this work. The authors are also grateful to Raymond G. Cox and Howard Brenner for advice and discussions and thank Hidenori Hasimoto for making the inertial lift integral data available [Equation 8b.].

Partial funding for this work was obtained from Dr. Francois Fiessinger, Societe Lyonnaise des Eaux et de l'Eclairage, Le Pecq, France.

Literature Cited

1. Belfort, G.; Altena, F.W. Desalination, 1983, 47, 105-127.
2. Altena, F.W.; Belfort, G.; Otis, J.; Fiessinger, F.; Rovel, J.M.; Nicoletti, J. Desalination, 1983, 47, 221-232.
3. Reihanian, H.; Robertson, C.R.; Michaels, A.S., J. Membrane Sci., 1983, 16, 237.
4. Malthiasson, E., Ph.D. Thesis, Lund University, Alnarp, Sweden, 1984.
5. Jonsson, G., J. Memb. Sci., 1984, submitted.
6. Vilker, V.L.; Colton, C.K.; Smith, K.A. AIChE J., 1981, 27(4), 632-645.
7. Fane, A.G.; Fell, C.J.D.; Suki, A. J. Memb. Sci., 1983, 16, 195-210.
8. Belfort, G.; Marx, B. Proc. 6th Int. Symp. Fresh Water From the Sea, 1978, 4, p. 183.
9. Trettin, D.R.; Doshi, M.R. Ind. Eng. Chem., 1980, 19, 80.
10. Michaels, A.S. Chem. Eng. Prog., 1968, 64, 31.
11. Blatt, W.F.; Dravid, A.; Michaels, A.D.; Nelson, L. In "Membrane Science and Technology"; J.E. Flinn, Ed.; Plenum Press: New York, 1970.
12. Trettin, D.R.; Doshi, M.R.; in ACS SYMPOSIUM SERIES, No. 154 ("Synthetic Membranes: Vol. 11 Hyper- and Ultrafiltration Uses), A.F. Turbak Ed.; ACS, Washington, 1981, p. 373.
13. Probstein, R.F.; Leung, W.F.; Alliance, Y. J. of Phys. Chem., 1979, 83(9), 1228.
14. Zydney, A.L.; Colton, C.K. "Continuous Flow Membrane Plasmapheresis: Theoretical Models for Flux and Hemolysis Prediction", ASAIO Annual Mtg., Chicago, 1982.
15. Green, G.; Belfort, G. Desalination, 1980, 129-147.
16. Belfort, G.; Chin, P.; Dziewulski, D. Proc. of World Filtration Congress III, Vol. II, 1982, pp. 548-555.
17. Altena, F.W.; Belfort, G. Chem. Engrg. Sci., 1984, 39(2), 343-55.
18. Weigand, R.J.; Altena, F.W.; Belfort, G. 1984, submitted for publication.
19. Otis, J.R.; Altena, F.W.; Mahar, J.T.; Belfort, G. 1984, submitted for publication.
20. Cox, R.G.; Brenner, H. Chem. Engrg. Sci., 1968, 23, 147.
21. Berman, A.S. J. Appl. Phys., 1953, 24, 1232-35.
22. Yuan, S.W.; Finkelstein, A.B. Trans. ASME, 1956, 78, 719-724.
23. Vasseur, P.; Cox, R.G. J. Fluid Mech., 1976, 78, 2, 385-443.
24. Ishii, K.; Hasimoto, H. J. Phys. Soc. Japan, 1980, 48(6), 2144.
25. Otis, J., MS Thesis, Rensselaer Polytechnic Institute, Troy,
26. Hasimoto, H., personal communication.
27. Abcor Inc., 850 Main St., Wilmington, MA 01887, Pamphlets GN/DB-80-4 and GN/DB-80-5.
28. Forschungsinstitut Berghof GmbH, P.O. Box 1523, Berghof, D-7400 Tübingen-1, West Germany.
29. De Danske Sukkerfabrikker, DDS-Nakskov, DK-4900 Nakskov, P.O. Box 149, Denmark. Pamphlets 1624-GB-0881-5OCT, 1643-GB/D/F-0681.
30. Dorr-Oliver, Inc., 77 Havemeyer Ln., P.O. Box 9312, Stamford, CT 06904, Bulletin No. 10-8, UF Systems-Instructions.

31. Romincon, Inc., 100 Cummings Pk., Woburn, MA 01801, UF cartridge and membrane pamphlet; Amicon Corp., 21 Hartwell Ave., Lexington, MA 02173, Publication No. 455.
32. Schleicher and Schüll GmbH, Postfach 4, D-3354 Dassel, West Germany, Catalogue E2/0580.
33. Enka AG/Membrana, Postfach 200916, Ohder St. 28, D-5600 Wuppertal 2, West Germany, Product bulletin (Membrana USA) C-21A, TB-37, TB-38, TB-39.
34. Klein, W; Hoelz, W. from Enka AG see Reference 33, Technical paper "Crossflow Microfiltration in Chemical Processes".
35. Pusch, W.; Walch, A. Angew. Chem. Int. Ed. Engl., 1982, 21, 660-685.
36. Nagata, N.; Belfort, G., unpublished data.

RECEIVED February 22, 1985

Reverse-Osmosis and Ultrafiltration Membrane Compaction and Fouling Studies Using Ultrafiltration Pretreatment

BRIAN J. RUDIE, TIM A. TORGRIMSON, and D. DEAN SPATZ

Osmonics, Inc., Minnetonka, MN 55343

Is it compaction or fouling that causes membrane flux to deteriorate over time? For years, compaction of the membrane itself has been considered a primary reason for flux decline. Recent findings indicate that compaction does not play the role suggested by many and is much less important than fouling, even on closed loop tests using purified water.

The data on numerous membranes run for a minimum of 1000 hours at temperatures varying from 25°C (77°F) to 50°C (122°F) and pressures from 25 psig (172 kPa) to 800 psig (5512 kPa) was collected. Comparison of membrane compaction rates indicates that fouling is a much greater cause of flux degradation than has previously been reported. The use of ultrafiltration prior to the membranes on test helped to substantiate the effect of fouling. The system used for these tests is a much improved method of obtaining compaction results and it is suggested that all compaction studies use UF pretreatment immediately prior to the membrane being tested.

The ability to predict membrane performance over time is a necessity to properly design any RO or UF system. Whether the RO or UF system membrane configuration employed is of the hollow fiber, tubular, plate and frame, or spiral wound configuration; system designers must consider changes in solute passage and permeate rate that occur with operating time to design a system that will meet the requirements of the end user.

Predicting system performance requires a thorough understanding of the physical and chemical properties of the solution being processed as well as a thorough understanding of the basic physical and chemical properties of the membrane employed. Essential to the understanding of how a membrane performs on any given solution is the understanding of how the membrane performs under controlled conditions. This paper discusses a better approach to

0097-6156/85/0281-0403$06.00/0

understand and predict membrane flux based on data obtained through closely controlled experiments.

Discussion

Initial attempts were made to collect meaningful data on membrane flux over time in a closed-loop test cell system. The objective was to get the non-fouling induced decline in flux, i.e. the compaction rate, of the membrane. This initial system contained a combination of brass, stainless steel and plastic components. Due to contamination of the feed solution from corrosion of the metal components, airborne particulates and microorganisms, meaningful data could not be obtained due to significant fouling. Two new test systems were designed and built at Osmonics in an attempt to reduce contamination of the feed solution and to collect meaningful data.

One test system was designed such that the feed solution to the test cells was continuously prefiltered through ultrafiltration spiral-wound membrane elements immediately prior to the test cells. The other test system did not employ the use of ultrafiltration of the feed solution. Both systems were constructed of all 316SS and plastic components.

The high pressure plumbing of both systems was constructed of 316 stainless steel components. The low pressure plumbing was constructed of either 316 SS or inert plastic components. Flow and pressure was generated by multi-stage centrifugal TONKAFLO pumps using all SS and plastic wetted parts. Flow schematics of the two test systems are shown in Figures 1 and 2.

The flow through the UF test system is as follows: the feed solution is contained in a 55 gallon polyethylene tank. The feed solution is fed by gravity to the inlet of the low pressure pump. A throttling valve at the discharge of the pump allows for flow and pressure adjustment.

Five micron cartridge filters are installed to remove any large particulate matter that may be present in the feed solution. The 316 SS filter housing is located at the discharge end of the first pump rather than ahead of the inlet to the pump to prevent cavitation of the pump. A 316 stainless steel heat exchanger with a tempering valve is used to maintain the desired temperature of the solution. In the system containing the ultrafiltration prefiltration (Figure 2), the pressurized feed solution passes through the ultrafiltration housing containing two OSMO-411-PT2(PS) spiral-wound UF elements. Note that this system contains a second multi-stage centrifugal pump. If low pressure testing ($\leq$400 psi) is desired, the second pump is by-passed. If high pressure testing (600-1000 psi) is desired, the flow is directed through the second pump to increase pressure to the UF housing.

The concentrate flow through the UF elements is controlled by a valve at the concentrate discharge side of the element housing. In this system, the recovery is established at 100% (i.e. no concentrate flow). Periodically, the concentrate valve is opened to increase concentrate flow over the membrane to flush away foulant materials. The permeate from the UF housing provides the direct feed to the test cell manifold. It is important to recognize that the UF runs with a transmembrane pressure of 40-50 psi and that

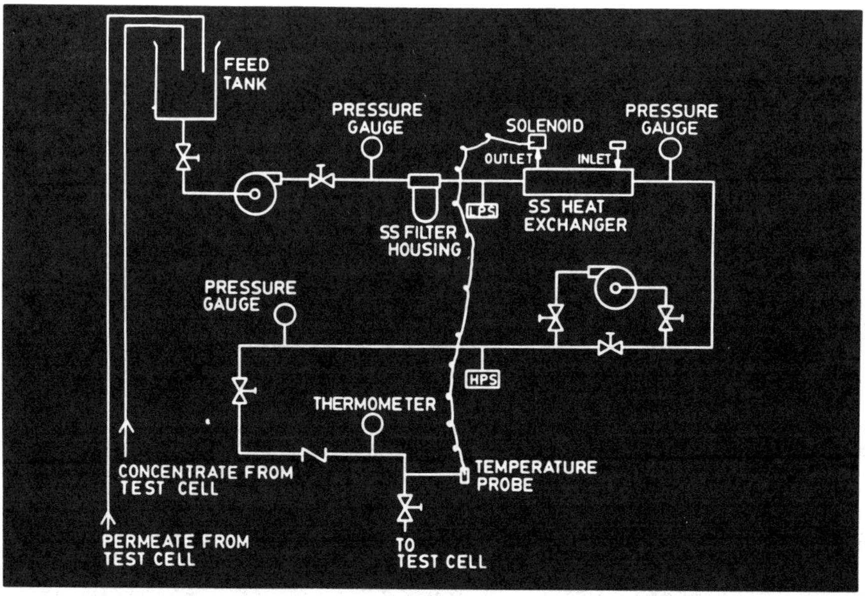

Figure 1a. Flow schematic of test cell loop without UF prefiltration.

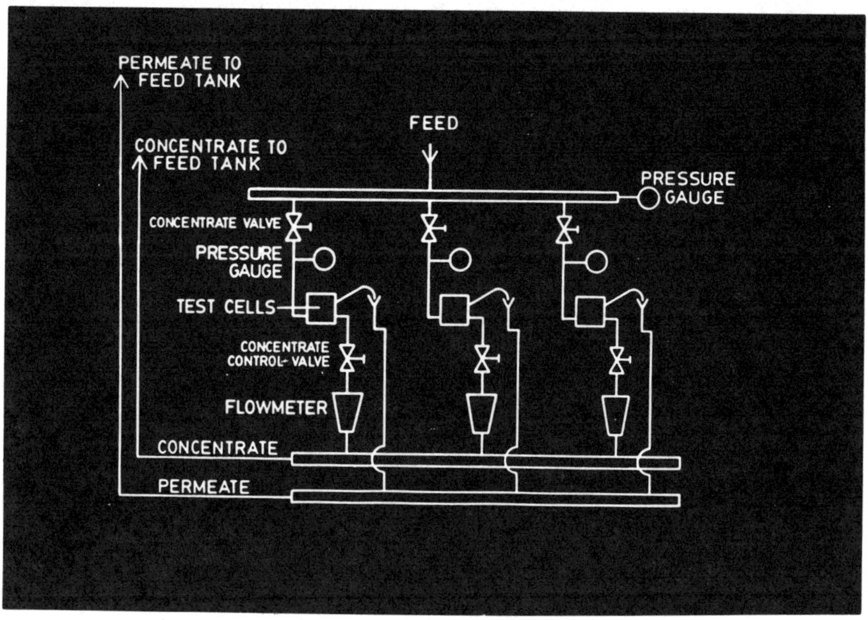

Figure 1b. Test cell.

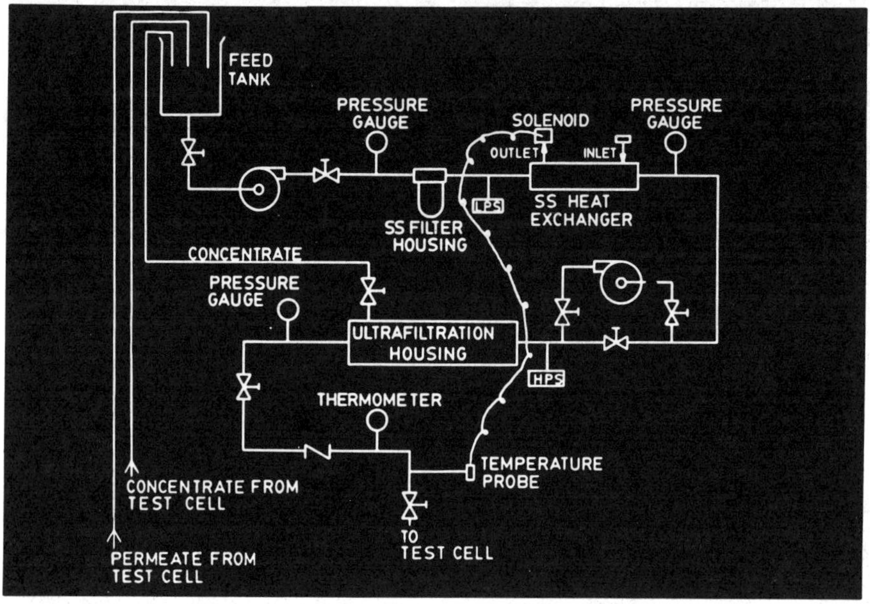

Figure 2. Flow schematic of test cell loop with UF prefiltration. Test cell is shown in Figure 1b.

the permeate is at the desired test cell operating pressure. Thus, there is no repressurization of the permeate as it feeds directly to the test cells with minimal contact of metals prior to contacting the membrane in the test cells.

The flow through each test cell is controlled by the 316 SS concentrate valve on the discharge side of each cell. The permeate and concentrate are returned to the feed tank, closing the loop.

Both systems were equipped with numerous control devices to protect the system components and to maintain the desired flows, temperatures, and pressures. In the event of reduced pressure due to blinding of the cartridge filters or a leak in the system, the low pressure switch automatically cuts off the power to the pumps. A high pressure switch automatically shuts off power to the pump in the event of over-pressurization due to a restriction in the plumbing. A temperature probe senses the temperature of the feed solution and controls the cooling water flow through the heat exchanger via a solenoid valve to maintain the desired temperature.

Data was collected on numerous CA and PS membranes run on a feed solution consisting of 2000 ppm NaCl and 0.5% formaldehyde in RO H_2O. Formaldehyde was used to eliminate potential bacteria mold or algae growth which can be a major problem on closed-loops without bactericides. Each membrane was tested for a minimum of 1000 hours continuous operation at various operating temperatures and applied pressures to determine the effects of temperature and pressure on membrane flux decline. New membrane samples were used for each test conducted. Table I reflects the battery of tests performed on both test loops.

Table I. Membranes Tested and Associated Operating Pressures and Temperatures

Membrane Type	Transmembrane Pressure (psig)	Operating Temperatures °C (°F)		
Polysulfone (UF)	25	25(77)		
Polysulfone (UF)	50	25(77)		
Cellulose Acetate (UF)	100	25(77)		
Cellulose Acetate (RO)	400	25(77),	40(105),	50(122)
Cellulose Acetate (RO)	600	25(77),	40(105)	
Cellulose Acetate (RO)	800	25(77)		50(122)

Visual examination of the membranes after long term testing indicated that both new all 316SS/plastic systems significantly reduced membrane fouling when compared to the first test loop which contained brass components. The membranes tested on the new test system without UF prefiltration still showed significantly more membrane fouling than membranes tested on the system employing UF prefiltration. Nevertheless, some membrane fouling was still observed on the membranes even with UF prefiltration, presumably, from inorganic materials that passed through the UF membrane.

Procedure

Multiple 2 inch diameter samples of each membrane were tested simultaneously to reduce experimental error. The temperature of the test solution was maintained at ±2°C of the desired test temperature throughout each test. Sodium chloride passage was monitored throughout the test as a measure of membrane integrity. Membrane flux was monitored as a measure of membrane compaction and/or fouling.

Feed flow rate was maintained at 0.3 to 0.5 gpm (1.3 to 1.9 lpm) for each cell. This flow rate range corresponded to a Δ P across the cell of 3-8 psid, respectively. Flow conditions across the cell were determined empirically by a plot of Δ P vs. flow rate. Non-laminar flow was found when the Δ P was greater than 1.25 psid. Operation with a Δ P of 3-8 psid across the cell, in conjunction with cell design, insured transitional to turbulent flow conditions across the membrane surface.

The test cells used for these tests are shown in Figure 3. The flow through the test cell is shown in Figure 4.

Feed enters the cell at the base and is directed to the center of the membrane sample. The concentrate then flows radially across the membrane surface to the concentrate outlet port located at the periphery of the cell. The concentrate then passes through a concentrate valve and flow meter. The concentrate flow is controlled by adjusting the concentrate valve.

The membrane is backed by a SS porous disc. The pure water (permeate) that passes through the membrane passes through the porous disc, and is collected through a port in the top of the cell. The permeate then exits the test cell and returns to the feed tank.

For each battery of tests conducted, initial membrane flux values were taken at 0.5 hours after the initiation of the test. Additional flux values were measured and recorded at two hour intervals during the first 24 hours of the test and then every other day throughout the duration of the test. Short intervals were used in the initiation of the test because most literature shows a sharp drop in flux at the beginning of a test.(1) Our data confirms that the most significant flux reduction occurs within the first 10 hours of operation.

The flux data obtained for each membrane was statistically analyzed via a linear regression program. The time and percent of original flux value was converted to log form and entered into the computer. A "least-squares"(2) linear regression was used to determine the best fit line and the resulting lines were displayed on log-log graphs.

By taking the values from the graphs, the slopes(m) were then calculated using the following equation:(3)

$$M = \frac{\text{Log } F_2 - \text{Log } F_1}{\text{Log } T_2 - \text{Log } T_1}$$

Where F_2 = Flux @ T_2 (Time)
and F_1 = Flux @ T_1 (Time)

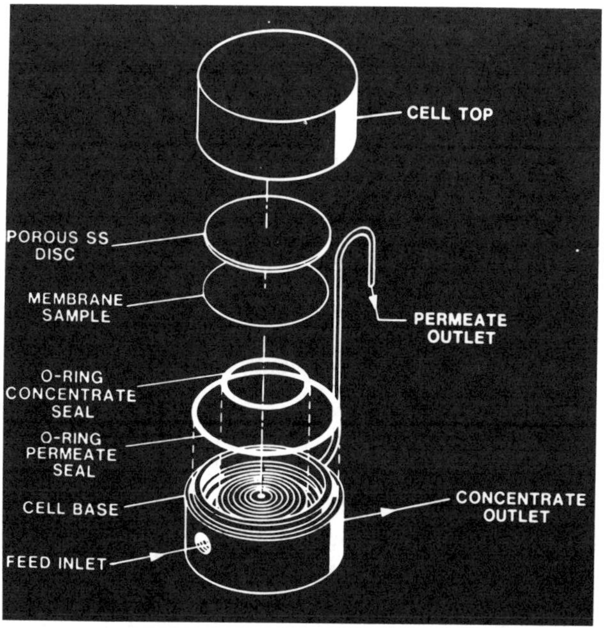

Figure 3. Test cell: exploded view.

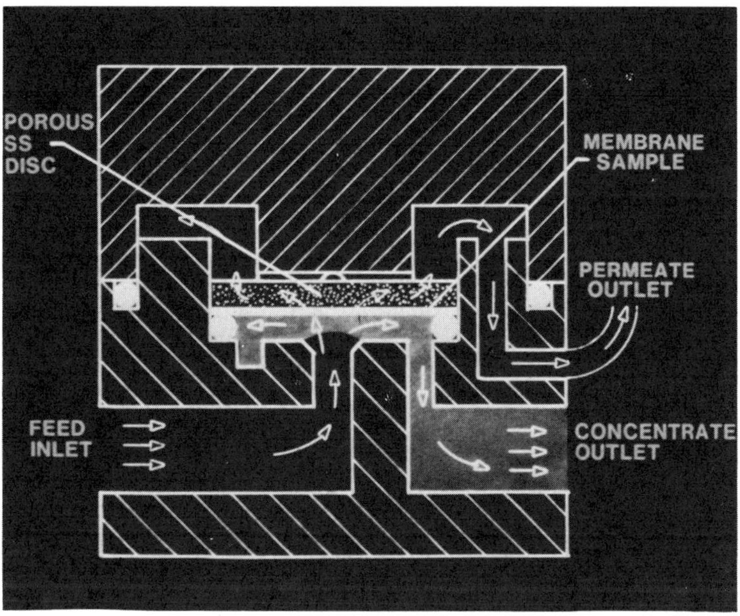

Figure 4. Test cell cross section and flow path.

As can be seen from the equation, for any situation where there is a membrane flux decline with increasing time, ($F_2<F_1$), the slope of the line will be a negative value. The larger the negative slope value, the more rapid the flux decline. Conversely, a flux increase with time would result in a positive slope.

Table II displays the calculated slopes of the UF membranes tested with and without UF prefiltration.

Note that the higher the initial flux, the more negative the slope value when tested without UF prefiltration. With UF prefiltration, the slopes are nearly all equal. The exception is the CA(UF) membrane which actually showed a flux increase with time when tested with UF prefiltration but a flux decline when tested without UF prefiltration. It is speculated that the increase in flux of the CA membrane w/UF prefiltration is due to physical changes in the membrane pore structure as a result of pressure. Without UF prefiltration, these physical changes are not evident due to the significant effect of fouling. In every case, the slope of the line for each membrane is more substantially negative without UF prefiltration than with UF prefiltration (approximately 10 times more negative).

Table III displays the calculated slopes of the RO membranes tested with and without UF prefiltration.

Again, the higher flux membranes tend to show the most significant flux decline. The slope of the line for each membrane is more negative without UF prefiltration, however the effect is more pronounced at lower temperatures and pressures than at higher temperatures and pressures.

Looking at the 50°C experiments (samples 3 and 8), the difference between slopes is not readily discernable between UF prefiltered and non-UF prefiltered membrane. This suggests that temperature-induced softening of the polymer had a more significant effect on flux loss than the effect of fouling. Further testing is planned on membranes with higher melting temperature polymers to substantiate this.

Observing the slope values of the 3% NaCl passage membranes at 25°C and various transmembrane pressures (Samples #1, 4 and 6), we see that increased pressure has a minimal effect on the rate of flux decline. However, there is still a significant decrease when the slope with UF prefiltration is compared to the slope without UF prefiltration.

The graphs of the data presented in Tables II and III, vividly show the effect that UF prefiltration has on reducing flux decline (Figures 5 and 6).

Summary

The results of the experiments to date at Osmonics graphically indicate that membrane flux reduction over time results primarily from fouling with compaction being a minor factor at 25°C(77°F), but a more important factor at 50°C. There is some evidence to suggest that membrane fouling may also contribute to accelerated membrane compaction. Additional testing on other membrane materials and at even higher temperatures will be undertaken to further identify the true cause for flux decline in RO/UF systems.

Table II. Comparison of Slope Values of Various UF Membranes With and Without UF Prefiltration

Sample	Type	Trans-Membrane Pressure (psi)	Temp. °C(°F)	Slope W/O UF Prefilt.	Slope With UF Prefilt.	Initial Flux (gfd)	MWCO
1	PS	25	25(77)	-0.160	-0.013	210	50,000
2	PS	50	25(77)	-0.140	-0.010	196	20,000
3	PS	100	25(77)	-0.180	-0.011	265	1,000
4	CA	100	25(77)	-0.056	+0.012	68	1,000

Table III. Comparison of Slope Values of Various RO Membranes With and Without UF Prefiltration

Sample	Type	Trans-membrane Pressure (psi)	Temp. °C(°F)	Slope W/O UF Prefilt.	Slope With UF Prefilt.	% Slope Reduc. w/UF Prefilt.	Init. Flux (gfd)	Init. Nom. NaCl Pass. (%)
1	CA	400	25(77)	-0.016	-0.005	69	14.8	3
2	CA	400	25(77)	-0.014	-0.006	57	19.7	8
3	CA	400	50(122)	-0.022	-0.021	4.5	12.2	3
4	CA	600	25(77)	-0.017	-0.009	53	26.7	3
5	CA	600	25(77)	-0.062	-0.008	87	36.3	8
6	CA	800	25(77)	-0.020	-0.014	30	28.0	3
7	CA	800	25(77)	-0.090	-0.021	77	31.3	8
8	CA	800	50(122)	-0.057	-0.054	5.3	38.8	3

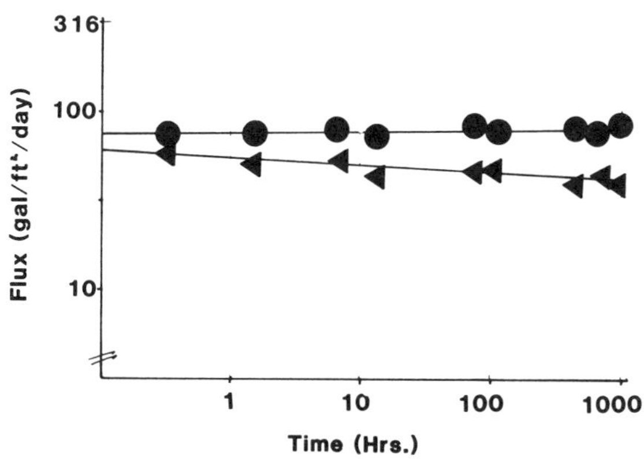

Figure 5a. Comparison of UF membrane flux decline with (●) and without (◀) UF prefiltration for a 1000 MWCO CA membrane (77 °F and 100 psi).

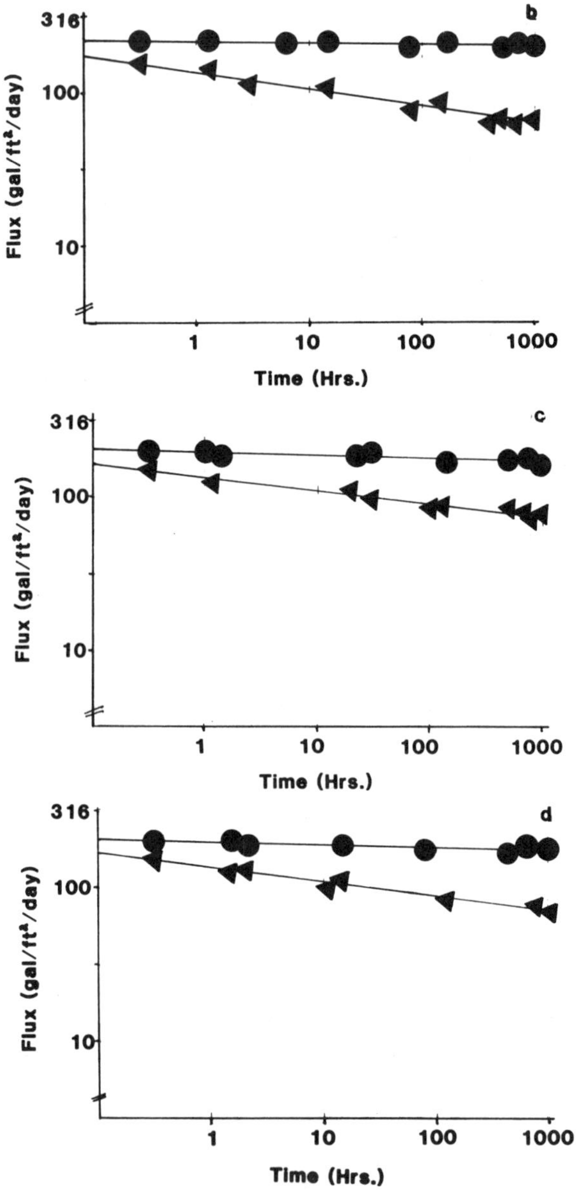

Figures 5b–5d. Comparison of UF membrane flux decline with (●) and without (◀) UF prefiltration for b, 1000 MWCO PS membrane (77 °F and 100 psi); c, 20,000 MWCO PS membrane (77 °F and 50 psi); and d, 50,000 MWCO PS membrane (77 °F and 35 psi).

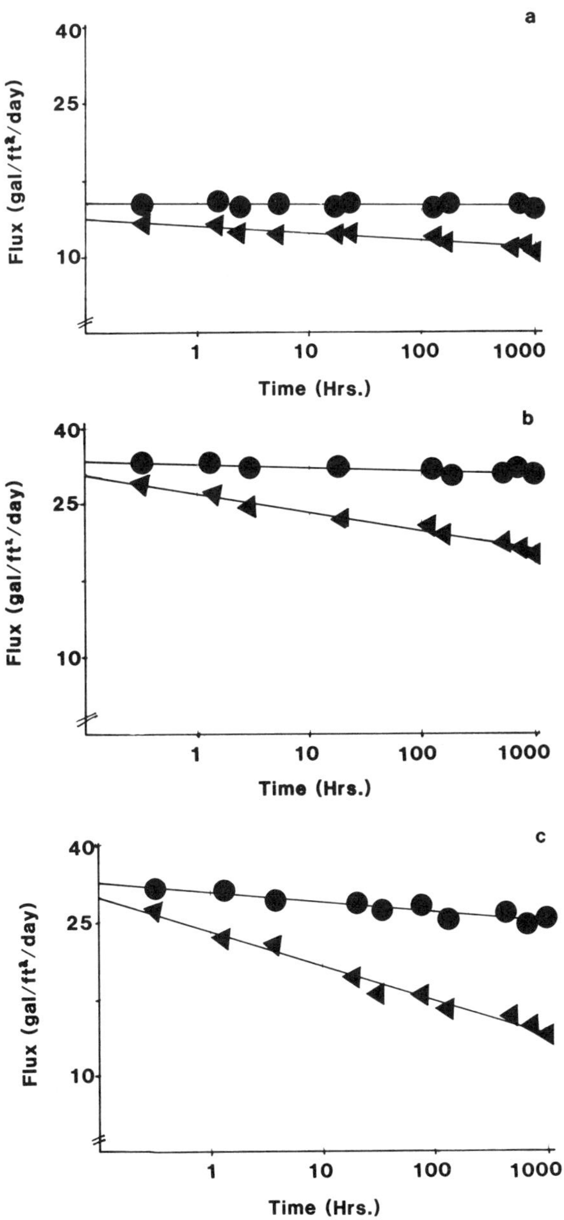

Figures 6a-6c. Comparison of RO membrane flux decline with (●) and without (◀) UF prefiltration for a, 3% NaCl passage CA membrane (77 °F and 400 psi); b, 8% NaCl passage CA membrane (77 °F and 600 psi); and c, 8% NaCl passage CA membrane (77 °F and 800 psi).

With UF prefiltration, membrane fouling and hence flux decline can be dramatically reduced. By employing ultrafiltration membranes as prefiltration ahead of RO membrane systems, significant improvements in RO system performance can be expected. As RO membrane becomes more expensive, the use of UF pretreatment should increase significantly to lengthen useful RO membrane life and give more predictable system performance.

Literature Cited

1. Lacey, R. E., and Loeb, S. "Industrial Processing with Membranes"; Wiley-Interscience: New York, 1972; p. 140-143.
2. Holman, "Experimental Methods for Engineers"; McGraw Hill: New York, 1978; p. 74-77.
3. King, W. M., Hoernschemeyer, and Saltonstall, C. W. "Reverse Osmosis Membrane Research"; Lonsdale, Podall, Ed.; Plenum Press: New York, 1972; p.154.

RECEIVED February 22, 1985

30

Determination of Gel Volume Deposited on Ultrafiltration Membranes

J. L. GADDIS[1], D. A. JERNIGAN[1,3], and H. G. SPENCER[2]

[1]Department of Mechanical Engineering, Clemson University, Clemson, SC 29631
[2]Department of Chemistry, Clemson University, Clemson, SC 29631

A method is developed to interpret the transient flux response of an ultrafiltration membrane to yield an estimate of gel volume deposited. Experiments of this type are presented for polyvinyl alcohol (PVA) ultrafiltration in step pressure changes up to pressures of 4.2 MPa. Thicknesses of 20 μm are typically indicated for this combination.

Any pressure-driven membrane process tends to produce permeated solvent more or less in proportion to the pressure applied to the membrane. As the pressure increases the permeate production at some point fails to maintain proportionality to the pressure increase — a form of diminishing return. Michaels (1) proposed for the common case of separation of large molecules that the retained molecules, being unable to diffuse sufficiently against the incoming solvent flow, reach a concentration at which the mixture provides hydraulic resistance to the solvent, termed as "gel". Blatt and coworkers (2) and later Shen and Probstein (3) studied the development of concentration near the membrane and the necessary distribution of flux along the channel subject to the gel concept of constant concentration. By simple extension of such concepts a finite membrane element may be shown to produce a flux-pressure result as shown in the solid line of Figure 1. Up to point "A" the permeate flux is limited by the membrane permeability to solvent flow. Beginning above point "A" there is at first a small, then growing from downstream to upstream, coating of gel on the membrane as the pressure is increased. Since the gel thickness increases along the channel, the upstream portions are more productive. The fraction of non-gel-affected membrane surface shrinks rather quickly as the pressure is raised. Therefore the average flux reflects rather strongly the tendency of the gel coating.

Some investigators suspected that the gel concept might be incorrect and that the effect might be only osmotic resistance. Trettin and Doshi (4) have illuminated the subject and show that

[3]Current address: CARRE, Inc., Seneca, SC 29678

0097-6156/85/0281-0415$06.00/0

both effects are possible. Observations of time-dependent fluxes by Reihanian et al (5) indicate that the filter cake layer or compressed gel layer tends to resist diffusive tendencies. The transients reported herein are all associated with increasing pressure steps and not decreasing steps which do fail to mirror the increasing pressure effect. The pressures are felt to be high enough to neglect osmotic pressure effects.

The Hypothesis

If a step change in pressure is applied to an ultrafiltration membrane operating at steady state it should eventually respond from point "O" to point "2" on Figure 1. Point "O" is typically in the non-permeable region so part of the membrane is covered with flux retarding, concentrated solute. Point "2" has a greater extent and thickness of concentrated solute, or gel. If the pressure step is applied suddenly, the gel does not form instantly and hence the flux increases along a line of temporarily constant permeability shown by the heavy dashed line. Because there is no absolutely perfect step, there is a rounding of the curve as the membrane-gel combined resistance builds. Then, while pressure is kept constant at P_2, the flux accomodates to the steady point 2.

This process is sketched on a time history of flux in Figure 2. Prior to time, t_o, the flux is constant at $\bar{J}_o$. (The overbar on the symbol denotes spatial average over the module). At time t_o the pressure is raised to P_o and maintained. The average flux increases along with the pressure until the growing resistance of gel coat forces it to diminish to the steady point 2.

The "wall" denotes herein the interface between the concentration boundary layer and membrane or gel. At the wall there is, due to bulk motion, a solute flux Jc_w passing toward the gel, where J is the local solvent flux and c_w is the concentration of solute at the wall, which in this case is the gel concentration. The diffusion of solute away from the gel may be described as $D\ \partial c/\partial y|_w$ where D is a suitably defined diffusion coefficient and $\partial c/\partial y|_w$ is the solute concentration derivative evaluated at the wall. The net solute flux between these two must be interpreted as a deposition of solute onto the gel surface.

$$\frac{d(m''_g)}{dt} = Jc_w - D\left.\frac{\partial c}{\partial y}\right|_w \qquad (1)$$

whenever steady state is reached the gel mass is constant and the right hand side is a familiar boundary condition. m''_g is the mass of gel per unit area. Values of m''_g, J, and c_w are both time and streamwise coordinate dependent, and the following simplifying assumptions are presumed.

Initially there is a region from 0 to x^* having no gel ($m''_g = 0$), $c_w < c_g$, and $J = J_p$ where J_p is the flux limited by permeability and pressure. Following this is a region $x > x^*$ having a layer of gel growing (3) according to $x^{+1/3}$, J decaying according to $x^{-1/3}$, and $c_w = c_g$. Here c_g is the supposed unique gel concentration. In steady state following the pressure step the gel region starts at $x = x_p$; $x_p < x^*$. The variations of parameters are similar to those before the pressure step.

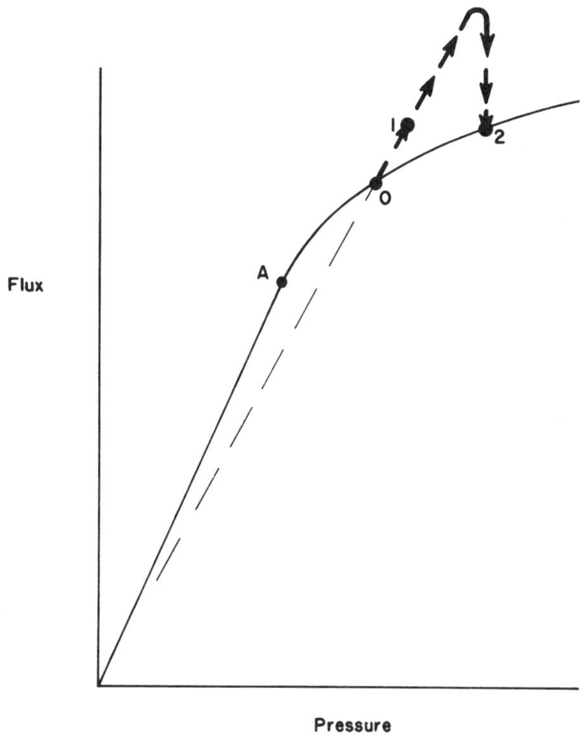

Figure 1. Flux Versus Pressure Characteristic of Gel Layers.

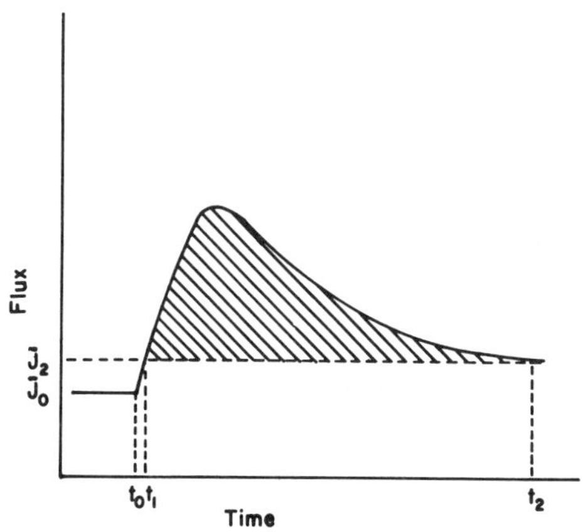

Figure 2. Flux History in Response to a Step in Pressure.

The transient between steady operating points is assumed to approximate the following. In the permeable region $x < x_p$ the volume flux and concentration field are assumed to be quasi steady so that as soon as the pressure reaches the steady final value no further changes occur. In the region aggressed by gel $x_p < x < x^*$ it is assumed[1] that the pressure rise forces a rapid rise in concentration at the membrane surface to $c = c_g$. It is assumed that for the ensuing period the gel thickness grows monotonically causing a reduction in flux to equilibrium. Also during this time, the diffusion remains time invariant by quasi-equilibrium assumption, though this is felt to be approximate only. Finally in the region $x > x^*$ the gel grows monotonically and the diffusion, by quasi-equilibrium arguments, is again time invariant.

The pressure cannot negotiate a true step change and so the event that signals the start of the assumed quasi-equilibrium period must be reckoned. We have used herein the time at which the average flux reaches the asymptotic value. This time is the t_1 of Figure 2. Probably the end of pressure rise would be better in keeping with the nature of the interpretive assumptions, since only at that point will the gel begin to cover the region to $x = x_p$. However, the actual pressure trace was not recorded in coordination with the flux trace. Only a few seconds, representing about 1 percent of the total active duration of experiment, separate the two events. Thus the timing errors are estimated to be of small consequence.

In equation (1) during the transient operating period we integrate from $x = 0$ to ℓ and from time t_1 to t_2. There results

$$\int_{t_1}^{t_2}\int_0^{\ell} \frac{d\ddot{m}_g}{dt}\,dxdt = \int_{t_1}^{t_2}\int_0^{x_p} Jc_w\,dxdt + \int_{t_1}^{t_2}\int_{x_p}^{\ell} Jc_w dxdt$$

$$- \int_{t_1}^{t_2}\int_0^{x_p} D\,\frac{\partial c}{\partial y}\Big|_w\,dxdt - \int_{t_1}^{t_2}\int_{x_p}^{\ell} D\,\frac{\partial c}{\partial y}\Big|_w\,dxdt \qquad (2)$$

The diffusion at quasi steady state must equal Jc_w in the region 0 to x_p, so the first and third integrals cancel each other. The value of $D\ \partial c/\partial y|_w$ equals J_2c_w at each position according to the assumption. This substitution is made in the fourth integral on the right. The left side is simply the mass of gel per unit of width, $\dot{m}_g$, accumulated during the time interval. Equation (2) may be rewritten

$$\dot{m}_g = c_g\int_{t_1}^{t_2}\int_{x_p}^{\ell} J(x,t)dxdt - c_g\int_{t_1}^{t_2}\int_{x_p}^{\ell} J_2(x)dxdt \qquad (3)$$

Further it is convenient to add and subtract a term $c_g \int_{t_1}^{t_2}\int_0^{x_p} J_p dxdt$ and divide the equation by c_g to produce $V_g' = \dot{m}_g/c_g$, the gel volume per unit of width.

---

[1]see the Appendix for a partial justification.

$$V_g' = \int_{t_1}^{t_2} \int_0^{\ell} [J(x,t) - J_2(x)]dx\, dt \tag{4}$$

Using $\bar{J}$ as the spatial mean value of flux $\ell\bar{J} = \int_0^{\ell} J dx$

gives

$$V_g'' = \int_{t_1}^{t_2} (\bar{J}(t) - \bar{J}_2)dt \tag{5}$$

This expression is useful because one only measures mean fluxes on channels. The V_g'' is the gel volume deposited per unit area in the transient from t_o to t_2. Graphically, Equation (5) represents the shaded area of Figure 2.

Resistances of the membrane, R_m, and gel, R_g, are additive in the form

$$R = R_m + R_g$$

Here R is the resistance to flow $P/\bar{J}$ and P is the pressure difference across the membrane; osmotic pressure is assumed negligible. With only increases in pressure considered the trend of Figure 1 shows clearly that $R_2 > R_1$ and R_m is not pressure dependent so $R_{g2} > R_{g1}$. Data for resistance increase are derived from steady state values only as

$$R_{g2} - R_{g1} = \Delta R = \frac{P_2}{\bar{J}_2} - \frac{P_1}{\bar{J}_1} \tag{6}$$

Upon the determination of resistance increase from Equation (6) and the volume from equation (5), it is possible to compute a resistivity, $\mathcal{R}$, as

$$\mathcal{R} = -\frac{dp}{dx}/J_x$$

This resistivity is μ/K where μ is the viscosity and K the specific permeability of Darcy's law (see Reference 6). To compute the restivity, the pressure on the newly formed gel layer should be $J_2\Delta R$ on the thickness V_g'', on average. The quotient of these is the average pressure gradient through the gel layer, so

$$\mathcal{R} = \frac{\Delta R}{V_g''} \tag{7}$$

This resistivity is approximate since it is the ratio of spatially averaged quantities. A more careful estimate can be computed by knowledge of the extent and thickness distribution of the gel layer. Some calculations of this type have been made for a range of hypothetical step changes. These calculations though not presented tend to show systematic shifts of 15 to 20 percent from the value given by equation (7). For the purpose of this discussion the cursory evaluation of Equation (7) is considered adequate.

Experimental Setup

A schematic of the ultrafiltration system employed for this study is illustrated in Figure 3. The reservoir maintained the feed solution and accepted the return of the membrane retent (concentrate) and permeate. A positive displacement pump with an attached pneumatic accumulator and by-pass loop pressurized the system. Bourdon tube gauges measured both the pump exit and module supply pressure. Temperature was monitored by a thermocouple placed on the concentrate line close to the module. Tap water flowing countercurrent through a concentric heat exchanger provided temperature control.

Permeate from the module flowed through a 0.34-mm diameter orifice across which pressure difference was measured with a 1 psi pressure transducer. The pressure transducer signal was recorded continuously by a strip chart recorder and by an x-y plotter during transient operation. The orifice-transducer-recorder system was calibrated from steady state data measured with a graduated cylinder and a stopwatch. The concentrate flow was measured with a float type flowmeter calibrated for each fluid concentration and temperature tested.

The membrane module consisted of a hydrous zirconium oxide (7) membrane formed on the 12.5-mm internal diameter of a 0.91 m long, sintered stainless steel support tube obtained from Mott Metallurgical Corporation in Farmington, Connecticut. The support tube had a 0.5-micron pore size (rating of the manufacturer). Buttwelded to each end of the support tube was a 0.3-m length, 15.8-mm internal diameter, stainless steel tube.

The permeate chamber surrounding the membrane tube was a 2-inch, schedule 40, stainless steel pipe sealed with tube fittings to the solid tubing on either end of the membrane tube. The permeate chamber was unusually large to accommodate some features not germane to this investigation. Other details of the equipment may be found in Reference 8.

Procedure

Solutions of 2 percent and 4 percent mass concentration of polyvinyl alcohol (PVA) at 60°C and 80°C were tested in the ultrafiltration loop with flow rates from 3.8 to 15.1 liters per minute. The PVA was DuPont Elvanol T-25 which is a highly hydrolyzed, high molecular weight blend commonly used for textile warp sizing. Solvent water was RO permeate from laboratory tap water. A typical procedure follows.

The pump was energized with the reservoir filled with 57 liters of demineralized water. The pump discharge pressure was adjusted to 4.8 MPa using the by-pass loop valve and the upstream module valve. Membrane pressure was maintained low until data was to be taken.

The solution to be tested was mixed in the feed reservoir. The 2 percent and 4 percent concentration of PVA solutions contained 1140 grams and 2280 grams, respectively, of dry powder per 57 liters of water. PVA was slowly added to the reservoir to prevent coalescence. Mixing the PVA was accomplished at 60°C or 80°C depending on the operating temperature of the upcoming run. The desired test temperature was maintained by adjusting the cooling water valve in order for the heat exchanger to remove excess heat generated by the pump. The by-pass loop valve and valves upstream and downstream of the module were adjusted simultaneously to maintain the 4.8 MPa pump discharge, pressure the desired flow rate, and initial module pressure. After steady state was reached these three valves were quickly coordinated to raise the pressure to the new level while maintaining the pump discharge pressure and throughflow rate. The temperature varied only slightly since the pump power was maintained constant. The permeate flux was recorded on an X-Y plotter versus time and also on a strip chart recorder. The plotter scale was used for flux record and the recorder maintained a record of sequence of runs.

When the data for the final pressure setting was recorded, the feed solution was pumped into a storage tank. The feed reservoir was filled with 20 liters of water to rinse the ultrafiltration membrane at 80°C for 2 hours. During this rinse a low module pressure and high recirculation rate were maintained. With rinsing complete and system drained, the feed solution was transferred back into the reservoir to start another run.

Results and Discussion

Figure 4 shows the steady state average flux obtained at various operating pressures and shear levels (indicated by flow rate). These data for 2 percent and 60°C are typical of those obtained. In each pressure excursion data were only taken in asending pressure to avoid hysteresis effects. The membrane was washed between series. Data obtained at such steady operating conditions were used to provide the orifice calibration of Figure 5. The strip chart R value was an arbitrary scale on the X-Y Plotter used to record transient flux data.

Figure 6 is a tracing of an actual X-Y (R-time) trace obtained when the pressure was raised from 2.1 MPa to 2.8 MPa at 7.6 dm^3/min flow at 2 percent concentration and 60°C. The transient period for the pressure rise was not measured but normally was consumated within 5 to 10 seconds. The transient period for the flux was much longer, usually 100 to 200 seconds. By using the steady state calibration, Figure 5, the arbitrary scale of the recorder was replaced by permeate flow rate to provide an average flux history.

A minor complication emerged due to temperature transients. The procedure described produced ideally constant pump power so that no temperature change would be expected. Indeed, it was found that the temperature was usually constant to within 1°C. However, the large permeate chamber, when subjected to temperature changes, would experience expansion or contraction of the fluid sufficient to cause excursions such as shown in Figure 7. In extreme cases, the runs were disqualified because their interpretation was deemed unreliable. For the most part, the temperature variations were 0.3°C or less, producing traces like Figure 7. Care was exerted to control the final temperature to within 0.1°C of its initial value and to include the temperature-induced wiggle completely within the

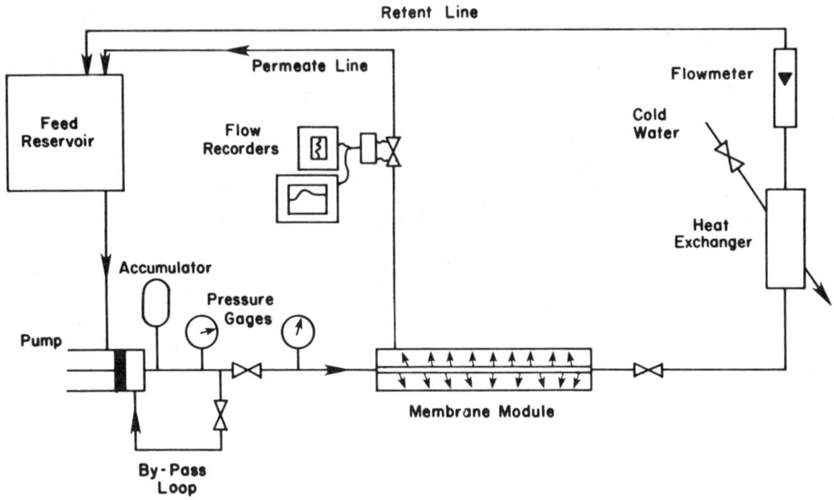

Figure 3. Schematic of Test Apparatus.

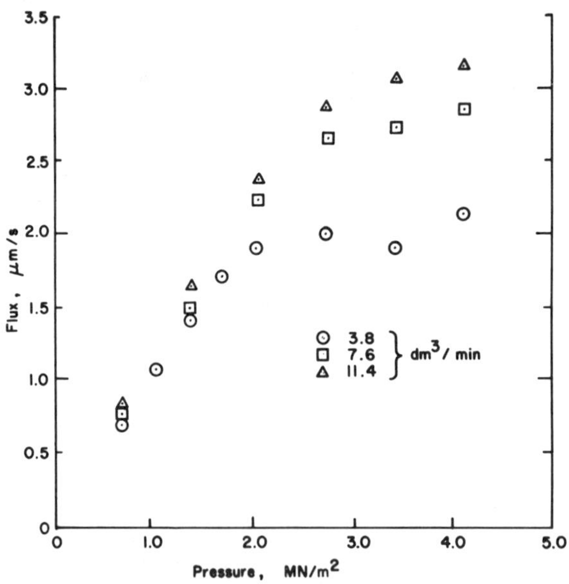

Figure 4. Steady State Flux Data for 2% PVA at 60°C in a Tubular Membrane.

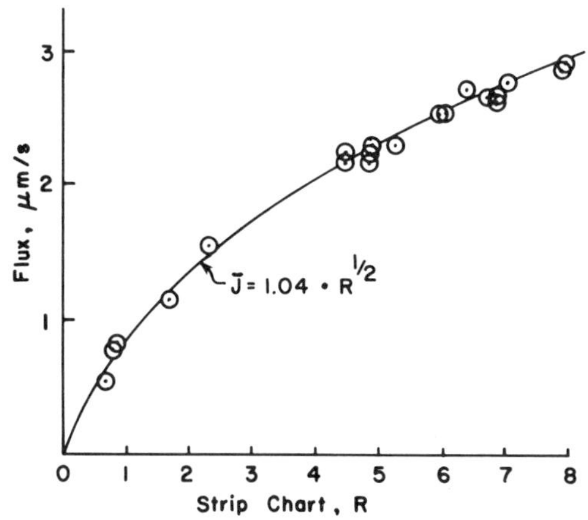

Figure 5. Permeate Flow Meter Calibration.

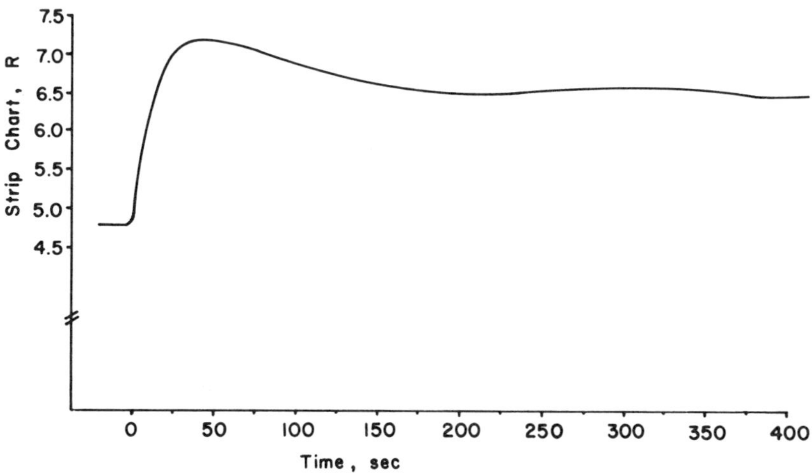

Figure 6. Recorded Flow Trace During a Step From 2.1 MPa to 2.7 MPa at 2%, 60°C.

analysis. Thus any rise was compensated by a corresponding fall in permeate flow.

In addition to runs having large temperature variations some runs were not analyzed because no gel was formed and others because of flux levels too low to measure with confidence. There remained 29 individual runs which were analyzed according to Equation (5) and Equation (6). The volume of gel deposited rather than its volume per unit area is presented in Figure 8. A considerable scatter is definitely noted to the data such that any results should be considered provisional in nature. The data isolated in the triangular region all belong to the subset of 4 percent concentration, 60°C. The flow rates under these conditions were the lowest of all the concentration-temperature pairs run and the flow rates are therefore subject to the highest influence of experimental uncertainty.

The line shown in Figure 8 is suggested as a possible trend for the data. Such a line passing through the origin has constant resistivity; the particular line has $R = 1.9 \times 10^{16}$ Pa-s/m². For an average viscosity of 4×10^{-4} Pa-s, this resistivity corresponds to a Darcy law permeability of 2.1×10^{-20} m². According to existing correlations for permeability of particles, for example Carman-Kozeny (5), and for a void fraction of 0.4, the mean particle diameter would be 4.5 nm. This does not mean that PVA forms a packed bed of 4.5 nm particles, but rather that the gel layer has the resistivity of such a packed bed.

The mean thicknesses of deposit shown on Figure 8 range generally up to 2.2×10^{-5} m or in one case to 3.8×10^{-5} m for the 0.0357 m² test section. The mass of gel is anticipated to be 0.2 to 0.6 g at a volume of about 1 cm³. Detection of such small mass amounts by analysis of mass depletion is clearly impractical. In principle one can analyze the residue washed off following deposition as was reported in Reference 8. However no attempt was made to determine gel mass and hence concentration.

The experimental uncertainty of the procedure is fairly significant compared with the scale of results. Coarsely, the flow rates may be determined to within ± 0.005 cm³/s. If this uncertainty is applied over the 100-second integration period the volume uncertainty would be = 0.5 cm³. This certainly would account for the scatter tendency shown in Figure 8. The increase in resistance is much more accurately determined. Most values should lie within ± 10% of the value indicated.

Conclusion

Subject to the approximations made, a method has been advanced which allows determination of gel volume in a membrane system. Experiments with polyvinyl alcohol solutions illustrate the use of the method. Transient flux responses to step pressure changes required hundreds of seconds to be consummated; volumes of gel of the order of cubic centimeters on a test section of 357 cm² were registered. The thickness of the gel layer was frequently 20 μm and produced resistances to flow of 5×10^{12} Pa-s/m.

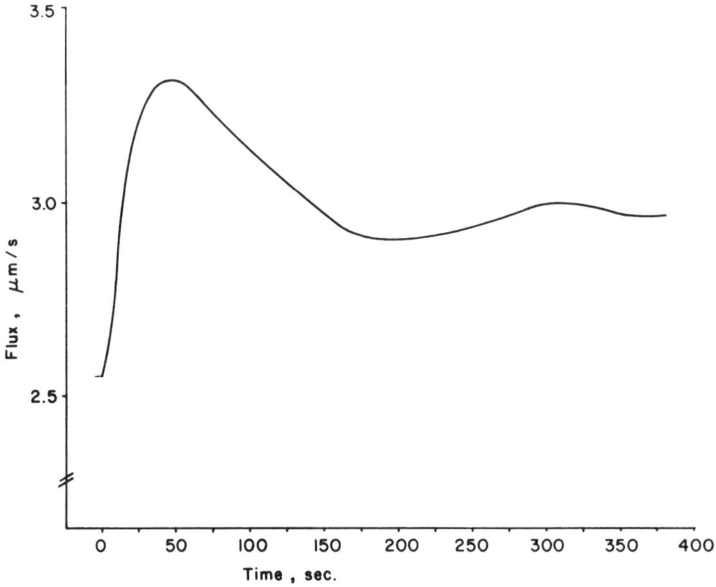

Figure 7. Flux History in Step From 2.7 MPa to 4.2 MPa Showing the Effect of Temperature Variation.

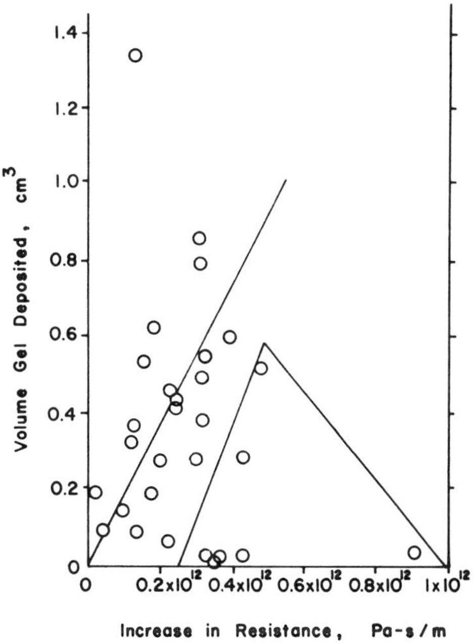

Figure 8. Plot of Gel Volume Versus Increase in Resistance.

Appendix

This appendix considers justification of the assumption that the concentration rises rapidly at the membrane to the gel concentration level in response to sudden flux increases. The slowest point to respond should be the most upstream point, x_p, which is included in treatment herein.

The accepted differential equation is, assuming constant diffusion coefficient D and assuming the velocity normal to the membrane surface is equal to permeate flux:

$$D \frac{\partial^2 \tilde{c}}{\partial \tilde{y}^2} + J \frac{\partial \tilde{c}}{\partial \tilde{y}} = \frac{\partial \tilde{c}}{\partial \tilde{t}} + u \frac{\partial \tilde{c}}{\partial \tilde{x}}$$

Auxiliary conditions are

$$D \frac{\partial \tilde{c}}{\partial \tilde{y}} = - J\tilde{c} \qquad \tilde{y} \to 0$$

$$\tilde{c} \to c_b \qquad \tilde{y} \to \infty$$

plus appropriate initial conditions. The dimensional variables are marked with a tilde and may be expressed in terms of non-dimensional variables by

$$\tilde{c} = c \cdot c_b; \; y = \frac{J\tilde{y}}{D}; \; x = \frac{J\tilde{x}}{D}; \; t = \frac{J^2}{D} \tilde{t}$$

Thus

$$D \frac{\partial^2 c}{\partial y^2} + \frac{\partial c}{\partial y} = \frac{\partial c}{\partial t} + \frac{u}{J} \frac{\partial c}{\partial x}$$

$$\frac{\partial c}{\partial y} = - c \qquad y = 0$$

$$c \to 1 \qquad y \to \infty$$

Introduce $\eta = 1 - e^{-y}$ such that

$$(1 - \eta)^2 \frac{\partial^2 c}{\partial \eta^2} = \frac{\partial c}{\partial t} + \frac{u}{J} \frac{\partial c}{\partial x}$$

$$\frac{\partial c}{\partial \eta} = - c \qquad \eta = 0$$

$$c = 1 \qquad \eta = 1$$

Break the solution into a sum $c^*(x,\eta) + c'(\eta,t)$, treating x as a parameter in c'. The transient response as $t \to \infty$ does not vanish but tends to $(1 - \eta)B$. Treating B as B(x) plausibly allows solution to $c^*(x, \eta)$ which together with $(1-\eta)B$ forms the steady state solution. The equation for c' and its boundary conditions are

$$(1 - \eta)^2 \frac{\partial^2 c'}{\partial \eta^2} = \frac{\partial c'}{\partial t}$$

$$c' = 0 \qquad \eta \to 1$$

$$c' = - \frac{\partial c'}{\partial \eta} \qquad \eta = 0$$

The solution should be related strongly to the solution of the simplified form

$$\frac{\partial^2 c'}{\partial \eta^2} = \frac{\partial c'}{\partial t}$$

This simplified form produces solutions which are lower in amplitude and almost twice as responsive than the precursive form, as judged from numerical studies. The solutions to diffusive slab problems with constant concentration boundary conditions by experience respond (to 95 percent) within $t \gtrless 1$. Numerical results indicate response for the system with $t \gtrless 0.4$. Here response means

$$\frac{c'(o,t) - c'(o,\infty)}{c'(o,o) - c'(o,\infty)} < 0.05$$

For the experiment herein, using $D = 10^{-10}$ m²/s and $J = 2 \times 10^{-5}$ m/s, the real time $t < 0.25$ sec corresponds to $t < 1$. Thus the response is assumed to be rapid.

Literature Cited

1. Michaels, A. S. Chemical Engineering Progress 1968, 64, 31-43.
2. Blatt, W. F.; David, A.; Michaels, A. S.; Nelson, L. in MEMBRANE SCIENCE AND TECHNOLOGY, J. E. Flinn, Ed; Plenum Press, N.Y., 1970, 47ff.
3. Shen, J. J. S.; Probstein, R. F.; Ind. Eng. Chem., Fundam. 1977, 16, 459-465.
4. Trettin, D. R.; Doshi, M. R.; in "Synthetic Membranes: Volume II Hyper- and Ultrafiltration Uses"; A. F. Turbak, Ed.; ACS Symposium Series No. 154, American Chemical Society; Washington, D.C., 1981; p. 373-409.
5. Reihanian, H.; Robertson, C. R.; Michaels, A. S., Journal of Membrane Science, 1983, 16, 237=258.
6. Dullien, F. A. L., The Chemical Engineering Journal, 1975, 10, 1-34.
7. Kuppers, J. R.; Marcinkowsky, A. E.; Kraus, K. A.; Johnson, J. S., SEPARATION SCIENCE, 1967, 2, 617-623.
8. Jernigan, D. A., M.S. Thesis, Clemson University, Clemson, S.C., 1981.
9. Nakao, Shin-ichi; Nomura, T.; Kimura, S., AIChE Journal, 1979, 25, 615-622.

RECEIVED February 22, 1985

31

Pretreatment, Fouling, and Cleaning in the Membrane Processing of Industrial Effluents

C. A. BUCKLEY, K. TREFFRY-GOATLEY, M. J. SIMPSON, A. L. BINDOFF, and G. R. GROVES

Pollution Research Group, Department of Chemical Engineering, University of Natal, Durban 4001, South Africa

The role and methodology of hyperfiltration pilot-plant investigations in the industrial effluent environment are examined. The objectives of a factory site investigation are enumerated and the importance of pretreatment and cleaning techniques for the long term maintenance of membrane flux is stressed. A method of correcting membrane flux for variations in operating pressure, temperature and feed osmotic pressure using specific membrane flux is presented together with a method of estimating osmotic pressure using batch concentration experimental results. An example of pulp mill effluent is used to illustrate two mechanisms of fouling viz: feed/membrane interaction and precipitation on the membrane. The first is apparent from total recycle experiments, while the second is apparent during serial batch concentrations. The specific membrane flux results from the operation of a hyperfiltration pilot-plant at a textile dyehouse were examined to determine the magnitude of long term fouling and the effectiveness of membrane cleaning procedures.

South Africa is a relatively arid country and it has been projected (1) that without substantial water reclamation and recycling the supply and demand will be equal by the year 2020. The effective use of fresh water supplies is being impaired through quality deterioration brought about by the discharge of partially treated waste waters. For this reason and also to prevent intractable, toxic or carcinogenic substances from entering water supplies or water reclamation systems, the closed loop recycling of factory effluent is necessary.

Hyperfiltration has several advantages as a treatment method for industrial effluents because of its inherent ability to remove a wide range of pollutants including total dissolved solids.

The problems associated with the disposal of the concentrates may be reduced by achieving a very high (95 percent or greater)

0097-6156/85/0281-0429$06.00/0

water recovery. In view of the unique nature of the effluent from any factory and often, the variability in the composition of the effluent, a pilot-plant investigation is necessary before a hyperfiltration treatment/water recycle plant can be designed.

This paper examines the role and the methodology of hyperfiltration pilot-plant investigations in the industrial environment.

Objectives of Pilot-Plant Investigation

The operation of a pilot-plant at the factory site is necessary to develop the following design information :-

(i) membrane selection, particularly on the susceptibility of different membranes to fouling,
(ii) membrane flux - water recovery relationship,
(iii) permeate quality - water recovery relationship,
(iv) module type and pretreatment requirements,
(v) formulation of cleaning solutions and cleaning frequency,
(vi) estimate of membrane life, and
(vii) the sensitivity of the system to upsets in the factory effluent.

Because, in general, hyperfiltration of industrial effluents will be used for the removal of both inorganic and organic materials, the following factors also have to be determined :-

(i) the lowest quality of water which is acceptable for reuse in the factory.
(ii) the nature of the concentrate.

Frequently some of the above points can be answered by small scale laboratory tests. However, for most industrial effluents it will be necessary to conduct a 3 - 6 month factory trial in order to design with confidence. On-site testing assists the factory management with the important aspect of technology transfer.

Assessment of Membrane Flux Performance

The long term maintenance of membrane flux by adequate pretreatment and the cleaning strategy is probably the most important design objective. Because industrial effluents vary in composition, the interpretation of membrane flux data is difficult. Factors to be considered include fouling, applied pressure, temperature, the concentration of compounds contributing to the osmotic pressure and water recovery.

The use of specific membrane flux, referenced to a particular temperature (e.g. 25°C), is often helpful in assessing module performances :-

$$J_s = \frac{J \cdot TCF}{(P_{ave} - \pi_{ave})} \qquad (1)$$

Where J_s is the specific membrane flux at the reference temperature (25°C),

J is the measured flux at temperature T,

TCF is the temperature correction factor,
P_{ave} is the average transmembrane pressure, and
π_{ave} is the average transmembrane osmotic pressure.

The temperature correction factor is usually given for a particular membrane by its manufacturer and often takes the following form :-

$$TCF = 1{,}03^{(T - 25)}$$

Where T is the temperature (°C).

Alternatively, a temperature correction factor based on viscosity may be applied.

Osmotic pressure measurements are not usually available and hence correlation with conductivity is needed. The conductivity measurements should be referenced to the correct temperature. The osmotic pressure - conductivity relationship is :-

$$\pi = a \cdot C \qquad (2)$$

Where a is constant,
C is concentration.

The constant, a, may be determined from data collected during batch concentration experiments using the following equation :-

$$J/P = J_S - C/P \cdot J_S \cdot a \qquad (3)$$

Several experiments are needed to obtain an average value of the constant, a because of the general variability of industrial effluents.

The use of a specific membrane flux against time plot will correlate the variables of temperature, applied pressure and concentration or water recovery and hence provide information on the degree of fouling. Membrane fouling alters the value of the specific membrane flux and thus information on the effectiveness of membrane cleaning or changes in pretreatment may be assessed.

Hyperfiltration Treatment of Industrial Effluents

Pulp Mill Effluent Treatment of Industrial Effluents. Initial laboratory results obtained using thin-film composite (TFC) membranes for the treatment of a pulp mill effluent indicated that the effluent without pretreatment was badly fouling and that standard cleaning methods only partially restored the permeate flux (2).

The effect of the two different pretreatment methods (A and B) are examined using the methodology presented above. A commercial TFC spiral module was used in these experiments.

Figure 1 gives the results of a series of batch concentrations for pretreatment method A. Between each batch concentration, the membrane module was exposed to the effluent for approximately 48 hours. This uncorrected membrane flux against feed conductivity plot indicates progressive membrane fouling because the slope ($-J_S$) of the plot for each batch is changing.

The data is replotted as specific membrane flux (45°C) versus conductivity in Figure 2. From examination of this, two fouling effects are noticable : firstly, fouling is occurring during the batch and secondly, fouling takes place from batch to batch during the total recycle mode. The first effect is thus concentration dependent and was determined to be due to insoluble salt precipitation caused by carryover from pretreatment. The second effect was probably due to feed-membrane interaction and the development of a better pretreatment system was necessary to eliminate this effect.

The specific membrane flux (45°C) - conductivity plot for the serial batch concentrations using pretreatment method B is given in Figure 3. Membrane fouling was minimal during this trial.

Pilot-Plant Hyperfiltration Treatment of Textile Dyehouse Effluents. Textile dyehouse effluents are highly variable in composition and are dependant mainly on the type of fibre being processed. For example, the dyeing of cotton produces an alkaline effluent high in soluble colour and salts whereas polyester dyeing effluent is acidic and contains colloidal dyestuffs and surfactants.

For mixed cotton/polyester dyeing effluent, a treatment system consisting of alum flocculation, microfiltration and spiral TFC membraned hyperfiltration was developed (3). A 40 m^3/d pilot - plant was used at a factory site to determine the long term membrane performance and to obtain the required design parameters (4).

The hyperfiltration pilot unit consisted of two stages and used brackish water membranes in the first stage and seawater membranes in the second. The feed to the pilot-plant was very variable and its composition is summarised in Table I. Each stage of the hyperfiltration unit was operated at a fixed feed flow rate and the reject flow was controlled using a process control valve. A by-pass on the process control valve was set such that the maximum inlet system pressure was achieved when the process control valve was fully closed.

The actual membrane flux - time plot for the first stage is given in Figure 4 and membrane performance is difficult to assess from this.

TABLE I. Dyehouse Effluent Characteristics

Parameter		Mean	Max	Min
Temperature	°C	35	40	20
pH		7,91	12,24	5,50
Conductivity	mS/cm	0,65	7,89	0,11
Total solids	g/l	0,61	5,3	0,13
Sodium	mg/l	152	495	45
COD	mg/l	757	5 461	16
Total carbon	mg/l	175	415	75
Colour ADMI	units	1 368	6 504	170

These flux results were corrected for temperature, osmotic pressure (using conductivity) and operating pressure and are replotted as specific membrane flux (25°C) in Figure 5. The specific water flux after membrane cleaning is plotted in Figure 6.

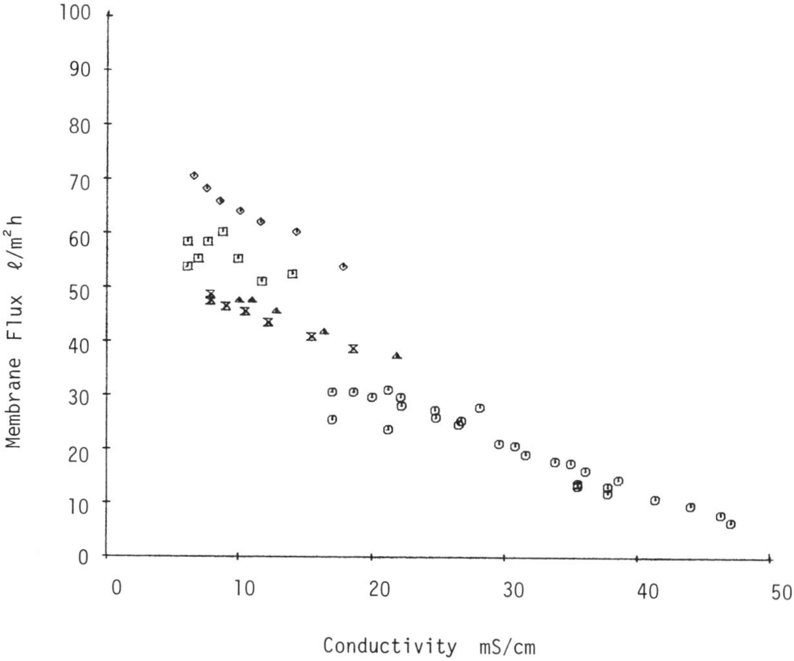

Figure 1. Membrane flux for serial batch concentrations: pretreatment A. Key: ◇, batch 1; □, batch 2; △, batch 3; ⧖, batch 4; and ⊙, batch 5.

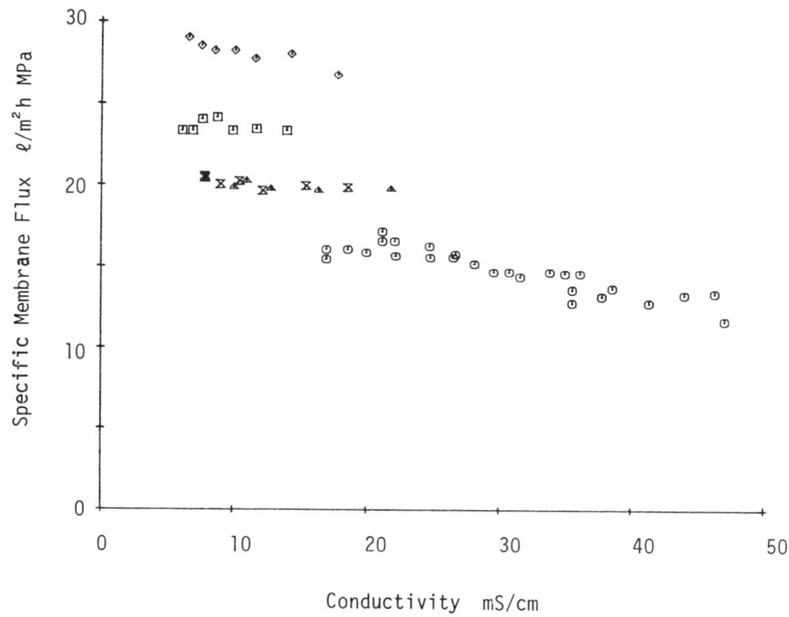

Figure 2. Specific membrane flux for serial batch concentrations: pretreatment A. Key: ◇, batch 1; □, batch 2; △, batch 3; ⧖, batch 4; ⊙, batch 5.

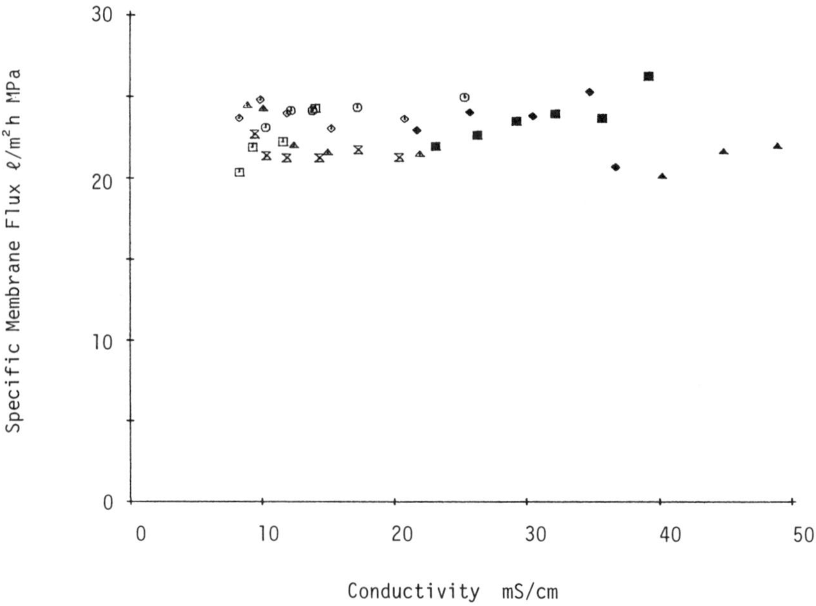

Figure 3. Specific membrane flux for serial batch concentrations: pretreatment B. Key: ◇, batch 1; □, batch 2; △, batch 3; ⧗, batch 4; ⊙, batch 5.

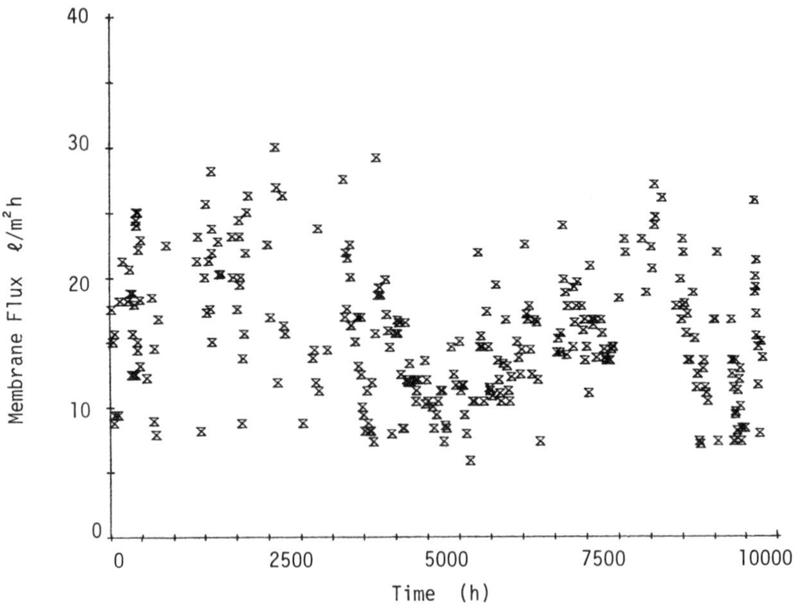

Figure 4. Membrane flux for the hyperfiltration (Stage 1) of dyehouse effluent.

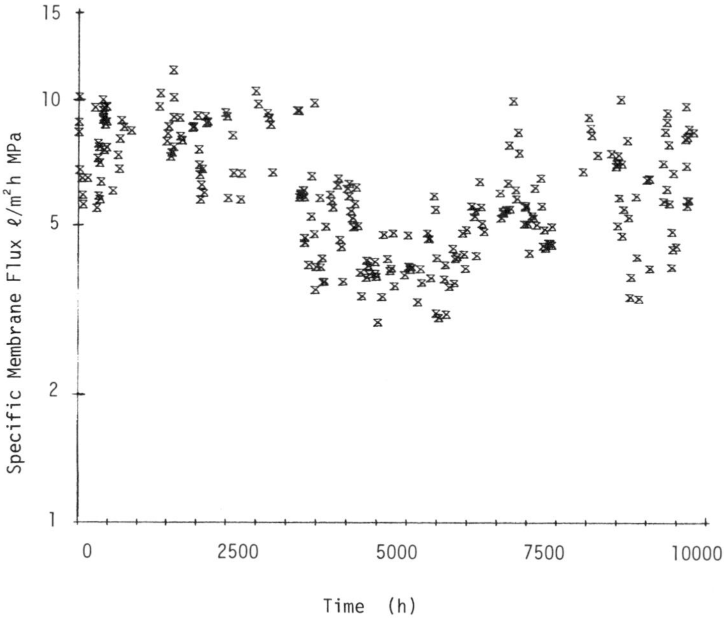

Figure 6. Specific water flux after membrane cleaning for the hyperfiltration (Stage 1) of dyehouse effluent.

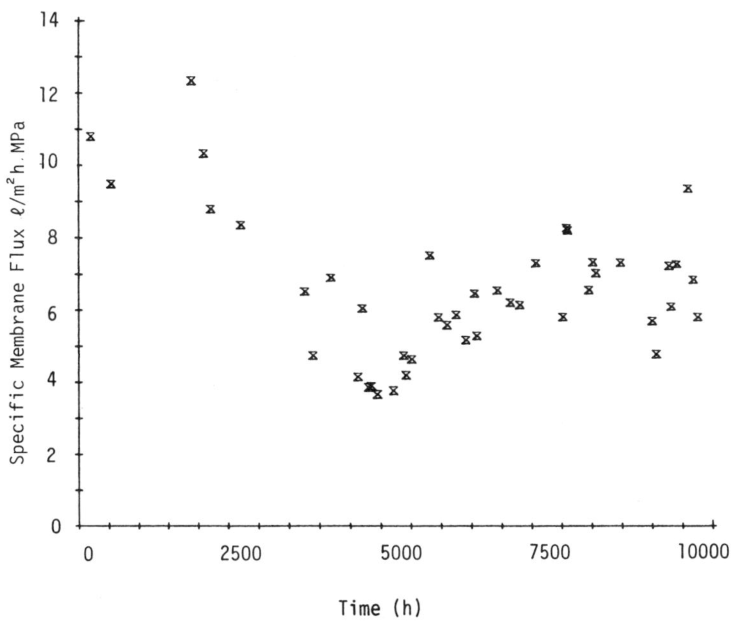

Figure 5. Specific membrane flux for the hyperfiltration (Stage 1) of dyehouse effluent.

It is evident from Figure 6 that the membrane set was becoming progressively fouled during the first 5 000 hours of operation. The progressive nature of this fouling is not evident from Figures 4 and 5.

After 5 000 hours of operation, a weekly cleaning strategy and a change in cleaning chemical formulation was instituted. In addition iron-free alum was substituted for the commercial grade in the flocculation stage. These changes gave a gradual improvement in both specific membrane flux and specific water flux after membrane cleaning. At 9 000 hours, the source of effluent was changed from the mixed factory effluent to cotton dyeing effluent only. The cotton dyeing effluent was not pretreated and some degree of fouling became evident.

The pilot-plant investigation indicates that difficult industrial effluents, as typified by mixed dyehouse effluents, may be treated successfully by hyperfiltration. A design specific membrane flux of 6 - 8 l/m^2h MPa is realistic for this particular membrane for this application duty.

Conclusions

The importance of adequate pretreatment to prevent serious feed-membrane interactions and the implementation of a cleaning strategy have been stressed for the successful implementation of hyperfiltration for the treatment of industrial effluents.

The interpretation of pilot-plant data is simplified by the adoption of specific membrane flux plots as described in the paper. The specific membrane flux correlates the effects of temperature, concentration and pressure and may be used for the assessment of long term performance and the effect of water recovery.

Membrane fouling is of obvious importance in the treatment of industrial effluents and the effect of different pretreatment techniques may be assessed both quickly and accurately using specific membrane flux plots.

Nomenclature

a	constant
C	concentration
J	the measured flux at temperature T
P	pressure
T	temperature
TCF	the temperature correction factor
π	osmotic pressure

Subscripts

ave	average
s	specific

Acknowledgments

This research work was carried out under a grant entitled "Water Management and Effluent Treatment in the Textile Industry" from the Water Research Commission.

Literature Cited

1. Kriel, J. P. Congress of the South African Association for the Advancement of Science, Stellenbosch, South Africa, 1975.

2. Simpson, M. S.; Groves, G. R.; Desalination 1983, 47, 327-3.

3. Treffry-Goatley, K.; Buckley, C. A.; Groves, G. R.; Desalination 1983, 47, 313-20.

4. Pollution Research Group, "A Guide for the Planning, Design and Implementation of Water Treatment Plants in the Textile Industry; Part 2 : Effluent Treatment/Water Recycle Systems for Textile Dyeing and Printing Effluents", University of Natal, Durban, South Africa (in press).

RECEIVED February 22, 1985

Gas Permeability of Polypeptide Membranes

Y. TSUJITA and A. TAKIZAWA

Department of Polymer Engineering, Nagoya Institute of Technology, Gokiso, Showa-ku, Nagoya 466, Japan

Permeation, diffusion, and sorption of carbon dioxide gas through or into the polypeptide membrane, poly(γ-benzyl-L-glutamate) forms A, B, and C, poly(β-benzyl-L-aspartate) α and ω forms, racemic form of poly(γ-benzyl glutamate), and poly(γ-methyl glutamate) C, D-D, and D films were studied in the light of the stacking of the benzene ring in the side chain, the first order transition due to the breakdown or formation of the stacking, and interhelix distance. Sorption was depressed by the presence of the intra-helix stacking, indicative of less sorption sites. Permeation and diffusion were enhanced by the breakdown of the benzene ring stacking at the transition temperature as observed for poly(γ-benzyl-L-glutamate) form A and racemic form of poly(γ-benzyl glutamate). The diffusion coefficient was calculated by the Cohen-Turnbull theory. The fractional free volume at the glass-like transition temperature was estimated to be 0.07 ~ 0.08, suggestive of a loosely packed side chain region, when a better agreement between the experimental and calculated diffusion coefficient was obtained by a trial and error method.

The side chain of polypeptides consisting of α helix is well-known to be coarse and loosely packed structure, compared to the main chain structure (1) . It is interesting to develop membrane performance utilizing the characteristic side chain region of polypeptides, functionality and coarse structure in the side chain region. Many polypeptide membrane studies performed by our group (2-6) and Nakagawa et al.(7-11) could elucidate that the permeation of small molecules through membrances takes place in the side chain regions, not in the α helix regions. The permeation depends to a major extent upon the side chain structure such as arrangement and packing of the side chain and also the primary structure of the side chain. A different side chain structure might be, in part, caused by a different array of a helical rod. Thus, it is considered that

0097-6156/85/0281-0439$06.00/0

permeation behavior depends also upon the helical array influencing the side chain structure.

In the present paper, the permeability coefficient ($\overline{P}$) of carbon dioxide gas, determined by the product of its diffusion coefficient ($\overline{D}$) and solubility coefficient (S) was discussed in terms of the side chain structure such as the benzene ring stacking and inter-helix distance. The samples used here are the solid state modification (forms A, B, and C) of poly(γ-benzyl-L-glutamate)(PBLG), α and ω forms of poly(β-benzyl-L-aspartate)(PBLAsp), racemic form of poly(γ-benzyl glutamate)(PBG) and poly(γ-methyl glutamate)(PMG). PBLG form A, PBLAsp ω form, and racemic form of PBG are able to form the intrahelix or interhelix benzene ring stacking. The first order transition occurs due to the breakdown or formation of the benzene ring stacking. The effect of the structural change of the side chain at the transition on the $\overline{P}$ was also discussed.

Experimental

PBG, PBLG, PBLAsp, and PMG were prepared by the polymerization of γ-benzyl glutamate N-carboxy anhydride, γ-benzyl-L-glutamate N-carboxy anhydride, β-benzyl-L-aspartate N-carboxy anhydride, and γ-methyl glutamate N-carboxy anhydride, respectively, at room temperature by using triethylamine as initiator.

PBLG forms A and B were cast at 45°C and at 60°C from 3 wt % PBLG solution in benzene, respectively. PBLG form C was cast at room temperature from 3 wt % PBLG solution in ethylene dichloride. PBLAsp α form was prepared by quick casting from 4 wt % solution in chloroform in several hours. The cast films thus obtained were dried under vacuum at room temperature for several days after immersion into a large amount of methyl alcohol for one day. PBLAsp ω form was obtained by annealing the α form at 150°C. PMG C, D-D, and D films, corresponding to used cast solvent, chloroform, dichloromethane-dimethyl formamide, and dimethyl formamide, respectively, were prepared at room temperature by casting method as described above. PMG cast from ethylene dichloride solution was treated in about 80 % formic acid aqueous solution for 66 hrs. and dried for two days after washing by water. The β form of PMG thus obtained was designated as PMG β film.

The sorption experiment with carbon dioxide gas was performed with a recording electrobalance (Model RG, Cahn Division, Ventron Instrument Corp., Paramount, Calif.). The film suspended in the electrobalance, of about 70 mg weight, was dried under vacuum until the weight of the film became constant. Then the weight sorbed under various pressures of carbon dioxide gas (70 ~ 350 torr) was measured at 10 ~ 40°C temperature range by the electrobalance.

The permeation experiment was performed over a wide temperature range covering the first order transition temperature, with a Rouse type apparatus, in which the pressure on the down-stream side was always zero using a liquid nitrogen trap. The permeability coefficient was evaluated from the linear slope, at steady state, of the permeation curve which is represented by a plot of the volume permeated at the standard state per unit area of the film against time.

Results and Discussion

Sorption. Sorption isotherms of carbon dioxide gas for three forms of PBLG including annealed forms A and B at 150°C, PBLAsp α and ω forms, and racemic form of PBG are shown in Figures 1, 2, and 3, where the volume sorbed (at the standard state) of carbon dioxide gas per residue of polypeptide is plotted against the fugacity ratio f_2/f_2° as activity. The sorption isotherms at various temperatures above glass-like transition were reduced to a single curve if no structural change occurs during the experiment. This indicates the temperature independence of the volume sorbed and the solubility coefficient in these temperature ranges. However, a different isotherm was obtained at 10°C for racemic form of PBG (Figure 3), the solubility coefficient being affected by the side chain mobility, i.e., glass-like transition temperature. The volume sorbed at 25°C was in the order:

racemic form of PBG ≳ PBLG form C > PBLAsp α form > PBLG form B > PBLAsp ω form > PBLG form A

It is noted that racemic form of PBG with interhelix stacking (12-14) could sorb nearly the same amount as PBLG form C without the stacking. Therefore it can be said that the interhelix stacking does not greatly influence, but the intrahelix stacking formed for PBLG form A (15,16) and PBLAsp ω form (17,18) depresses the solubility remarkably, suggesting the effective sorption site of unstacked benzene rings.

It was found that the following Langmuir sorption isotherm equation might apply to the sorption isotherm data shown in Figures 1, 2, and 3:

$$\frac{p}{n} = \frac{1}{ab} + \frac{p}{b} \qquad (1)$$

where n is the sorbed amount under pressure p, a the sorption energy factor, and b the saturation value of n (sorption capacity).
The Langmuir constant a and b calculated from a Langmuir plot were tabulated in Table I. The sorption capacity b indicates that PBLAsp

Table I. Langmuir Constants (Sorption Energy Factor a and Sorption Capacity b) for the Polypeptide-Carbon Dioxide Gas System

Polypeptide	Temperature	a	b
	°C	$cmHg^{-1}$	cc STP / residue
Racemic form of PBG	20	0.065	250
PBLG form A	25	0.052	148
PBLG form B	25	0.056	225
PBLG form C	25	0.066	264
PBLAsp α form	35	0.043	268
PBLAsp ω form	35	0.039	183

α form and PBLG forms B and C with unstacked benzene ring have a higher sorption capacity site than PBLAsp ω form and PBLG form A with the intrahelix benzene ring stacking and racemic form of PBG has a remarkably high sorption capacity, although it can form the interhelix benzene ring stacking.

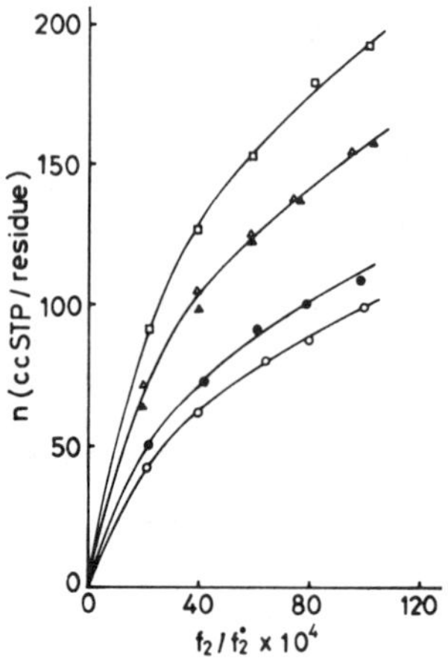

Figure 1. Sorption isotherms of carbon dioxide gas for PBLG forms A, B, and C including annealed A and B: (○) form A, (●) annealed form A, (△) form B, (▲) annealed form B, (□) form C. (Reproduced with permission from Ref. 4. Copyright 1978, John Wiley & Sons, Inc.)

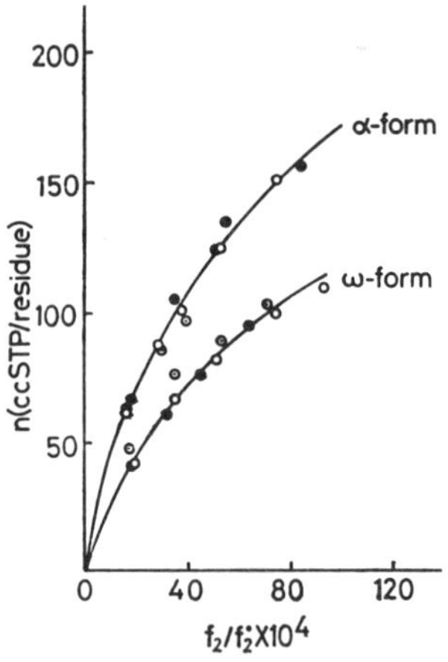

Figure 2. Sorption isotherms of carbon dioxide gas for PBLAsp α and ω forms at various temperatures: (●) 30°C, (○) 35°C, (⊙) 40°C. (Reproduced with permission from Ref. 5. Copyright 1980, Marcel Dekker, Inc.)

Permeation. The temperature dependence of the $\overline{P}$ of PBLG forms A, B, and C, PBLAsp α and ω forms, and racemic form of PBG is shown in Figures 4, 5, and 6, respectively. Of three kinds of solid state modification of PBLG, the $\overline{P}$ of form A was lower, reflecting the presence of the benzene ring stacking. The form A exhibited an abrupt increase of the $\overline{P}$ around 130-135°C, where the breakdown of the stacking occurred, in accordance with the endothermic peak at 135°C and the decrease in modulus (4,19). Such a result shows explicitly that the breakdown of the stacking increases the number of sorption sites or molecular motion in the side chain region, and that carbon dioxide gas diffuses through the side chain region between α helices. The $\overline{P}$ of the form B changed remarkably around 110°C, corresponding to the transformation of α helical hexagonal array reported by McKinnon and Tobolsky (20, 21). On the other hand, PBLG form C with the absence of the stacking had no abrupt increase of the $\overline{P}$ in the high temperature region. The plot of the $\overline{P}$ versus reciprocal of temperature showed inflections at 20 ~ 30°C, coinsident with the glass-like transition which is due to the onset of micro-brownian motion of the side chain.

The $\overline{P}$ of PBLAsp α form increased up to 110°C as shown in Figure 5, however it did not vary significantly in the temperature range from 110 to 135°C. The increasing tendency of the $\overline{P}$ owing to the increasing temperature is considered to be balanced with the decreasing tendency due to the α-ω transition, which is observed irreversibly around 135°C for PBLAsp α form. Therefore, it can be said that the $\overline{P}$ decreases considerably at the α-ω transition temperature, where the loose packing in the α form transformed into the dense packing in the ω form (22), accompanying by a decrease of sorption sites and a narrowing of the interhelix distance. The $\overline{P}$ of the ω form is smaller than that of the α form in the temperature range studied. This is due to the lesser amount sorbed of the ω form and also to the smaller diffusion coefficient of the ω form, causing by the presence of the intrahelix stacking.

A broad inflection appeared at about 75°C for the α form and 50°C for the ω form, which are ascribed to the onset of the side chain microbrownian motion of the α and ω forms, respectively. This might correspond to the dielectric dispersion at a frequency of 1 KHz observed at 76°C for the α form and at 59°C for the ω form(23).

The $\overline{P}$ of racemic form of PBG shown in Figure 6 showed an abrupt increase at about 90°C, being due to the breakdown of the interhelix benzene ring stacking. The breakdown would increase the interhelix distance and the molecular motion in the side chain region (24), both of which would enhance the permeation of carbon dioxide gas through racemic form of PBG. The $\overline{P}$ decreased abruptly at 65°C during cooling process. This indicates the rearrangement of the side chain to form the stacking and the confirmation of the first order and reversible transition (25).

The $\overline{P}$ of the polypeptide accompanying by the transition is shown together in Figure 7. The abrupt change of the $\overline{P}$ was observed at the transition, i.e., the breakdown of the benzene ring stacking of PBLG form A and racemic form of PBG and α-ω transition of PBLAsp. Below the transition temperature of each polypeptide, the $\overline{P}$ is in the following order:

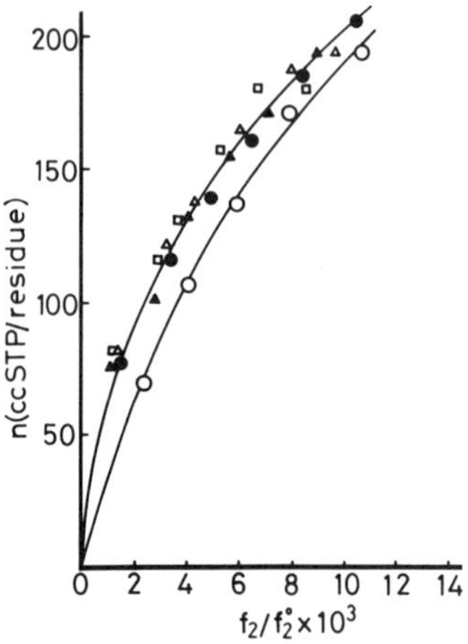

Figure 3. Sorption isotherms of carbon dioxide gas for racemic form of PBG at various temperatures: (○) 10°C, (●) 20°C, (△) 25°C, (▲) 30°C, (□) 40°C. (Reproduced with permission from Ref. 6. Copyright 1981, Hüthig & Wepf Verlag Basel.)

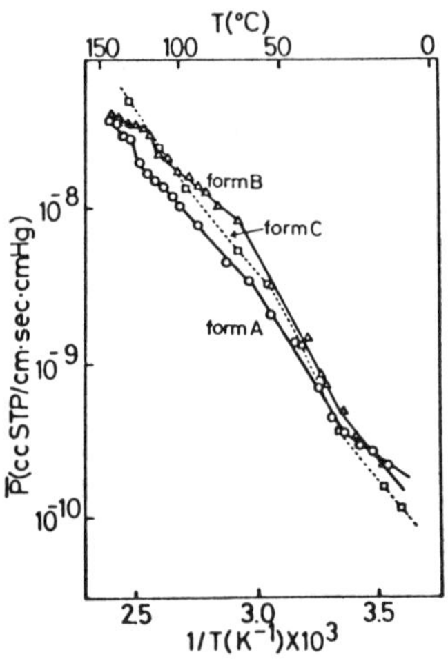

Figure 4. Temperature dependence of the permeability coefficient of carbon dioxide gas through PBLG forms A, B, and C: (○) form A, (△) form B, (□) form C. (Reproduced with permission from Ref 4. Copyright 1978, John Wiley & Sons, Inc.)

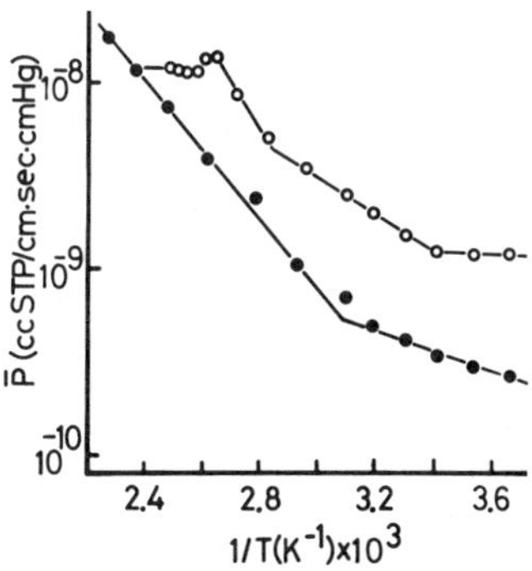

Figure 5. Temperature dependence of the permeability coefficient of carbon dioxide gas through PBLAsp α and ω forms: (○) α form, (●) ω form. (Reproduced with permission from Ref. 5. Copyright 1980, Marcel Dekker, Inc.)

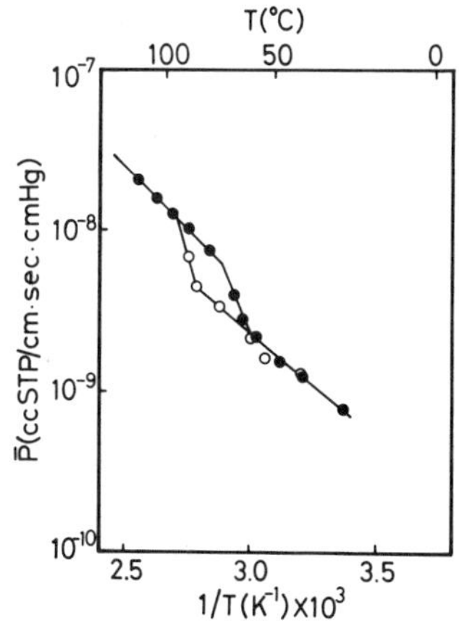

Figure 6. Temperature dependence of the permeability coefficient of carbon dioxide gas through racemic form of PBG: (○) heating curve, (●) cooling curve. (Reproduced with permission from Ref. 6. Copyright 1981, Hüthig & Wepf Verlag Basel.)

PBLG form A > racemic form of PBG > PBLAsp ω form

This is considered to depend upon both the type of the stacking (inter or intrahelix stacking) and the degree of the stacking. Above the transition temperature the $\overline{P}$ of PBLG form A and racemic form of PBG appears to converge to the same level at higher temperatures. This is reasonable, since the side chain of these PBGs is in a completely disordered state and possesses the same degree of molecular motion at higher temperatures.

The temperature dependence of the $\overline{P}$ of PMG C, D-D, D and β films is shown in Figure 8. The $\overline{P}$ of PMG β film consisting of the β form conformation was lower than that of various PMG films of the α helix conformation over an entire temperature range studied, supposing the presence of the β form sheet perpendicular to permeation direction (26). PMG film of the α helix exhibited no remarkable difference of the $\overline{P}$. A considerable decrease of the $\overline{P}$ was observed for all the PMG films studied around 100°C, where there appeared no characteristic structural change by the method such as thermal analysis, X-ray diffraction, and viscoelasticity (27). The permeation experiment is considered to very sensitive to structural change of the membrane,which is not detected by conventional techniques by molecular motion and structural analysis. Such sensitive permeation behaviors observed also in Figures 4 and 5 will be valuable to analyze a detailed structure of polypeptides.

Diffusion. Assuming that the temperature independence of the solubility coefficient aforementioned can be extended to the higher temperature range, one could obtain the temperature dependence of the diffusion coefficient as shown in Figure 9 for PBLAsp α and ω forms and in Figure 10 for racemic form of PBG. The $\overline{D}$ of the α form is larger than that of the ω form in the temperature range 0-50°C as shown in Figure 9. This is expected because of the large interhelix distance of the α form. The experimental solid line in Figure 10 showed a similarity to permeation behavior in Figure 6. The $\overline{D}$ of racemic form of PBG increased gradually with temperature; the breakdown of the stacking at about 90°C results in a remarkable increase in the $\overline{D}$.

The side chain region of polypeptide consisting of α helix may be relatively loosely packed irrespective of the presence of the benzene ring stacking. Thus the side chain region may be treated as being in the liquid state. Hence, it is relevant to examine the diffusion behavior in the side chain region of polypeptide using the Cohen-Turnbull theory (28) for diffusion in the liquid state. The diffusion coefficient is formulated according to the Cohen-Turnbull theory as follows:

$$\overline{D} = g\ a^*\ u\ \exp\left(-\frac{\gamma v^*}{v_f}\right) \qquad (2)$$

where g is a geometric factor, a* approximately equal to a molecular diameter of the penetrant gas, u gas kinetic velocity, γ an overlapping factor between 1 and 0.5, v_f free volume. The values of g, a*, and u for carbon dioxide gas are known (29,30). Free volume of polypeptide is evaluated by a trial and error method using the experimentally obtained $\overline{D}$ value and Equation 2. Namely the $\overline{D}$

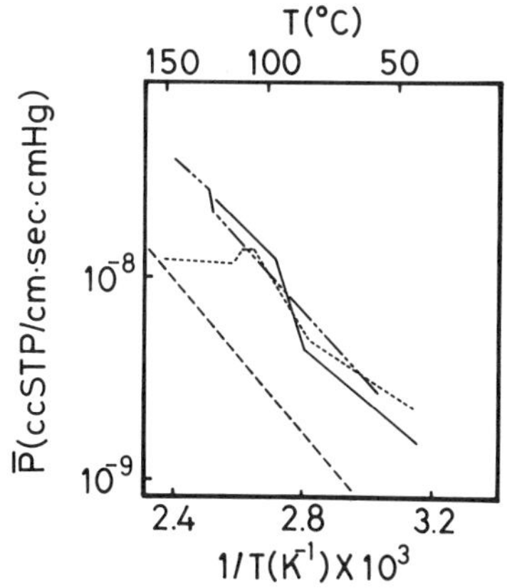

Figure 7. Temperature dependence of the permeability coefficient of carbon dioxide gas through polypeptide membrane with the transition due to the breakdown or formation of the benzene ring stacking: (——··——) PBLG form A, (-----) and (––––) PBLAsp α and ω forms, (————) racemic form of PBG. (Reproduced with permission from Ref. 6. Copyright 1981, Hüthig & Wepf Verlag Basel.)

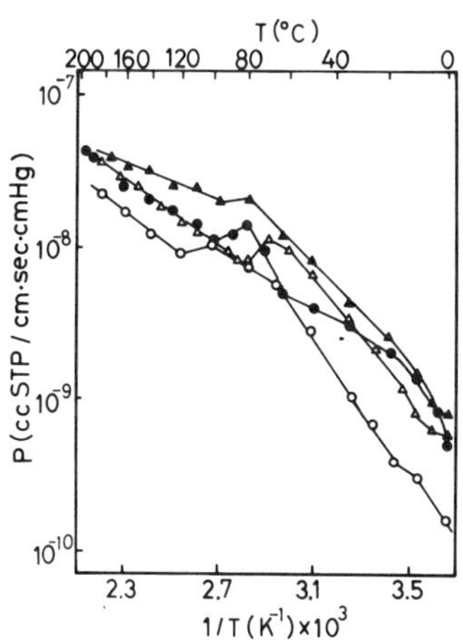

Figure 8. Temperature dependence of the permeability coefficient of carbon dioxide gas through PMG C, D-D, D, and β films: (●) C film, (▲) D-D film, (△) D film, (○) β film.

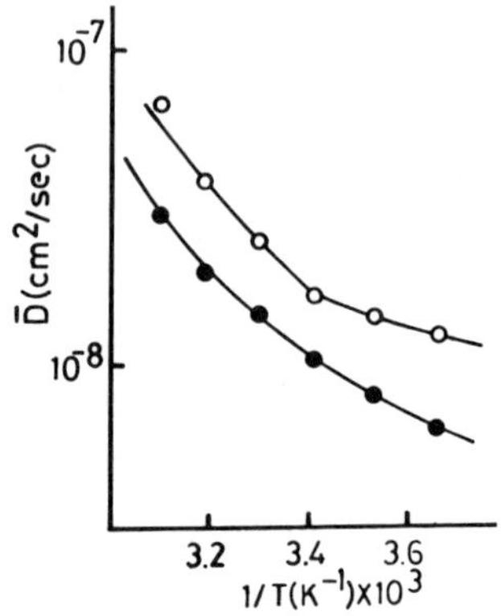

Figure 9. Temperature dependence of the diffusion coefficient of carbon dioxide gas through PBLAsp α and ω forms: (○) α form, (●) ω form. (Reproduced with permission from Ref. 5. Copyright 1980, Marcel Dekker, Inc.)

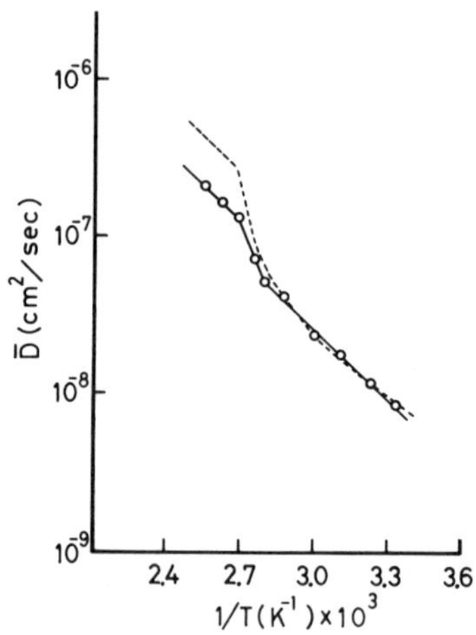

Figure 10. Experimental and calculated temperature dependence of the diffusion coefficient of carbon dioxide gas through racemic form of PBG: (—○—○—) experimental curve, (--------) calculated curve with the fractional free volume of 0.068 and $\gamma v^* = 120$ cm^3/mol. (Reproduced with permission from Ref. 6. Copyright 1981, Hüthig & Wepf Verlag Basel.)

calculated with the assumed values of γv^* and fractional free volume ($f_{T_g'}$) at the glass-like transition temperature (T_g') instead of free volume was fitted to the experimentally obtained $\overline{D}$. The better agreement is also shown, as an example, in Figure 10, where the solid and broken lines ($f_{T_g'} = 0.068$ and $\gamma v^* = 120$ cm^3/mol) represent the experimental and calculated curves, respectively. The calculated value of $\overline{D}$ also demonstrates the transition behavior at about 95°C, which is due to the abrupt increase in free volume on the transition based on the breakdown of the benzene ring stacking. The value of $f_{T_g'}$ of 0.068 was remarkably high. In addition, the value of $f_{T_g'}$ of PBLAsp α and ω forms was very high and in the vicinity of 0.08 (5).

Rigorously speaking, the application to the Cohen-Turnbull theory to polypeptide may not be accurate in regard to the determination of the absolute value of $f_{T_g'}$ since such polypeptides are composed of a main chain and side chain with or without the stacking and thus is a heterogeneous system. However, a comparison of the fractional free volume may have a physical significance. These $f_{T_g'}$ values of polypeptide are about three times larger than the fractional free volume at glass transition temperature for conventional amorphous polymers, which is known to be 0.025 according to the iso-free volume theory. According to the two-dimensional molecular array of racemic form of PBG, the fractional free volume is estimated to be 0.093, assuming that the cylindrical α helix has a close packed structure and unoccupied volume by α helix corresponds to free volume (6). $f_{T_g'}$ obtained is ranging between 0.025 and 0.093. This indicates that the side chain region of polypeptide is packed much more loosely than amorphous polymers and plays a very important role in diffusion.

Acknowledgments

The authors are grateful to the Ministry of Education, Japan, for a partial support of this research through Grant-in-Aid for Scientific Research.

Literature Cited

1. Elliott, A. In "Poly-α-Amino Acids Protein Models for Conformational Studies" Fasman, G. D. Ed.; Marcel Dekker: New York, 1967; Chap. 1
2. Takizawa, A.; Okada, H.; Kadota, S.; Nonoyama, H. J. Appl. Polym. Sci., 1974, 18, 1443
3. Takizawa, A.; Hamada. T.; Okada, H.; Imai. S.; Kadota, S. Polymer, 1974, 15, 157
4. Oohachi, Y.; Hamano, H.; Tsujita, Y.; Takizawa, A. J. Appl. Polym. Sci., 1978, 22, 1469
5. Kodama, H.; Tsujita, Y.; Takizawa, A. J. Macromol. Sci., Phys., 1980, 17, 57
6. Vivatpanachart, S.; Tsujita, Y.; Takizawa, A. Makromol. Chem., 1981, 182, 1197

7. Minoura, N.; Fujiwara, Y.; Nakagawa, T. J. Appl. Polym. Sci., 1978, 22, 1593
8. Minoura, N.; Nakagawa, T. ibid, 1979, 23, 815
9. Minoura, N.; Nakagawa, T. ibid, 1979, 23, 2729
10. Minoura, N.; Fujiwara, Y.; Nakagawa, T. ibid, 1979, 24, 965
11. Minoura, N.; Fujiwara, Y.; Nakagawa, T. ibid, 1981, 26, 1301
12. Tsuboi, M.; Wada, A.; Nagashima, N. J. Mol. Biol., 1961, 3, 705
13. Mitsui, Y.; Iitake, Y.; Tsuboi, M. ibid, 1967, 24, 15
14. Squire, J. M.; Elliott, A. ibid, 1972, 65, 291
15. Parry, D. A. D.; Elliott, A. J. ibid, 1967, 25, 1
16. Yoshikawa, N.; Tsujita, Y.; Uematsu, I. Polym. J., 1975, 7, 96
17. Bradbury, E. M.; Downie, A. R.; Elliott, A.; Hanby, W. E. Proc. Roy. Soc. (London), 1960, A259, 110
18. Bradbury, E. M.; Brown, L.; Downie, A. R.; Elliott, A.; Fraser, R. D. B.; Hanby, W. E. J. Mol. Biol., 1962, 5, 230
19. Watanabe, T.; Tsujita, Y.; Uematsu, I. Polym. J., 1975, 7, 181
20. McKinnon, A. J.; Tobolsky, A. V. J. Phys. Chem., 1966, 70, 1453
21. McKinnon, A. J.; Tobolsky, A. V. ibid, 1968, 72, 1157
22. Tsujita, Y. Sen-i to Kogyo, 1977, 33, 45
23. Tsujita, Y.; Uematsu, I. Polym. J., 1974, 6, 274
24. Fukuzawa, T.; Uematsu, Y.; Uematsu, I. ibid, 1974, 6, 537
25. Yoshikawa, M.; Tsujita, Y.; Uematsu, I.; Uematsu, Y. ibid, 1975, 7, 96
26. Tsujita, Y.; Sumitomo, T.; Nomura, M.; Takizawa, A. Submitted to J. Appl. Polym. Sci.
27. Watanabe, J.; Sasaki, S.; Uematsu, I. Polym. J., 1977, 9, 451
28. Cohen, M. H.; Turnbull, D. J. Chem. Phys., 1959, 31, 1164
29. Kumins, C. A.; Roteman, J. J. Polym. Sci., 1961, 55, 683
30. Crank, J.; Park, J. S. In "Diffusion in Polymers": Academic Press: London and New York, 1968; pp. 120-125

RECEIVED February 22, 1985

Solvent-Exchange Drying of Cellulose Acetate Membranes for Separation of Hydrogen–Methane Gas Mixtures

B. S. MINHAS, TAKESHI MATSUURA, and S. SOURIRAJAN

Division of Chemistry, National Research Council of Canada, Ottawa, Ontario, Canada K1A 0R9

Cellulose acetate reverse osmosis membranes of different surface porosities were prepared using different solvents in the solvent exchange drying method. These membranes showed a wide variation in the permeation rate and separation of hydrogen-methane gas mixtures. An attempt was made to develop a correlation between the membrane performance and the solvent used for drying. The membranes were characterized and their performance was predicted on the basis of the pore flow model developed in the previous work. The results obtained are discussed.

Though the history of gas separations by membranes can be traced back to 1831 (1) when ivestigations were reported for the enrichment of oxygen in air by rubber membranes, membranes for industrial gas separations could not be used until recently. Low permeation rates and poor separations by earlier membranes made them no match for conventional processes such as cryogenics and adsorption (2). Breakthrough in the formation of appropriate membranes with higher flux came with the development of asymmetric cellulose acetate membranes for reverse osmosis water desalination (3). The reverse osmosis membranes, when dried in a manner to preserve their porosity and the surface pore structure, showed higher permeation rates and significant separations for gaseous mixtures (4, 5).

Fundamental studies have been reported on the permeation of different gases through dry reverse osmosis membranes of different polymeric materials (6, 7) and the separation of hydrogen-methane gas mixtures by cellulose acetate membranes (8, 9) establishing a valid means of treating gas permeations and separations through reverse osmosis membranes. In the above studies the gaseous flow through the pores in the surface layer of the asymmetric porous membrane is considered to be governed by Knudsen, slip, viscous and surface flow mechanisms. The contributions of different flows, defined by different flow mechanisms, to the total flow through the membrane are dependent upon the pore size and the pore size distribution in the

0097–6156/85/0281–0451$06.00/0
Published 1985, American Chemical Society

surface layer of the membrane. For achieving a membrane of appropriate pore size and pore size distribution which gives high separation factor and high permeation rate, a firm cause and effect relationship has to be established between the variables involved in the membrane formation process and the performance data of membrane produced. Among the many variables involved in the formation of cellulose acetate membranes, particularly, we have identified that the evaporation period, shrinkage temperature and the solvent used for replacement of water in the membrane during the drying process are some of the important factors affecting the ultimate pore size and pore size distribution of the membrane and consequently the membrane performance data (9). The objective of this investigation is primarily to study the effect of various solvents used in the solvent exchange drying process of membrane on its subsequent performance in the separation of hydrogen-methane gas mixtures. The effects of process variables such as operating pressure and hydrogen mole fraction in the gas mixture on the separation factor and the product permeation rate were also studied. An attempt was made to correlate the membrane performance with the solvents used for drying the membrane. The predictability of membrane performance is also attempted on the basis of gas transport mechanism referred above.

Gas Transport Through Porous Reverse Osmosis Membranes

Transport equations for the gaseous flow through pores on the surface of an asymmetric porous membrane have been reported earlier by Rangarajan et al (7). These equations have been developed assuming that the pores can be represented by a bundle of cylinderical capillary tubes running through the membrane surface layer. The gaseous flow through these capillaries is governed by four flow mechanisms, namely, Knudsen, slip, viscous and surface flows. The contributions from the first three mechanisms are categorized as pore flow which is free of any gas-polymer interaction. Surface flow, on the other hand, involves interaction forces between the gas and the membrane material. The surface flow mechanism is applicable to all pores irrespective of their sizes, whereas pore flow is governed by one of the first three flow mechanisms depending upon the pore size, pore size distribution and the mean free path of the gas. The mean free path of a gas can be calculated from the expression (10),

$$\lambda = \left(\frac{\underset{\sim}{R}\ T}{\sqrt{2}\ \pi\ d^2\ N\ \bar{P}}\right) \qquad (1)$$

It is assumed that Knudsen flow occurs in the pores of radii > 0 to 0.05λ, slip flow occurs in the pores of radii 0.05λ to 50λ, and viscous flow occurs in the pores of radii larger than 50λ. Further, it is assumed that the variation of the pore sizes on the surface of an asymmetric porous membrane is represented by a single equivalent normal distribution

$$N(R) = \left(\frac{N_t}{\sqrt{2\pi}\ \sigma}\right) \exp\left\{-\tfrac{1}{2}\left(\frac{R-\bar{R}}{\sigma}\right)^2\right\} \qquad (2)$$

The development of a transport equation, for the calculation of gas permeability coefficient, based on the four flow mechanisms described above and the single equivalent normal distribution of pores on the surface of a membrane has been reported by Rangarajan et al (7). According to this theoretical development a membrane can be characterized by the average pore radius, $\bar{R}$, the standard deviation, σ, and by two additional parameters A_1 and A_2, all of which can be determined from the permeability data of a reference gas. The quantity A_1 is related to the total number of pores and the effective membrane thickness, and should remain constant for a given membrane irrespective of the gas. On the other hand, the quantity A_2 is related to the adsorption equilibrium, the mobility of the sorbed gas species and the membrane pore structure. The value of A_2 for various gases has been related to that of a reference gas under the assumption that different gases do not affect the pore structure of a given membrane.

$$(A_2)_i = (A_2)_{ref}\ \phi_i \qquad (3)$$

The parameter ϕ_i, called relative surface transport coefficient, for various gases has been reported by Rangarajan et al (7) taking nitrogen as the reference gas. As stated by Rangarajan et al (7) the effective mean pore radius $\bar{R}$ is distinct from the geometrical mean pore radius $\bar{R}^*$. The former mean pore radius is expected to depend upon the gas-membrane interaction and the mobility of the adsorbed gas molecule. According to the paper referred above, for any gas i, $\bar{R}_i$ is computed from the following equation:

$$\bar{R}_i = \bar{R}_{ref} + \Delta_i \qquad (4)$$

The quantities Δ_i called the radius correction factor, for several gases are reported in literature (7) with respect to the reference gas, nitrogen. The standard deviation, σ, is expected to remain constant for a given membrane irrespective of the gas.

These four characteristic parameters can be used to compute the permeation rate and the separation factor of a gas mixture with respect to the characterized membrane. The method for the calculation of product composition have been reported by Mazid et al (8). According to this method the separation factor, S_{12}, defined as:

$$S_{12} = \frac{X_{13}/X_{23}}{X_{12}\ /X_{22}} \qquad (5)$$

can be obtained from the calculated compositions of product gases. The product permeation rate can be calculated from the fluxes, J_i, of each component of the gas mixture. The equations for the evaluation of flux, J_i, has been reported in the literature (8).

$$[PR] = \sum J_i \qquad (6)$$

Thus to predict the separation factor and the permeation rate first the membrane is characterized by evaluating $\bar{R}$, σ, A_1 and A_2 using the permeability of a reference gas through the membrane. These reference parameters are then used to evaluate the characteristic parameters with respect to each component of the gas mixture. These evaluated parameters are subsequently used to predict the separation factor and the permeation rate for a gas mixture for the characterized membrane.

Experimental

The membranes used in the present investigation were cast using a solution of the following composition (wt. %): cellulose acetate (Eastman 398-3) 17, acetone 69.2, magnesium perchlorate 1.45 and water 12.35 (11). All membranes were cast to equal nominal thickness. The casting solution and the casting atmosphere temperatures were kept constant at 10° and 30°C, respectively. The relative humidity of the casting atmosphere was maintained at 65%. Membranes were gelled in ice cold water after 60 s of solvent evaporation time and then shrunk in hot water at 80°C. These membranes were then dried, in order to use them for gas separation experiments by a multiple stage solvent exchange drying technique. In this technique the water in the membrane is first replaced by a water miscible solvent (called the "first solvent") which is a nonsolvent for the membrane material. Then, the first solvent is replaced by a second solvent which is volatile. The second solvent is subsequently air evaporated to obtain the dry membrane. A number of different solvents were used both as the first solvent and as the second solvent. First solvents used include isopropyl alcohol, tertiary butyl alcohol, ethylene glycol, diethylene glycol, triethylene glycol and ethylene glycol monoethyl ether. The solvents used as the second solvent include pentane, hexane, cyclohexane, benzene, toluene, carbondisulfide, triethylamine and isopropyl ether. The replacement of water in the membrane by the first solvent was done by successive immersion in first solvent-water solutions which were progressively more concentrated in the first solvent. For example in four stage replacement, which was the most common in our experiments, 25, 50, 75 and 100 vol. % aqueous solutions of the first solvent were used. When isopropyl alcohol was employed as the first solvent the replacement was carried out also in one, two and three stages in order to study the effect of number of stages on the membrane performance. In the cases where the first solvent used was any one of the glycols an intermediate solvent was also used, for membrane drying, between the first and the second solvents, as the glycols are not miscible with the second solvents. Intermediate solvents used were ethyl alcohol and n-butyl alcohol which are miscible with both the first solvents and the second solvents. Usage of different solvents in the drying procedure resulted in membranes of different average pore sizes and pore size distributions.

Membranes dried by the combination of different solvents were numbered in the form of CA(K)-mn for the purpose of membrane identification, where CA indicates cellulose acetate material, K denotes the number of stages involved in the replacement of water in the membrane by the first solvent, and m and n are the numbers given to the first and to the second solvents, respectively, used in the membrane drying process. Numbers given to the first solvents are as follows: tertiary butyl alcohol-1, isopropyl alcohol-2, ethylene glycol monoethyl ether-3, ethylene glycol-4, diethylene glycol-5, triethylene glycol-6; whereas the second solvents were numbered as follows: pentane-1, hexane-2, cyclohexane-3, benzene-4, toluene-5, triethylamine-6, isopropyl ether-7 and carbondisulfide-8.

The equipment used in the present investigation and the details of the experimental procedure have been reported previously (4, 7). Air in the reverse osmosis cells and in the feed gas line was removed by flushing them with the feed gas mixture of hydrogen and methane. The mole fraction of hydrogen in the feed gas was changed from 0.883 to 0.116. All the experiments were conducted at room temperature and the feed pressure was varied in the range of 400 to 2400 kPa abs. The permeate flow rate was measured by a soap bubble meter. The composition of the gas was measured by gas chromatography using Tracor MT 160/220 model equipped with a Porapak Q column. The accuracy involved in gas composition analysis was ±1%. All the gases (pure and mixed) were obtained from Matheson Canada with a specified purity of 99.9%.

The liquid chromatography experiments for measuring the retention volume of various second solvents, in the presence of first solvent was conducted using the equipment model ALC 202 from Waters Associates. The column used was 0.16 cm inner diameter and 60 cm in length and packed with commercially available cellulose acetate powder. The solutions of various second solvents (1 vol. %) in a first solvent were injected into the stream of the first solvent flowing through the column as a carrier solvent in order to measure the retention volume of the second solvent. Either tertiary butyl alcohol or isopropyl alcohol was used as the first solvent. When tertiary butyl alcohol was used as the carrier solvent, the entire chromatography system including the column was maintained at 27°C in order to avoid the freezing of tertiary butyl alcohol. All the solvents used were of reagent grade.

Results and Discussion

Membrane Performance. Cellulose acetate membranes dried by solvent exchange technique using various combinations of first, second and intermediate solvents showed a wide variation in separation factor ranging from 1 to 28 for a feed mixture of hydrogen and methane. A wide variation in permeation rate was also found ranging from 1.25×10^{-4} to 9.71×10^{-7} kmol/m^{2}s, for the feed containing 0.883 mole fraction of hydrogen at the operating pressure about 2200 kPa abs. These wide variations in the separation factor and in the permeation rate emphasize the importance of solvent used in the solvent exchange technique of membrane drying. The membranes dried by the combination of isopropyl alcohol and hexane as the first and the second solvent, respectively, usually gave higher separation factors and permeation rates compared to the membranes dried using other combinations of solvents. It was also found that when membranes were dried using the combination of isopropyl alcohol and hexane solvents, the number of stages involved in the replacement of water in the membrane by isopropyl alcohol significantly affected the membrane performance. It may be seen in Table I that the membrane (CA(2)-22) dried by two stage process (replacing water with 50 vol. % aqueous solution of isopropyl alcohol which was then replaced by

Table I. Effect of Number of Stages Involved in the Replacement of Water in the Membranes by the First Solvent on the Separation Factor and the Permeation Rate

X_{12}	Membrane	Pressure, kPa	S_{12}	[PR], kmol/m^2.s
0.883	CA(1)-22	487.4	9.3	0.144×10^{-5}
		1411.3	4.3	0.261×10^{-5}
		1756.1	4.2	0.296×10^{-5}
	CA(2)-22	446.1	9.0	0.330×10^{-5}
		652.9	15.0	0.340×10^{-5}
		1135.5	21.0	0.700×10^{-5}
		1494.1	25.0	0.960×10^{-5}
		2224.9	28.0	0.150×10^{-4}
	CA(3)-22	487.4	6.1	0.254×10^{-6}
		914.9	7.0	0.497×10^{-6}
		1411.3	8.6	0.793×10^{-6}
		1756.1	13.3	0.990×10^{-6}
		2169.7	19.1	0.606×10^{-5}
	CA(4)-22	515.0	9.4	-
		1410.0	14.5	-
		2310.0	17.5	0.947×10^{-6}
0.781	CA(1)-22	452.9	6.8	0.104×10^{-5}
		1349.3	2.2	0.172×10^{-5}
		1756.1	1.6	0.205×10^{-5}
	CA(2)-22	446.1	15.0	0.170×10^{-5}
		790.8	20.0	0.360×10^{-5}
		1101.1	21.0	0.530×10^{-5}
		1549.2	18.5	0.780×10^{-5}
		2204.2	17.0	0.120×10^{-4}
	CA(3)-22	452.9	13.0	0.225×10^{-6}
		859.7	11.0	0.411×10^{-6}
		1756.1	7.2	0.832×10^{-6}
	CA(4)-22	515.0	7.3	-
		1410.0	10.0	0.110×10^{-6}
		2310.0	12.7	0.120×10^{-6}

Note: The first solvent is isopropyl alcohol; the second solvent is hexane.

nonaqueous isopropyl alcohol) gave higher separation factor as well as the permeation rate for hydrogen methane gas mixtures compared to the membranes dried by one (CA(1)-22), three (CA(3)-22) and four (CA(4)-22) stage processes. The highest separation factor obtained for the membrane CA(2)-22 was 28 at the operating pressure of 2225 kPa abs for a feed gas mixture containing 0.883 mole fraction of hydrogen. This result corresponds to the hydrogen permeation rate of 99 relative to methane, which is significantly higher than the values reported in the literature for various membranes. These results indicate that significant separation factors are obtainable by gas permeation under pressure through porous cellulose acetate reverse osmosis membranes if dried by suitable combination and sequence of solvents.

The present investigation also revealed that a membrane with a lower permeation rate does not necessarily yield a higher separation factor. The permeation rates and the separation factors obtained in the present study for some of the membranes dried using different combinations of solvents are presented in Table II. It may be seen in Tables I and II that the permeation rates and the separation factors obtained at all feed compositions and at all feed gas pressures were lower for the membrane dried using tertiary butyl alcohol as the first solvent and benzene as the second solvent (CA(4)-14) than those for the membrane dried using isopropyl alcohol as the first solvent and hexane as the second solvent (CA(2)-22) at corresponding feed compositions and feed gas pressures.

Correlation Between Membrane Performance and Drying Solvent. An attempt was made to develop a correlation between the properties of the solvent used for membrane drying and the separation factor. In order to develop such a correlation the retention volumes of various second solvents were measured in a liquid chromatography system in which the column was packed with cellulose acetate powder. The retention volume is considered to be a measure of the magnitude of the interaction force working between the second solvent and the cellulose acetate polymer in the presence of the first solvent. As described in the experimental section two first solvents, namely isopropyl alcohol and tertiary butyl alcohol, were used as the carrier solvent. Figure 1 shows the results of the chromatography experiments. In the figure retention volumes of various second solvents are plotted against the separation factor obtained from the membrane dried using the corresponding second solvent. It may be seen in Figure 1 that the membranes which were dried using isopropyl alcohol as the first solvent gave the highest separation factor at the retention volume corresponding to the second solvent hexane, irrespective of the operating pressure. On the other hand, when tertiary butyl alcohol was used as the first solvent, the second solvent which gave the maximum separation factor, shifted towards the higher retention volume, suggesting a stronger interaction force between the second solvent and the membrane material is needed. The second solvent corresponding to the maximum separation factor was found to be isopropyl ether. Though Figure 1 includes points for the membranes dried using different stages when isopropanol was used as the first solvent, the correlation emerging between the separation factor and the retention volume of the second solvent is quite clear. The above results indicate that for a given first drying

Table II. Some Experimental Data on Permeation Rate and Separation Factor for Different Membranes at Various Pressures and Feed Compositions

X_{12}	Membrane	Pressure, kPa	S_{12}	[PR], $kmol/m^2.s$
0.883	CA(4)-14	790.8	1.9	0.19×10^{-6}
		1114.8	1.74	0.33×10^{-6}
		1818.1	1.7	0.72×10^{-6}
		2176.6	1.7	0.97×10^{-6}
	CA(4)-17	466.7	4.6	0.47×10^{-5}
		790.8	3.3	0.87×10^{-5}
		1114.8	2.5	0.14×10^{-4}
		1818.1	1.5	0.27×10^{-4}
		2176.6	1.1	0.35×10^{-4}
	CA(4)-27	466.7	3.2	0.43×10^{-5}
		790.8	2.5	0.78×10^{-5}
		1114.8	2.1	0.12×10^{-4}
		1818.1	1.6	0.23×10^{-4}
		2176.6	1.3	0.28×10^{-4}
	CA(4)-32	466.7	1.2	0.15×10^{-4}
		790.8	1.0	0.42×10^{-4}
		1114.8	1.0	0.61×10^{-4}
		1818.1	1.0	0.10×10^{-3}
		2169.7	1.0	0.13×10^{-3}
	CA(4)-47[a]	446.1	1.9	0.19×10^{-4}
		859.7	1.2	0.40×10^{-4}
		1342.4	1.0	0.73×10^{-4}
		1811.2	1.0	0.11×10^{-3}
		2169.7	1.0	0.14×10^{-3}
	CA(4)-62[b]	446.1	1.9	0.24×10^{-5}
		859.7	1.9	0.57×10^{-5}
		1342.4	1.3	0.12×10^{-4}
		1811.2	1.2	0.17×10^{-4}
		2169.7	1.2	0.21×10^{-4}
0.781	CA(4)-14	1342.4	1.8	0.39×10^{-6}
		1825.0	1.8	0.63×10^{-6}
		2169.7	1.84	0.84×10^{-6}
	CA(4)-17	859.7	3.5	0.78×10^{-5}
		1342.4	2.5	0.14×10^{-4}
		1825.0	1.5	0.22×10^{-4}
		2169.7	1.2	0.36×10^{-4}
	CA(4)-27	459.9	2.8	0.33×10^{-5}
		859.7	2.3	0.72×10^{-5}
		1342.4	1.8	0.13×10^{-4}
		1825.0	1.6	0.19×10^{-4}
		2169.7	1.4	0.22×10^{-4}
	CA(4)-32	459.9	1.2	0.19×10^{-4}
		859.7	1.1	0.37×10^{-4}
		1342.4	1.1	0.61×10^{-4}
		1825.0	1.0	0.84×10^{-4}
		2169.7	1.0	0.99×10^{-4}

Table II. Continued

X_{12}	Membrane	Pressure, kPa	S_{12}	[PR], $kmol/m^2.s$
0.781	CA(4)-47[a]	473.6	1.5	0.16×10^{-4}
		859.7	1.2	0.30×10^{-4}
		1342.4	1.0	0.50×10^{-4}
		1818.1	1.0	0.78×10^{-4}
		2238.7	1.0	0.10×10^{-3}
	CA(4)-62[b]	473.6	2.3	0.25×10^{-5}
		859.7	2.0	0.54×10^{-5}
		1342.4	1.5	0.10×10^{-4}
		1818.1	1.3	0.15×10^{-4}
		2238.7	1.3	0.19×10^{-4}
0.484	CA(4)-14	859.7	1.8	0.13×10^{-6}
		1342.4	1.8	0.24×10^{-6}
		1804.3	2.0	0.38×10^{-6}
		2204.2	1.9	0.52×10^{-6}
	CA(4)-17	446.1	4.2	0.15×10^{-5}
		859.7	3.3	0.38×10^{-5}
		1342.4	2.2	0.69×10^{-5}
		1804.3	1.6	0.12×10^{-4}
		2204.2	1.3	0.18×10^{-4}
	CA(2)-22	1135.5	14.7	0.20×10^{-5}
		1549.2	13.7	0.27×10^{-5}
		1997.4	12.8	0.36×10^{-5}
	CA(4)-27	446.1	2.8	0.16×10^{-5}
		859.7	2.5	0.38×10^{-5}
		1342.4	1.9	0.68×10^{-5}
		1804.3	1.8	0.96×10^{-5}
		2204.2	1.6	0.12×10^{-4}
	CA(4)-32	446.1	1.1	0.65×10^{-5}
		859.7	1.1	0.24×10^{-4}
		1342.4	1.0	0.38×10^{-4}
		1804.3	1.0	0.51×10^{-4}
		2190.4	1.0	0.63×10^{-4}
	CA(4)-47[a]	501.2	1.8	0.86×10^{-5}
		873.5	1.1	0.14×10^{-4}
	CA(4)-62[b]	501.2	2.4	0.16×10^{-5}
		873.5	1.8	0.32×10^{-5}
0.238	CA(4)-14	446.1	2.0	0.35×10^{-7}
		1342.4	2.0	0.13×10^{-6}
		1790.5	1.9	0.24×10^{-6}
		2204.2	1.8	0.34×10^{-6}
	CA(4)-17	473.6	3.4	0.74×10^{-6}
		859.7	2.7	0.20×10^{-5}
		1342.4	2.2	0.42×10^{-5}
		1790.5	1.6	0.94×10^{-5}
		2238.7	1.3	0.14×10^{-4}
	CA(2)-22	446.1	11.0	0.18×10^{-6}
		790.8	10.5	0.33×10^{-6}
		1135.5	10.2	0.51×10^{-6}
		1549.2	10.0	0.78×10^{-6}

Continued on next page.

Table II. Continued

X_{12}	Membrane	Pressure, kPa	S_{12}	[PR], $kmol/m^2.s$
		1997.4	10.9	0.12×10^{-5}
0.238	CA(4)-27	473.6	2.4	0.96×10^{-6}
		859.7	2.2	0.21×10^{-5}
		1342.4	2.0	0.43×10^{-5}
		1790.5	1.73	0.61×10^{-5}
		2238.7	1.7	0.77×10^{-5}
	CA(4)-32	473.6	1.1	0.45×10^{-5}
		859.7	1.0	0.12×10^{-4}
		1342.4	1.0	0.22×10^{-4}
		1790.5	1.0	0.38×10^{-4}
		2204.2	1.0	0.48×10^{-4}
	CA(4)-47[a]	459.9	1.9	0.44×10^{-5}
		825.3	1.6	0.10×10^{-4}
		1328.6	1.3	0.17×10^{-4}
	CA(4)-62[b]	459.9	2.2	0.89×10^{-6}
		825.3	2.2	0.22×10^{-5}
		1328.6	1.9	0.45×10^{-5}
0.116	CA(4)-17	563.3	3.7	0.76×10^{-6}
		914.9	2.6	0.16×10^{-5}
		1411.3	2.2	0.38×10^{-5}
		1825.0	1.6	0.69×10^{-5}
	CA(2)-22	480.5	11.7	0.98×10^{-7}
		859.7	11.7	0.19×10^{-6}
		1342.4	11.3	0.36×10^{-6}
	CA(4)-27	563.3	3.0	0.98×10^{-6}
		914.9	2.2	0.18×10^{-5}
		1411.3	2.2	0.37×10^{-5}
		1825.0	1.9	0.49×10^{-5}
	CA(4)-32	563.3	1.3	0.48×10^{-5}
		914.9	1.1	0.10×10^{-4}
		1411.3	1.0	0.21×10^{-4}
	CA(4)-47[a]	452.9	2.0	0.38×10^{-5}
		804.6	1.4	0.74×10^{-5}
		1328.6	1.1	0.14×10^{-4}
	CA(4)-62[b]	452.9	2.1	0.73×10^{-6}
		804.6	2.0	0.16×10^{-5}
		1328.6	1.73	0.35×10^{-5}
		1756.1	1.6	0.52×10^{-5}
		2204.2	1.53	0.72×10^{-5}

a) ethylene glycol followed by ethyl alcohol
b) triethylene glycol followed by n-butyl alcohol

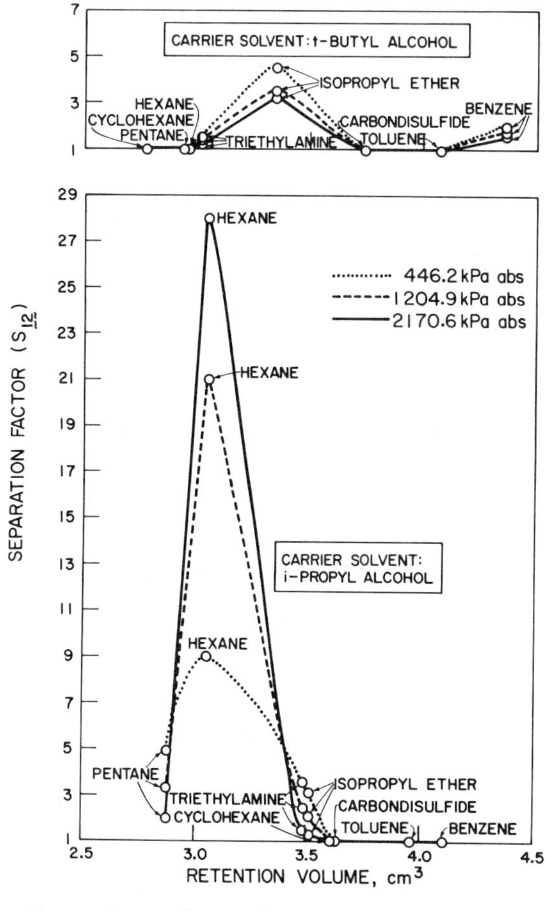

Figure 1. Retention volume of various second solvents versus separation factor of H_2/CH_4 gas mixture. Liquid chromatography column, packed with cellulose acetate powder. Membrane gas separation conducted at the hydrogen mole fraction of 0.883 in feed gas and various operating pressures.

solvent (carrier solvent in liquid chromatography), there is an optimum retention volume for the second solvent which corresponds to the highest separation factor for the gas mixture studied; and the above retention volume is independent of the operating pressure used in the gas separation experiment. Thus Figure 1 shows a useful correlation between the retention volume of the second solvent and the resulting porous structure of the membrane surface in terms of the gas separation experiment involved. Consequently, liquid chromatography data on retention volumes for different second solvents for a given first solvent offer a basis for the choice of the appropriate second solvent for obtaining the desirable porous structure on the resulting membrane capable of giving the highest separation factor for a given gas mixture.

Prediction of Membrane Performance. All the membranes formed in the present investigation were characterized in terms of $\bar{R}$, σ, A_1 and A_2 by using the transport equation described by Rangarajan et al (7) together with the permeability data of helium through the membranes. The results are shown in Table III. The table shows that there are over 8 fold variation in $(\bar{R})_{He}$ and 11 fold variation in σ. Furthermore, there are 380 fold variation in the value of A_1 and 500 fold variation in $(A_2)_{He}$. These variations indicate that the membranes tested have a wide range of porosity and also a wide range in surface flow contribution, which should result in a wide variation in permeation rates and separation factors of a gas mixture.

The parameters evaluated in membrane characterization were used to predict the permeation rate, [PR], for a given membrane under a given set of operating conditions such as hydrogen mole fraction in the feed gas and the operating pressure. The details of the prediction method were described previously (8). The comparison of calculated and experimental permeation rates is shown in Figure 2. The agreement between calculated and experimental values is excellent in some cases and unsatisfactory in many cases. Further, the product compositions and the separation factors were also predicted, using the approach developed previously (8), for all the membranes studied under all the combinations of operating pressures and feed compositions employed in the experiment using the membrane parameters given in Table III. The comparison of experimental and calculated values of product compositions is shown in Figure 3. These results also show excellent agreement in some cases, and unsatisfactory agreement in many cases. Such observed agreements and disagreements between the calculated and experimental data are similar to those reported earlier (8). These results indicate that while the transport equations employed are basically valid, the analytical expressions involved need improvement based on the physical chemistry of the gas adsorption process taking place in the system, and the porous structure of the membranes employed. In particular, it is necessary to examine the need for (i) alternative expressions for the individual and competitive gas adsorption processes, and (ii) bimodal pore size distribution on the membrane surface, for incorporation in the gas transport equations.

Table III. Membrane Characterization by Using Helium Permeation Data

Membrane	$(\bar{R})_{He} \times 10^{10}$,m	$\sigma \times 10^{10}$,m	$(A_1)_{He}$,m^{-3}	$(A_2)_{He}$,$kmol/m^3sPa^2$
CA(4)-11	8.0	4.3	6.24×10^{19}	5.15×10^{-8}
CA(4)-12	8.0	5.5	6.45×10^{18}	1.46×10^{-9}
CA(4)-13	12.0	3.3	1.01×10^{19}	4.38×10^{-9}
CA(4)-14	4.0	1.7	3.40×10^{17}	1.02×10^{-10}
CA(4)-15	7.0	0.8	3.15×10^{17}	3.15×10^{-10}
CA(4)-16	4.0	2.7	5.63×10^{18}	1.24×10^{-9}
CA(4)-17	24.0	0.7	1.64×10^{17}	5.55×10^{-10}
CA(4)-18	7.0	0.9	9.21×10^{17}	6.26×10^{-10}
CA(4)-21	26.0	0.5	1.85×10^{17}	4.18×10^{-10}
CA(1)-22	7.0	1.1	1.36×10^{18}	9.83×10^{-10}
CA(2)-22	5.0	3.7	2.47×10^{18}	1.30×10^{-9}
CA(3)-22	6.4	1.3	2.25×10^{17}	1.18×10^{-10}
CA(4)-22	3.2	2.3	4.24×10^{17}	1.15×10^{-9}
CA(4)-24	5.0	1.0	7.87×10^{17}	1.53×10^{-10}
CA(4)-25	7.0	0.7	6.99×10^{17}	5.83×10^{-10}
CA(4)-27	5.0	1.1	6.79×10^{18}	2.26×10^{-9}
CA(4)-28	24.0	0.7	5.37×10^{17}	2.53×10^{-10}
CA(4)-32	7.0	0.9	1.22×10^{19}	7.07×10^{-9}
CA(4)-42[a]	5.8	0.6	6.15×10^{17}	1.68×10^{-9}
CA(4)-47[a]	5.0	3.0	3.15×10^{19}	8.01×10^{-9}
CA(4)-52[b]	5.0	1.5	6.31×10^{19}	1.99×10^{-8}
CA(4)-62[c]	24.0	0.7	2.15×10^{18}	1.59×10^{-9}

a) ethylene glycol followed by ethyl alcohol
b) diethylene glycol followed by n-butyl alcohol
c) triethylene glycol followed by n-butyl alcohol

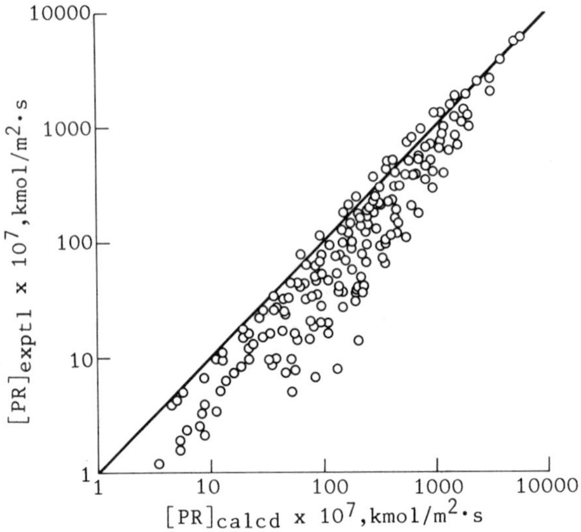

Figure 2. Comparison of experimental and calculated permeation rates of H_2/CH_4 gas mixture. Hydrogen mole fraction in feed gas mixture, 0.116–0.883; operating pressure, 450–2300 kPa abs; membrane material, cellulose acetate; membrane porosity given in Table III; operating temperature, room.

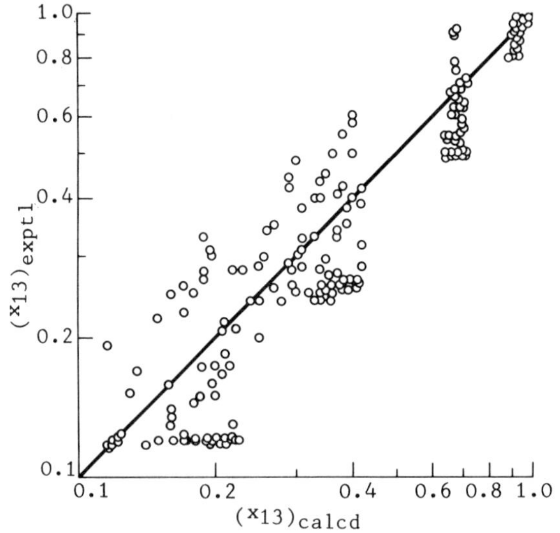

Figure 3. Comparison of experimental and calculated product compositions of H_2/CH_4 gas mixture. Hydrogen mole fraction in feed gas mixture, 0.116–0.883; operating pressure, 450–2300 kPa abs; membrane material, cellulose acetate; membrane porosity given in Table III; operating temperature, room.

Conclusion

It may be concluded from the present study that the performance of reverse osmosis cellulose acetate membranes for the separation of hydrogen-methane gas mixtures is strongly influenced by the solvent systems and operational sequence used with solvent exchange drying of the membrane. By selecting a proper combination of solvents and proper operational sequence a cellulose acetate membrane of desired pore size and pore size distribution can be made to give higher separation factor and permeation rate. The selection of a solvent system can be facilitated by measuring the retention volume of the solvent in a column packed with the membrane material in liquid chromatographic study where the carrier is the first solvent. It is also concluded from this study that further refinements in the prediction method are necessary in order to improve the agreement between calculated and experimental gas permeation and separation data in a wide range of operating conditions.

Nomenclature

A_1, $(A_1)_{\underset{\sim}{i}}$	= constant for a given membrane related to the porous structure, A_1 for gas i, m^{-3}
A_2, $(A_2)_{ref}$, $(A_2)_{\underset{\sim}{i}}$	= constant related to surface transport, A_2 for reference gas and gas i, respectively, $kmol/m^3.s.\ Pa^2$
d	= collision diameter, m
$J_{\underset{\sim}{i}}$	= flux of gas i, $kmol/m^2.s$
K	= number of stages involved in the replacement of water in the membrane by the first solvent
m	= number given to the first solvent
n	= number given to the second solvent
N	= Avogadro number
N(R)	= number of pores having a radius R, m^{-1}
N_t	= total number of pores
P	= pressure, Pa
$\bar{P}$	= mean pressure, Pa
[PR]	= total product permeation rate of the gas mixture, $kmol/m^2.s$
R	= pore radius
$\bar{R}$, $\bar{R}_{ref}$, $\bar{R}_{\underset{\sim}{i}}$	= mean pore radius, $\bar{R}$ for reference gas and gas i, respectively, m
$\bar{R}^*$	= geometrical mean pore radius, m
$\underset{\sim}{R}$	= gas constant $m^3.Pa/K.kmol$
$S_{\underset{\sim}{12}}$	= separation factor for gas mixture
T	= absolute temperature, K
$X_{\underset{\sim}{i}}$, $X_{\underset{\sim}{i}2}$, $X_{\underset{\sim}{i}3}$	= mole fraction of gas i, $X_{\underset{\sim}{i}}$ on the high pressure side, and on the permeate side, respectively

Greek Letters

$\Delta_{\underset{\sim}{i}}$	= radius correction factor, a constant for gas i for a given membrane material, m
λ	= mean free path of gas, m

σ = standard deviation for the pore size distribution, m

ϕ_i = characteristic parameter, called the relative surface transport coefficient, related to gas membrane interaction

Acknowledgments

The authors are grateful to the NRC Bioenergy Project for supporting this work. One of the authors (BSM) thanks the NSERC for a visiting fellowship. This paper was issued as NRC No. 24035.

Literature Cited

1. Mitchell, J.V. J. Roy. Inst. 1831, 2, 101, 307.
2. Rain, C. High Technol. 1983, Nov., 69-76.
3. Loeb, S.; Sourirajan, S., Department of Engineering, University of California, Los Angeles, Report No. 60-60, 1960, July.
4. Agrawal, J.P.; Sourirajan, S. J. Appl. Polym. Sci. 1969, 13, 1065.
5. Agrawal, J.P.; Sourirajan, S. J. Appl. Polym. Sci. 1970, 14, 1303.
6. Rangarajan, R.; Mazid, M.A.; Matsuura, T.; Sourirajan, S. Proc. Fourth Bioenergy R & D Seminar, National Research Council of Canada, Ottawa, 1982, pp. 435-40.
7. Rangarajan, R.; Mazid, M.A.; Matsuura, T.; Sourirajan, S. Ind. Eng. Chem. Process Des. Dev. 1984, 23, 79-87.
8. Mazid, M.A.; Rangarajan, R.; Matsuura, T.; Sourirajan, S. Ind. Eng. Chem. Process Des. Dev., in press.
9. Minhas, B.S.; Mazid, M.A.; Matsuura, T.; Sourirajan, S. Proc. Fifth Bioenergy R & D Seminar, National Research Council of Canada, Ottawa, 1984.
10. Metz, C.R. "Physical Chemistry"; McGraw-Hill: New York, 1976; p. 11.
11. Pageau, L.; Sourirajan, S. J. Appl. Polym. Sci., 1972, 16, 3185.

RECEIVED April 22, 1985

Pervaporation Membranes

Application in the Chemical Process Industry

H. E. A. BRÜSCHKE[1], G. F. TUSEL[1], and R. RAUTENBACH[2]

[1]GFT Ingenieurbüro, Gerberstrasse 48, 6650 Homburg, Federal Republic of Germany
[2]Rhein Westfaren Technische Hochschule, Turmstrasse 46, 5100 Aachen, Federal Republic of Germany

Data for the separation of alcohol-water mixtures by means of a newly developed pervaporation membrane are reported. The membrane is of the composite type, comprising polyvinylalcohol as the main polymer of the separating layer. Critical process parameters and the application of the pervaporation process in an industrial scale are discussed.

Membrane pervaporation is a very effective separation process, especially in the chemical industry. Various membranes of the composite type have been developed and tested for their pervaporation capabilities. These newly developed composite membranes exhibit excellent performance in the removal of water from mixtures with organic solvents, the organic solvent being a simple alcohol, such as ethanol or isopropanol; a ketone, such as acetone or methyl-ethyl-ketone; an ether, such as diethyl ether or dioxane or an ester, such as ethyl acetate. Starting with sub-azeotropic compositions, purities of the organic components of 99.8% or higher can be achieved. Especially feasible is the pervaporation process for multi-component mixtures. Since, even at high organic feed concentrations, the permeate streams contain approx. 90% water, evaporation heat for these permeate streams has to be supplied at a temperature level of only 50 - 100°C, making the use of waste heat possible. Different membranes still under development show extremely promising results with regard to the separation of organic-organic mixtures.

Pervaporation and its potentials were studied fairly extensively by a small number of researchers 20 - 30 years ago (1-2). However, pervaporation processes did not find their way into industrial application because of the lack of suitable membranes and because interest in membrane research was shifted to more promising process such as RO and UF.

The energy crisis and environmental pollution, however, forced engineers and scientists to look into possibilities of reducing

0097-6156/85/0281-0467$06.00/0

energy consumption and avoiding environmental pollution, especially in the chemical industry.

During recent years, scientists have developed a new and strong interest in pervaporation because of its high separation potential for such organic or aqueous-organic mixtures, which are difficult and costly to separate by conventional separation techniques (3-5). Pervaporation can - for example - successfully replace azeotropic distillation or extractive distillation; it is simpler and safer because addition of a third component is unnecessary and it is superior from an economical point of view.

Fundamentals of Pervaporation Processes

The difference between pervaporation and all other membrane processes is the phase change of the substance permeating through the membrane from a liquid feed to the vaporous permeate. The liquid "pervaporates" when passing through the membrane. Driving force for this process is the difference in the chemical potential of the species on both sides of the membrane. As non-porous membranes have to be used, the transport mechanism can be best described by a solution-diffusion model. In order to keep the difference in chemical potential sufficiently high, condensation on the permeate side of the membrane must be avoided. This is effected by continuous removal of the vapour by either sweeping with an inert gas or by a vacuum. Figure 1 shows the principal arrangement of a pervaporation system.

Contrary to the well-known processes of RO and UF, the performance of the pervaporation system is practically not influenced by the feed-side pressure. However, both selectivity and flux of a pervaporation membrane are highly dependent on the ratio of the total pressure at the permeate side of the membrane to the saturation pressure of the permeating components. In Figure 2 the dependence of flux on this ratio is presented for the benzene-cyclohexan system. This figure indicates clearly that in the practical range of application at values of $p/p°$ smaller than 0.4, flux and selectivity are fairly independent of this pressure ratio. This necessary range of pressure ratios corresponds to an absolute pressure between 2 and 50 mbar, which can easily be achieved by standard vacuum pumps. Like in vacuum distillation, it is only necessary to compress the inert gases from vacuum to atmospheric pressure if the vapours are condensed in the vacuum.

A normal Arrhenius-type dependence of flux on the operation temperature is found with all types of pervaporation membranes (8), whereas no simple dependence of selectivity on temperature can be observed (9).

Process Design

Process design in pervaporation has obviously to concentrate on two problems:

1. The design of a system with low pressure losses on the permeate side, since the actual pressure at the membrane surface determines the process rather than the suction pressure of the pump.
2. The low cost supply of the necessary evaporation enthalpy for the permeating components.

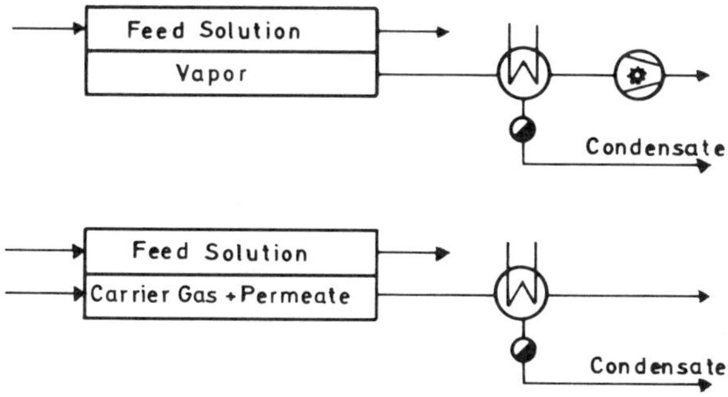

Figure 1. Schematic of Pervaporation.

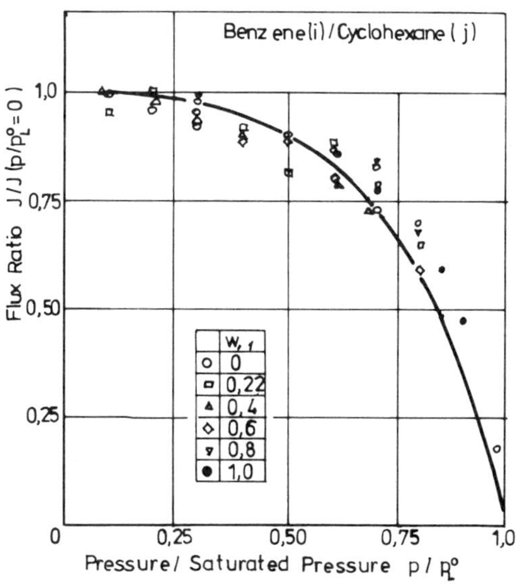

Figure 2. Influence of permeate pressure on flux of benzene through a polyethylene membrane.

In principle, the evaporation enthalpy can be supplied by the heating of one side of the feed channel. Such a design is expensive and it is a far better solution to draw the energy from the sensible heat of the feed. As a consequence, the temperature of the feed will decrease in the module, i.e. a marked difference exists between feed-entrance and feed-exit temperature. Since the flux depends very much on temperature (8), the feed must be reheated and, therefore, the combination of membrane modules and heat exchangers in series (Figure 3) is typical for pervaporation.

Membranes

Obviously membranes for pervaporation, which could be operated at elevated temperatures (100°C or more) would be highly desirable.

This is a very demanding task for membrane development in a field where the systems to be separated are mostly solvents by nature and where the transport mechanisms in the membrane are sorption and diffusion (6).

Recently new membranes have been developed for a certain group of separation problems - the dehydration of aqueous organic mixtures. These asymmetric composite membranes, comprising poly-vinyl-alcohol as the separating polymer layer, can be operated up to temperatures of 130°C. An important practical feature of these membranes is the simplicity of operation: shut-downs of the unit at weekends for example - and consequently a drying of the membranes - will not damage the membranes.

The membranes have proven their reliability for the separation of water-ethanol, water-isopropanol, water-methanol mixtures and systems such as methanol-ethanol and methanol-acetone.

Nevertheless, the development of membranes for pervaporation is still at an early stage. Promising results have been obtained in our laboratory with small membrane samples capable of separating olefines from paraffines, for example.

Processes and Economics

The basics of pervaporation will be explained using the ethanol-water system as an example.

In Figure 4 a diagram is presented, which shows on the x-axis the composition of an ethanol/water mixture and on the y-axis the composition of the permeate, as well as the composition of the vapours in equilibrium with the liquid. The upper curve is the equilibrium curve, which is well-known in distillation techniques. As can be seen, ethanol is the more volatile component and, therefore, enriched in the vapour phase compared to its liquid concentration. However, the higher the concentration of the liquid mixture in ethanol, the lower the ratio of its volatility over water volatility. In the ethanol-rich range a point called the azeotropic point is found, where liquid and vapour have the same composition.

The lower curve in Figure 4 shows the permeate composition versus feed composition, obtained with the new PVA-based membrane. As can be seen, water is by far the better permeating component and is, therefore, enriched in the permeate. It has to be stressed that

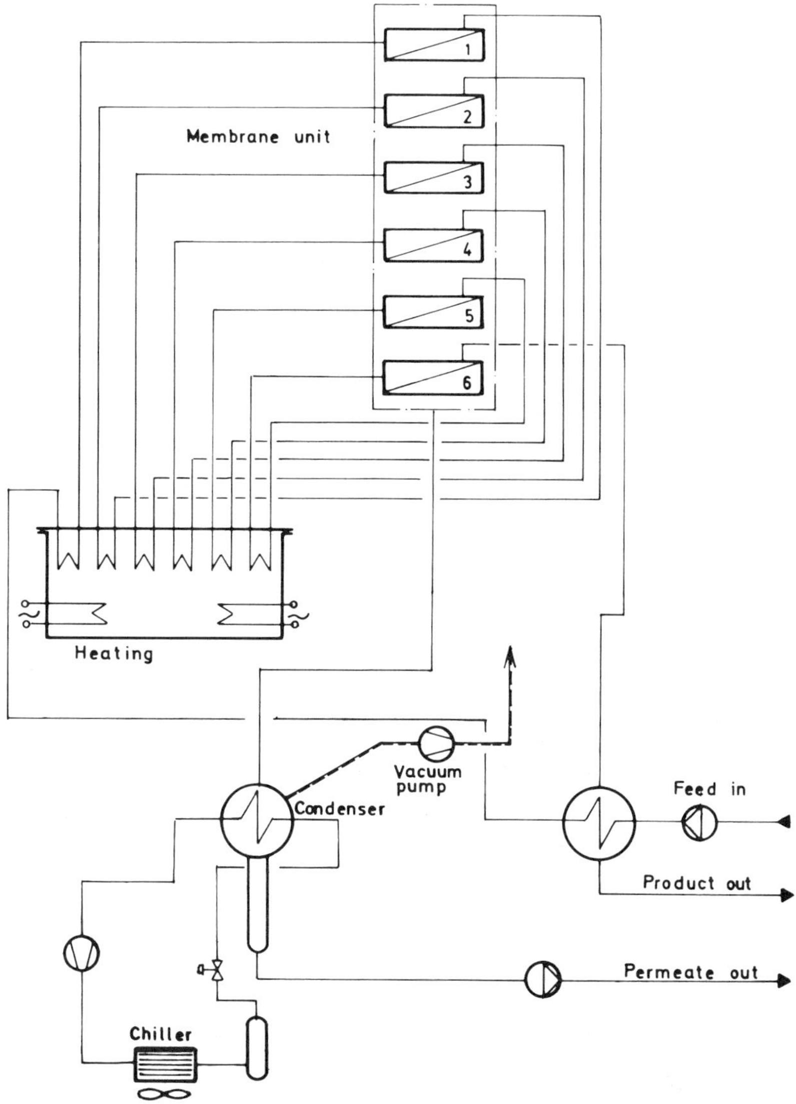

Figure 3. Typical flow sheet for a pervaporation unit.

the lower curve is no equilibrium curve, but obtained in a steady-state kinetic process. Starting at low water concentrations, water is constantly enriched in the permeate and it can be seen that over a broad concentration range, the composition of the permeate is nearly constant, independent of the feed composition, and only a very small portion of ethanol is present at all in the permeate. With increasing water concentration, the water content of the permeate goes through a minimum and in the water-rich range we observe a point where the membrane does not separate and which looks very similar to an azeotropic point. This diagram also clearly shows why and where pervaporation processes offer advantages over distillation. It shows as well that distillation and pervaporation possess optimal separation characteristics at different concentration ranges. A hybrid process, where each process is used in its most effective concentration range should give better results than each process alone.

In the water-rich area ethanol can be effectively separated from the mixture by distillation. At a concentration range of 60 - 80% (b.w.) of ethanol, however, the selectivity of the membranes with the now minor quantities of water becomes more effective. At high ethanol concentration, where distillation is no longer effective, pervaporation shows the highest selectivity. Furthermore, mainly the water is removed from the mixture. Therefore, only the aqueous portion has to be evaporated. The necessary heat input is, therefore, several times smaller than in distillation, where a multiple of the whole mixture (including the reflux) has to be evaporated.

It has to be pointed out that the characteristics shown in Figure 4 are only valid for a specific membrane, i.e. material of the active layer based on PVA, and for a specific range of temperature and permeate pressure. Figure 5 shows the separation behaviour of membranes with cellulose triacetate as the main polymer of the active layer. Figures 4 and 5 also clearly demonstrate that the interactions between feed and permeate and active layer polymer play a more important role in pervaporation than in other membrane processes.

Figure 6 shows a schematic diagram of a hybrid process for the production of dehydrated ethanol from a fermentation broth (7). In a simple distillation column, ethanol with a concentration of approx. 80% b.w. is produced and is further dehydrated in a pervaporation system. Waste heat from the distillation column is used to supply the necessary energy for pervaporation. The permeate containing a small amount of ethanol is condensed and recycled to the distillation column. Figure 7 shows a similar system, but for mixtures available at higher concentrations. This might be the case at central collection points collecting the product from smaller distilleries. The permeate is fed to a stripping column, which is used at the same time for preheating the main feed stream. In this way, ethanol losses are close to zero and environmental pollution, occurring in conventional dehydration units, does not exist. Depending on the plant size, the investment costs of systems described above, including vacuum system and stripping column, are between 40 and 80% of those of a conventional (azeotropic distillation) system. Savings in operation costs are in an even higher range. (For details in investment and operation costs refer to (8) and (9)). Figure 8 shows a unit supplied to the

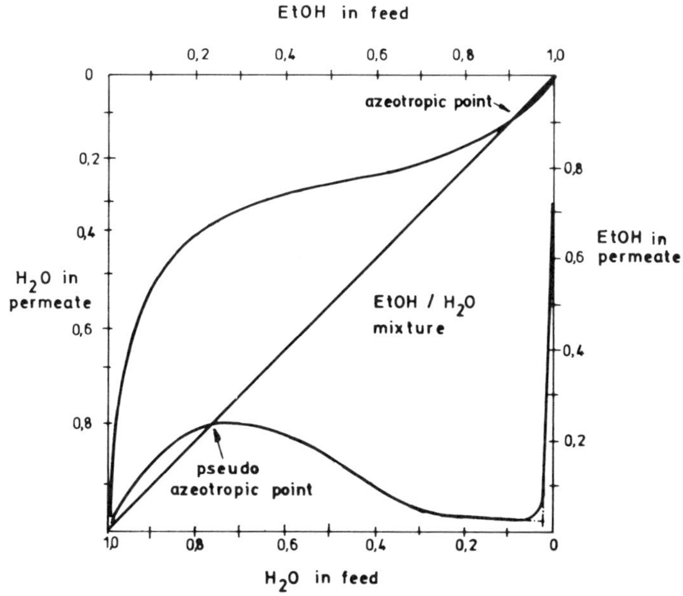

Figure 4. Permeate composition versus feed composition (composite membrane).
Feed temperature 90 - 100°C
P_p/P sat. >0.1.

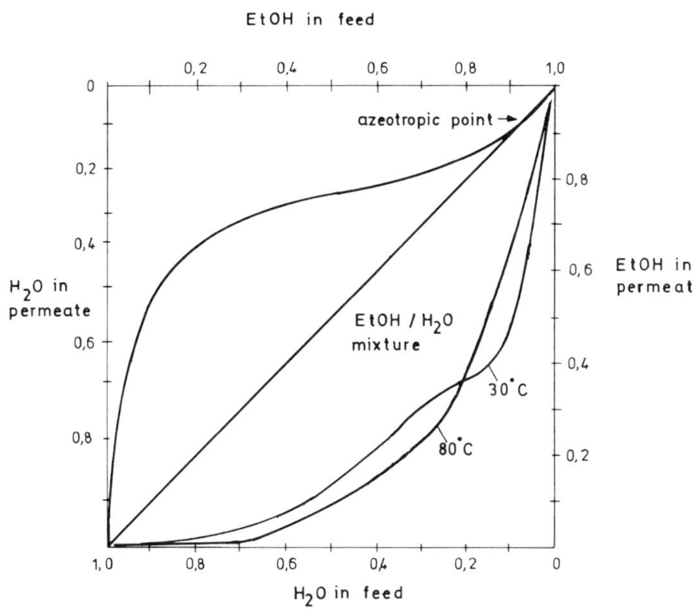

Figure 5. Permeate composition versus feed composition.
Asymmetric cellulose triacetate membrane
Permeate pressure 20 mbar.

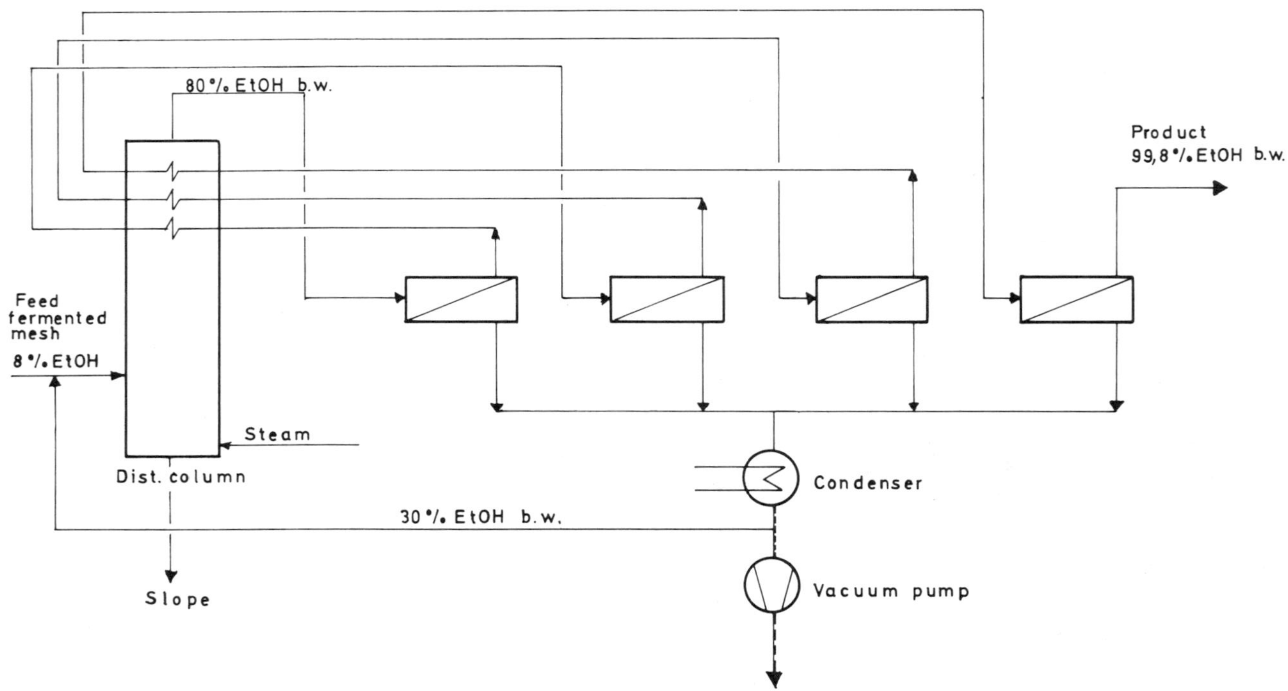

Figure 6. Combination distillation/pervaporation for EtOH dehydration.

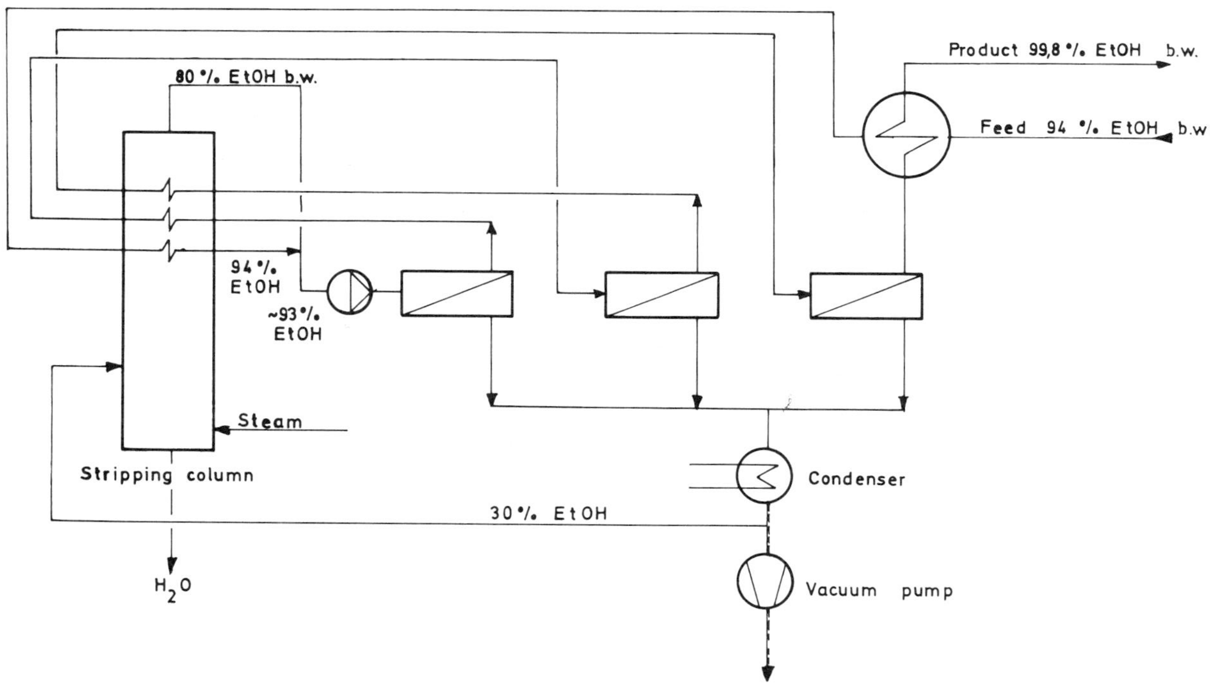

Figure 7. Combination of a separate pervaporation plant with a stripping column.

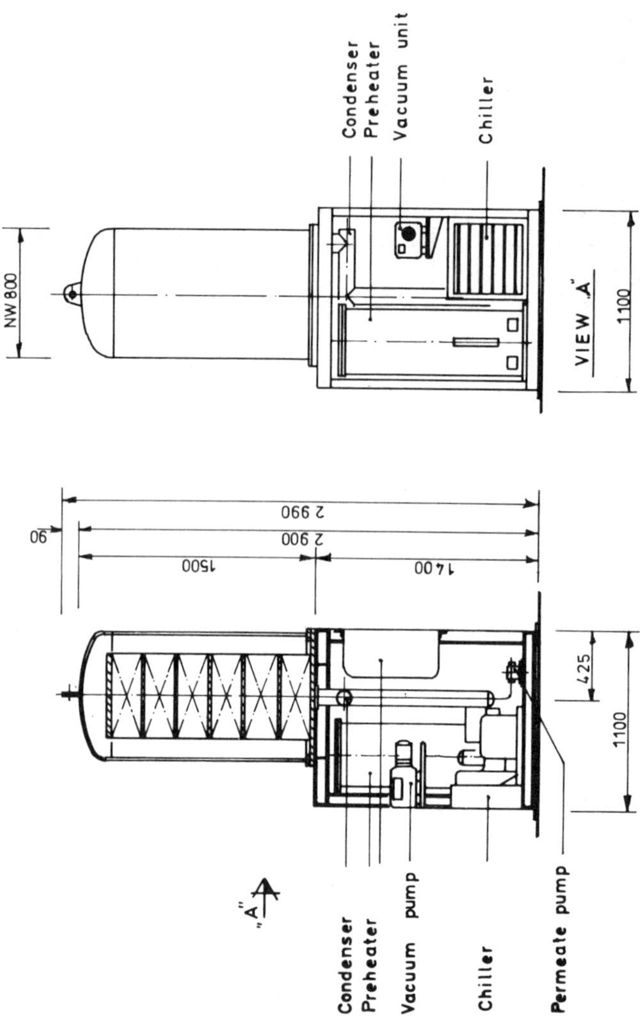

Figure 8. Pervaporation unit for the pharmaceutical industry.

pharmaceutical industries for the dehydration of ethanol for medical purposes.

Very similar results have been obtained for the following systems, using the same membranes:

a) Isopronal/water
b) Acetone/water
c) Methyl-ethyl-ketone/water
d) Dioxan/water
e) Ethyl-acetate/water
f) Diethylen glycol/water

The main differences between these systems and the ethanol-water mixture are found in slightly increased selectivities and fluxes in the organic-rich concentration ranges and shifting of the pseudo-azeotropic points. All these systems have been examined in the laboratory and partly on small technical scales. Small-size production units for the dehydration of several of the above-mentioned systems are under construction and will be in operation in early 1985.

Pervaporation processes exhibit extreme advantages in the dehydration of azeotrope-forming multi-component mixtures of organic solvents with water. In many cases, water removal from these systems is very difficult, if not impossible, irrespective of the type of thermal process. As the amount of water to be removed from such a mixture is normally not more than 10%, pervaporation offers a very simple and inexpensive solution.

The membranes under discussion can even be used to separate mixtures of methanol/ethanol and methanol/acetone, as methanol permeation rates are high compared to those of other organic solvents. In several cases, these separations are already applicable on a large scale and are more economical than distillation processes.

Future Trends

Development of pervaporation membranes is still in its early stages. Promising results have been obtained in the laboratory with special membranes, developed not for the removal of water, but for the separation of organic fluids. For different types of separation, different types of membranes are needed, as the solubilities of the permeating components seem to govern the overall separation characteristics. It will be possible to separate chemicals, such as olefines from paraffines by pervaporation processes.

Furthermore, it is possible as well to remove small amounts of organic substances (concentration below 1%) from water, down to levels below 1 ppm, using especially developed membranes.

Literature Cited

1. Heisler, E.G. "Solute and Temperature Effects in the Pervaporation of Aqueous Alcoholic Solutions"; Science 124, 1956, 77/78.
2. Binning, R.C.; James, F.E. "Membrane Permeation"; Petroleum Refiner 37, 1958, 5, 214-216.
3. Aptel, P.; Cuny, J.; Challard, N.; Neel, J. "Application of the Pervaporation Process to Separate Azeotropic Mixtures"; Journal of Membrane Science 1, 1976, 3, 271-287.

4. Nagy, E.; Borlai, O.; Kjhidy, A. "Membrane Permeation of Water Alcohol Binary Mixtures"; Journal of Membrane Science 7, 1980, 109/118.
5. Rautenbach, R.; Albrecht, R. "Separation of Organic Binary Mixtures by Pervaporation"; Journal of Membrane Science 7, 1980, 203-223.
6. Albrecht, R. "Pervaporation - Beiträge zur Verfahrensentwicklung"; Ph.D Thesis, RWTH Aachen (FRG), 1983.
7. Ballweg, A.H.; Brüschke, H.E.A.; Schneider, W.; Tusel, G.F. etal. "Pervaporation Membranes - An Economical Method to Replace Conventional Distillation and Rectification Columns in Ethanol Distilleries", 5th Int. Symp. on Alcohol Fuel Technology, Auckland, New Zealand, 1982.
8. Mokhtari-Nejad, E.; Schneider, W. Europe-Japan Congress on Membrane and Membrane Processes, 1984, Stresa, Italy.
9. Soukup, P.B. Diploma Thesis, University of Munich, 1983.

RECEIVED February 22, 1985

Dehydration of Alcohol-Water Mixtures Through Composite Membranes by Pervaporation

YUTAKA TAKETANI and HIROYOSHI MINEMATSU

Central Research Laboratories, Teijin Ltd., 4-3-2 Asahigaoka, Hino City, Tokyo 191, Japan

Three types of composite membranes from polyethyleneimine (PEI) with isophthaloyl chloride (IPC), trimesoyl chloride (TMC) and 5-chlorosulfonyl isophthaloyl chloride (CSIPC) were studied for dehydration of aqueous azeotropes of ethanol or isopropanol by pervaporation. The water selectivities of the composite membranes from PEI and CSIPC were the best of the three, and were 365 and 2069 for aqueous azeotropes of ethanol and isopropanol, respectively. The water selectivities decreased in the order of CSIPC≥TMC>IPC. The membranes from the PEIs modified with monochloroacetic acid or polyacrylic acid, and crosslinked with IPC improved the water selectivity at a cost of decreased flux rate. The performance of the composite membranes was analyzed and discussed from a chemical structural view point.

The reverse osmosis membrane process is well established both in the science and technology for the desalination of sea water and brackish water. The wide applications of the reverse osmosis membrane for the various processes are well studied (1). The high osmotic pressure required may have limited the application of reverse osmosis membrane technology (2,3). To avoid this, the pervaporation process is thought to be one of the alternatives which separates the liquid mixtures through membranes, where there are some variations in the ways of reducing the permeate pressure (4-6). The pervaporation is one membrane process that may provide an economical alternative to the distillation process. To be so, the membranes should have excellent selectivities and flux rates. Several works were done for the separation of an ethanol and water mixture by the pervaporation with various membranes (7-9). The recent work showed the results of an application of reverse osmosis composite membranes to the pervaporation process (10). In case of alcohol azeotropes such as an ethanol-water and isopropanol-water mixtures, the water is the minor component and, therefore, it would be more practical to remove

0097-6156/85/0281-0479$06.00/0

water than to extract alcohols from the mixture. For this purpose, pervaporation membranes should have high water selectivities. In order to prepare membranes of this type, one may apply the concept of preferentially absorbed water proposed by Sourirajan for the reverse osmosis membranes (1). If a membrane has the preferentially absorbed water layer at the surface of the membrane, then this layer of water might repel organic components, resulting in higher selectivity of water even in the pervaporation process.

In this report, three types of composite membranes were prepared by dipping microporous polysulfone membranes into a polyethyleneimine (PEI) solution, and then crosslinking the PEI with three different aromatic acid halides. The pervaporation performance of the three membranes was examined in the dehydration of ethanol or isopropanol azeotropic mixtures. The performance characteristics of the membranes were discussed with regard to their chemical structures.

Experimental

Pervaporation. The pervaporation apparatus consists of a constant temperature bath and a pump that circulates the feed solution through a radial flow cell (membrane surface area 11.1 cm^2) at a rate of about 1.2 l/min. and with the bath temperature controlled to ±0.2°C. The downstream compartment consists of two parallel pumping station that allows alternate sampling from the cold traps of ethanol and dry ice. The membrane was supported by a sintered metal plate. The composite membrane was put into the cell so that the thin layer of the composite membrane faced the feed solution. Before the membranes performance was measured, the preliminary sampling was made in the first two or three hours. The feed and permeate concentrations were determined by gas chromatography using Hitachi model 163 with stainless steel Porapak Q columns (oven temperature of 170°C and 190°C for ethanol and isopropanol, respectively, under a He stream of 20 ml/min.) equipped with Shimadzu CR-1 recorder. The concentration was determined by the calibration curves prepared with use of standard solutions. The flux was determined from the amount of material captured in the trap, and selectivity was calculated by the following equation.

$$\alpha^{H_2O}_{ROH} = (P_{H_2O}/P_{R\text{-}OH})/(F_{H_2O}/F_{R\text{-}OH})$$

where P and F denote the permeate and the feed weight fractions, respectively.

Materials. PEI was supplied in a 30% aqueous solution, manufactured by Nippon Shokubai Chemicals. Isophthaloyl chloride (IPC) and trimesoyl chloride (TMC) were of reagent grade from Wako Chemicals. 5-Chlorosulfonyl-isophthaloyl chloride (CSIPC) was prepared from the sodium 3,5-dicarboxy benzene sulfonate by the reaction with thionyl chloride and a small amount of DMF. CSIPC was recrystallized from CCl_4 and $CHCl_3$. All other materials were of reagent grade, and used without further purification.

Modification of PEI. Aqueous PEI solution was prepared by diluting 30% PEI solution (20 g) with 18.68 g of water. Monochloroacetic acid (1.32 g) was added to the solution and the mixture was stirred for 3 h at room temperature. The solution thus obtained was diluted to 2% with water, and was used for the composite membranes preparations.

Composite membranes. The composite membranes were prepared by the procedure described elsewhere (11,12). Throughout the experiment, polysulfone microporous membranes (pure water permeability $3.0–7.0 \times 10^5$ kg/m^2s kPa) were used as the support for the composite membranes. In order to isolate the active layer of the membrane, the composite membranes thus formed were immersed in $CHCl_3$ for 15–20 min. The polysulfone microporous membranes were dissolved, and crosslinked thin layers were floated off. An individual piece of the thin film was scooped with a preweighed filter paper, and was transferred to another $CHCl_3$ bath to rinse off the polysulfone completely. After this procedure, the membranes with filter papers were dried and weighed. The weight was approximately between 2–6 mg for a 5 cm diameter disk sample. These samples were divided into two parts and were used for chemical analyses.

Titration of composite membranes. The weighed thin layer films were immersed in deionized water overnight with stirring. The rinsed and filtered polymers were immersed again in deionized water. This procedure was repeated three times. The rinsed and dried films were then immersed in a 1/50 N NaOH aqueous solution and the mixture was stirred overnight. The titration of an aliquot of 10 ml of the filtrate with $AgNO_3$ aqueous solution gave the Cl content of B meq/g dried polymer, which would be equal to the amount of the salt of amines hydrochloride. The remaining filtrate was then titrated for the NaOH consumption measurement with a 1/100 N HCl aqueous solution to give A meq/g polymer. The value A meq/g would determine the sum of all acidic components such as -COOH, $-SO_3H$ and amines hydrochloride. The other run was conducted using the second portion of the weighed thin films. Instead of a 1/50 N NaOH solution, a 1/50 N HCl aqueous solution was used for the immersion, and the value of C meq/g polymer, was obtained by the measurement of the consumed amount of HCl with a 1/100 N NaOH aqueous solution. The C meq/g polymer was thought to be equivalent to the amount of free amine in the composite membranes. The blank experiment was also done by using the filter papers.

Instruments. The Nippon Denshi JIR-40X and JESCA-4 were used for the FT-IR and XPA measurements.

Results and Discussion

Pervaporation performance of composite membranes. The pervaporation experiments were done for the separation of azeotropic mixtures of two alcohols, that is, ethanol and isopropanol, in which water is the minor component, containing 5 wt% and 12 wt%, respectively. Three types of crosslinking reagents, IPC, TMC and CSIPC, were used to obtain *in situ* composite membranes on polysulfone microporous supports. The performances features are shown in Table I. It was reported that a high molecular weight PEI itself affords the insoluble thin layer, which showed rather high salt rejection

Table I. The Pervaporation Performance of Composite Membranes

Feed Composition wt/wt%	XLR[a)]	M.W of PEI							
		300				70,000			
		30°C		60°C		30°C		60°C	
		Flux (kg/m^2h)	$\alpha^{H_2O}_{ROH}$	Flux (kg/m^2h)	$\alpha^{H_2O}_{ROH}$	Flux (kg/m^2h)	$\alpha^{H_2O}_{ROH}$	Flux (kg/m^2h)	$\alpha^{H_2O}_{ROH}$
	none	-	-	-	-	12.5	1.7	-	-
H_2O : EtOH	IPC	1.35	13.2	4.73	21.0	1.46	9.6	5.16	11.4
5 : 95	TMC	0.94	18.6	3.00	33.8	1.06	16.9	4.85	15.9
	CSIPC	0.33 (0.20,	171.0 364.6)	1.35	273.3	0.37	78.4	1.33	112.0
	IPC	1.38	158.6	4.00	167.3	1.19	155.0	3.33	176.0
H_2O : i-PrOH 12 : 88	TMC	1.02	142.0	2.80	196.4	1.41	269.4	3.68	424.0
	CSIPC	1.02	2068.8	2.71	1873.0	1.33	763.0	3.31	845.4

a) Cross-linking reagents.

IPC: Isophthaloyl chloride. TMC: Trimesoyl chloride. CSIPC: 5-Chlorosulfonyl isophthaloyl chloride.

Operating pressure 26.6 Pa.

when it was operated under reverse osmosis conditions (11). So the results of PEI cured without crosslinking reagents is also shown in Table I for comparison. The heat cured PEI shows high flux and low selectivity, contrary to the other composite membranes when PEI was crosslinked with various aromatic acid halides. With the change of crosslinking reagents, the selectivity of membranes for both ethanol and isopropanol varies simultaneously regardless of the molecular weight of PEI as is shown in Table I. The polyfunctionality of the crosslinking reagent seemed to increase the water selectivity. The highest water selectivity was obtained from the lower molecular weight PEI crosslinked with CSIPC. The presence of chlorosulfonyl group was found more effective than an extra chlorocarbonyl group as was the case in TMC. It is well known that TMC forms trialkyl ester or amide when it is exposed to alcohols or amines without HCl acceptors. To the contrary, the chlorosulfonyl group of CSIPC has a lower reactivity to alcohols or amines, and remains unreacted and was then hydrolyzed by exposure to water. Table I seems to show that the presence of carboxylic acid or sulfonic acid in the composite membrane was essential to obtain good water selectivity. In case of reverse osmosis, the introduction of carboxylic or sulfonic groups increases the water flux rate and decreases the salt rejection, as is well known with the sulfonated polysulfone or sulfonated polyphenylene oxide. The present results show the reverse tendency. This is thought to be derived from the big differences in the feed composition, noting that water is the minor component in the pervaporation as in our case. Thinking of the reactivity difference between TMC and CSIPC, it seems conceivable that the density of covalent crosslinkage is not a very important factor for high water selectivity of the pervaporation membranes. However, crosslinking in the membranes is needed to some extent, as the secondary amide is thought to be one of the important structures for higher water flux rate in the reverse osmosis membrane study (13). When the pervaporation experiment was conducted at a higher temperature, both the flux and the selectivity increased. The water selectivity was higher for the isopropanol-water mixture than the ethanol-water mixture. This is partly due to the difference in the bulkiness of organic component as well as in the water content of the feed composition.

Structures of composite membranes. The surface analyses were done by FT-ATR IR spectroscopy and ESCA. The three types of composite membranes on polysulfone microporous membranes were examined by the FT-ATR IR spectrum method. In case of the FT-ATR IR method, the absorbance due to supporting polysulfone overlapped the spectra of composite membranes surface layers as the latter were so thin. The spectrum subtracted of polysulfone absorbances is shown in Figure 1. The absorbance due to the sulfo group were observed at 1200 cm^{-1}, 1040 cm^{-1} and 630 cm^{-1} in addition to the amide I band at 1660 cm^{-1} and amide II band at 1560 cm^{-1}. Between composite membranes from PEI with TMC and those with IPC, there was no significant difference in the IR spectra. In order to examine the chemical structure, the surface layers of the composite membranes were treated with the perfluoroalcohol to form a fluorine containing ester of the carboxylic groups left unreacted. The amount of F on the modified composite membranes was determined by XPS. The ratio of F/N is 0.252

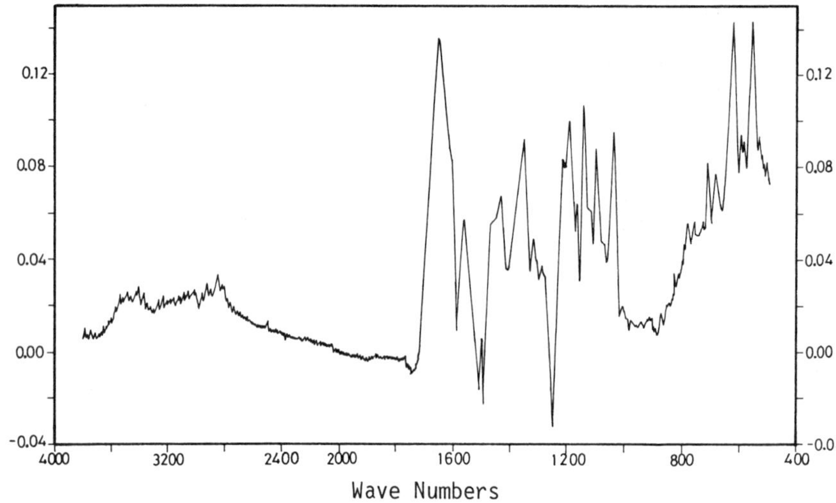

Figure 1. FT-ATR IR spectrum of PEI-CSIPC composite membrane.

and 0.316 for IPC and TMC crosslinked membranes, respectively, indicating that the TMC crosslinked ones had higher values of free carboxylic acid. Since the esterification with perfluoroalcohol was not applicable for sulfo group, the amount of SO_3H was not determined by this procedure.

The composite membrane thin layer was isolated and analyzed by the titration method shown in Experimental Section. Three values were obtained: A meq/g; -COOH, $-SO_3H$ and amines hydrochloride, B meq/g; amines hydrochloride, and C meq/g; amines.

Thus, the value (A-B) would give the total acids, namely -COOH and $-SO_3H$. For the total amines, the (B+C) value seemed to be appropriate. Both values are listed in Table II, which were supported by the XPS measurement. The composite membranes of PEI and CSIPC showed the smallest B value and the highest acid content.

The acid content is in the same increasing order as the water selectivity of the membrane. These results indicate that a high acid content is important to obtain a good water selective membrane for the alcohol azeotropic mixtures.

The remaining free acid would form a polyion complex both on and within the membranes with PEI. To examine the existence of polyion complex formation, the following experiments were conducted.

Table II. Titration Values for Composite Membranes

Composite Membrane	A value (meq/g)	B value (meq/g)	C value (meq/g)	Weak acid content[a)] (meq/)	Weak base content[b)] (meq/g)
PEI + IPC	3.46	1.28	1.18	2.18	2.46
PEI + TMC	3.18	0.61	1.02	2.57	1.63
PEI + CSIPC	3.27	0.12	1.83	3.15	1.95

a) (A - B), b) (B + C)

Composite membranes from modified PEI with IPC. Aqueous solutions of PEI was treated with monochloroacetic acid (MCA). The resulting solution of MCA-PEI was crosslinked with IPC to form composite membranes. After the reaction of PEI with MCA, no MCA could be detected in the mixture by gas chromatography. Figure 2 shows the performance results of the composite membranes when the molar ratio of MCA to PEI was changed. The water selectivity of the membranes crosslinked with IPC was greatly increased until the MCA/PEI ratio reached the value of 0.30, where no crosslinked membranes seemed to be formed. This was due to the reaction of MCA to primary amines, causing less reaction sites for the amide formation with the acid halide to form membranes. According to catalogue data, PEI was not linear and consisted of 25% primary amine, 50% secondary amine and 25% tertiary amine. The result indicates that primary amine and/or secondary amine were converted into the iminoacetic acid, resulting in the introduction of acidic components before crosslinking with acid halides. The absorbance increment of IR spectrum at 1650 cm^{-1} of the evaporated reaction mixtures were related with the amount of MCA introduced to PEI.

The similar results were obtained for the composite membranes from another modified PEI and IPC combination, in which polysulfone microporous membranes were treated beforehand with diluted aqueous polyacrylic acid (PAA) solution. The selectivities and fluxes of the membranes thus obtained are shown in Figure 3. The higher the PAA concentration, higher selectivities and lower fluxes were exhibited by the membranes. When PAA was coated afterward on the PEI dipped polysulfone microporous membranes, the composite membranes did not show any improvement of selectivities even after crosslinking with IPC. These two results indicated that the membranes of PEI crosslinked with IPC, which would be thought to have the least ionic character of the three crosslinking reagents, could improve their water selectivity to the feed mixtures by introducing the carboxylic acid group beforehand. These results also suggested that the contribution of polyion complex formation in the membranes is an important factor to obtain membranes with higher water selectivity.

The hydrochloric acid originating from the aromatic acid halides might also improve the membranes affinity to water since it would provide hydrophilic sites of amines hydrochloride. The performance of membranes from PEI neutralized with hydrochloric acid was not promising as shown in Figure 4. The neutralization by hydrochloric acid lowered the crosslinking density, which in turn made the membranes too loose.

Permeation behavior. Observed values for the activation energy of permeation are listed in Table III for each membrane. For the

Table III. Apparent Activation Energy of Permeation in Pervaporation

Composite Membranes	Activation Energy ($kJ \cdot mol^{-1}$)	
	Feed Azeotropes	
	H_2O/EtOH	H_2O/i-PrOH
PEI + IPC	25.5	40.2
PEI + TMC	33.1	40.2
PEI + CSIPC	35.6	40.2

Note: corresponding activation energy of permeation determined with PEI + IPC membrane for H_2O: 49.8 $kJ \cdot mol^{-1}$; for EtOH: 37.3 $kJ \cdot mol^{-1}$; for i-PrOH: 13.0 $kJ \cdot mol^{-1}$

permeation of isopropanol azeotrope, the observed activation energies were the same for the three types of membrane, and were very close to the that for the water permeation, 49.8 $kJ \cdot mol^{-1}$. This was indicative that the water permeation was dominant through the membranes. In case of ethanol azeotrope, the less water selective membranes of PEI and IPC showed the least activation energy. With the value of 37.3 $kJ \cdot mol^{-1}$ for the activation energy of EtOH, it seems to indicate that permeation of EtOH is predominant in these membranes.

The influence of the operating pressure on the permeate side was examined for the PEI-CSIPC composite membranes. The result is shown

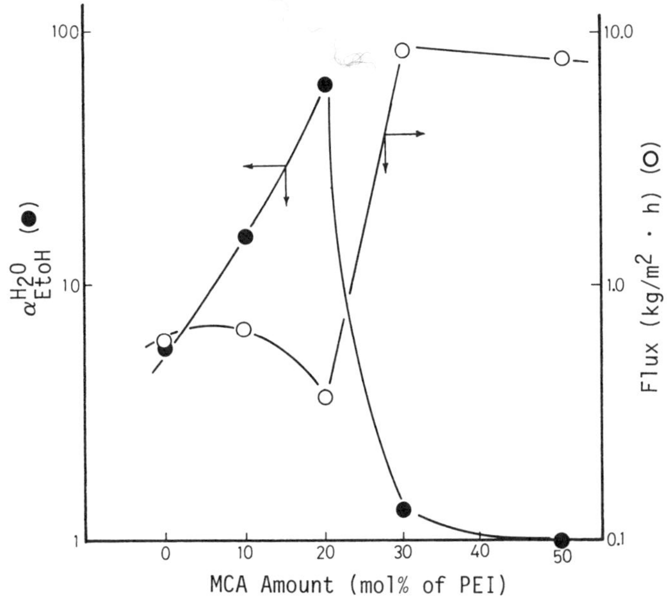

Figure 2. Flux and selectivity of MCA modified PEI compsite membranes with IPC. Feed solution, $H_2O/EtOH = 5/95$, 30°C

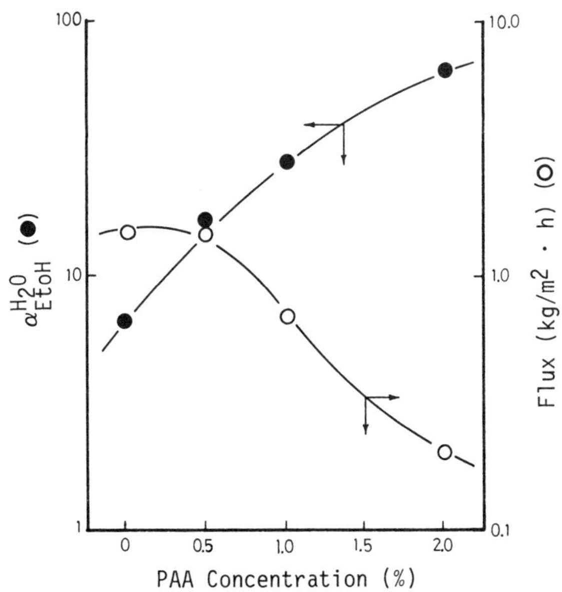

Figure 3. Flux and selectivity of PAA pretreated PEI composite membranes with IPC. Feed solution, $H_2O/EtOH = 5/95$, 50°C

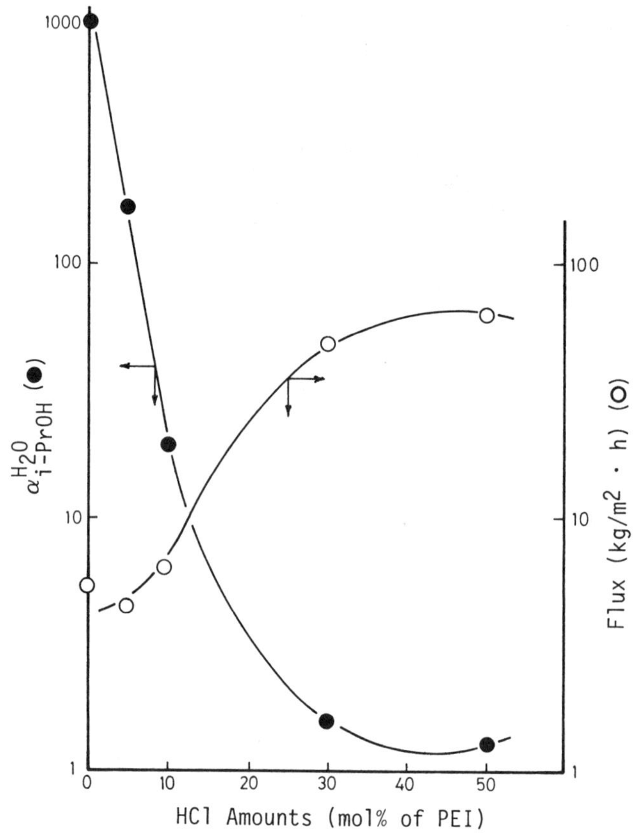

Figure 4. Flux and selectivity of PEI neutralized by HCl· composite membranes with CSIPC. Feed solution, H_2O/i-PrOH = 12/88, 50°C

in Fig. 5. Both flux and selectivity began to decrease rather abruptly at a pressure of 66.5 Pa. This seems to reflect the evaporation equilibrium of water under the conditions, as was predicted by Shelden (14).

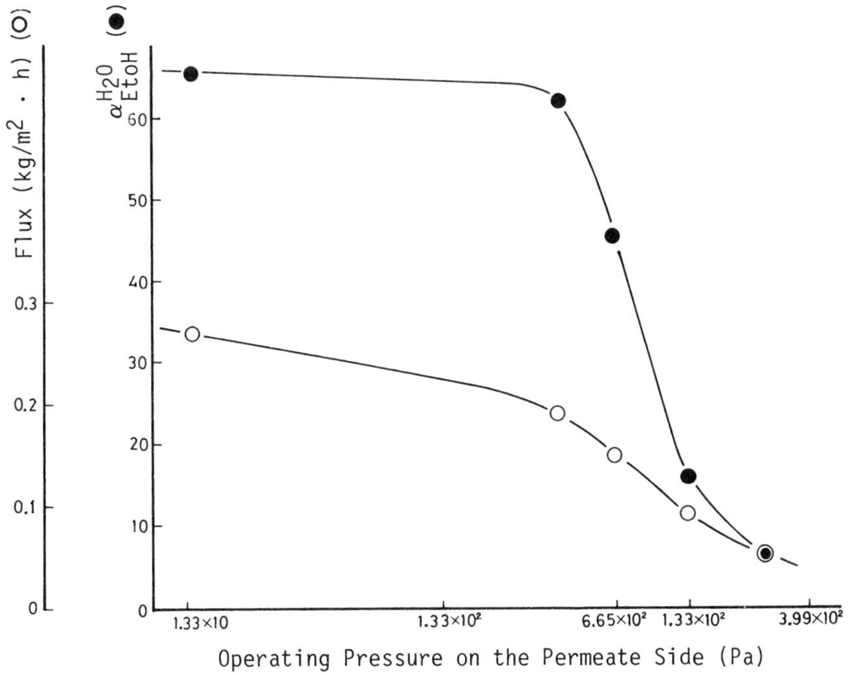

Figure 5. Influence of operating pressure at the permeate side on flux and selectivity of PEI composite membranes with CSIPC. Feed solution, H_2O/EtOH = 5/95, 30°C

Conclusion

Three kinds of crosslinking reagents were used for the preparation of PEI composite membranes. The pervaporation performance of the membranes was examined for the dehydration of aqueous azeotropes of ethanol and isopropanol. The water selectivity was greatly enhanced by applying CSIPC as a crosslinking reagent. The lower reactivity of the chlorosulfonyl group resulted in a larger amount of free sulfo groups in the composite membranes. The formation of polyion complex from the sulfonic acid and the amines probably enhanced the water uptake from the feed solution. The existence of covalent bond also seems to be necessary to obtain high separation factors. The mingled structure of strong hydrophilic parts of polyion complex and weak hydrophilic parts of aromatic amides appears to be essential to obtain a good membrane for extracting the less abundant water from alcohols azeotropes. The preferentially absorbed water in the polyion complex parts both at the surface and in the membranes may explain the low solubility of the abundant alcohol components. Further work is necessary to investigate better membranes having both high selectivity and flux rate.

Acknowledgment

This work was performed under the management of the Research Association for Basic Polymer Technology as a part of a project on Basic Technology for Future Industries sponsored by Agency of Industrial Science and Technology, Ministry of International Trade and Industry.

The authors are indebed to Dr. T. Doi, General Manager of the Research Association for Basic Polymer Technology for his guidance throughout the work. They also wish to thank Mr. H. Itagaki, Director of the Department of Research & Development of TEIJIN Ltd., and Dr. S. Ozawa, Director of Central Research Lab. of TEIJIN Ltd., for their encouragement and valuable suggestions.

Literature Cited

1. Sourirajan, S.; Matsuura, T. In "Reverse Osmosis and Synthetic Membranes"; Sourirajan, S., Ed.; National Research Council Canada: Ottawa, 1977; Chap. 1—3.
2. Ohya, H. Kagaku Kogyo Ronbunshu 1983, 9, 154.
3. Ohya, H.; Kazama, E. Kagaku Kogyo Ronbunshu 1982, 8; 144.
4. Binning, R. C.; Lee, R. J. U.S. Patent 2,953,502, 1960.
5. Yuan, S.; Schwartzbarg, H. G. AIChE Sumposium Series No.120, American Institute of Chemical Engineering: Washington, D.C., 1972; p. 41.
6. Aptel, P.; Challard, J.; Neel, J. J. Memb. Sci., 1976, 1, 271.
7. Sweeny, R. F.; Rose, A. Ind. Eng. Chem., Prod. Res. Dev., 1965, 4, 248.
8. Huang, R. Y. M.; Jarvis, N. R. J. Appl. Polym. Sci., 1970, 14, 2341.
9. Shantora, V.; Huang, R. Y. M. J. Appl. Polym. Sci., 1981, 26 3223.
10. Schissel, P.; Orth, R. A. J. Memb. Sci., 1984, 17, 109.
11. Rozelle, L. T.; Cadotte, J. E.; Cobian, K. E.; Kopp, C. V. Jr. In "Reverse Osmosis and Synthetic Membranes"; Sourirajan, S., Ed.; National Research Council Canada: Ottawa, 1977; pp. 249—261.
12. Kawaguchi, T.; Taketani, Y.; Hayashi, Y.; Ono, T.; Mori, K. U.S. Patent 4,265,745, 1981.
13. Taketani, Y.; Matsuura, T.; Sourirajan, S. Sep. Sci. Technol., 1982, 17, 821.
14. Shelden, R. A.; Thompson, E. V. J. Memb. Sci., 1978, 4, 115.

RECEIVED February 22, 1985

INDEXES

Author Index

Subject Index

Production by Meg Marshall
Indexing by Karen McCeney
Jacket design by Pamela Lewis

Elements typeset by Hot Type Ltd., Washington, D.C.
Printed and bound by Maple Press Co., York, Pa.